Venera Khoromskaia, Boris N. Khoromskij

Tensor Numerical Methods in Quantum Chemistry

Also of Interest

Tensor Numerical Methods in Scientific Computing
Boris N. Khoromskij, 2018
ISBN 978-3-11-037013-3, e-ISBN (PDF) 978-3-11-036591-7,
e-ISBN (EPUB) 978-3-11-039139-8

Numerical Tensor Methods.
Tensor Trains in Mathematics and Computer Science
Ivan Oseledets, 2018
ISBN 978-3-11-046162-6, e-ISBN (PDF) 978-3-11-046163-3,
e-ISBN (EPUB) 978-3-11-046169-5

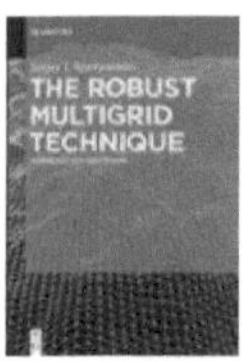

The Robust Multigrid Technique. For Black-Box Software
Sergey I. Martynenko, 2017
ISBN 978-3-11-053755-0, e-ISBN (PDF) 978-3-11-053926-4,
e-ISBN (EPUB) 978-3-11-053762-8

Direct and Large-Eddy Simulation
Bernard J. Geurts, 2018
ISBN 978-3-11-051621-0, e-ISBN (PDF) 978-3-11-053236-4,
e-ISBN (EPUB) 978-3-11-053182-4

Richardson Extrapolation. Practical Aspects and Applications
Zahari Zlatev, Ivan Dimov, István Faragó, Ágnes Havasi, 2017
ISBN 978-3-11-051649-4, e-ISBN (PDF) 978-3-11-053300-2,
e-ISBN (EPUB) 978-3-11-053198-5

Venera Khoromskaia, Boris N. Khoromskij

Tensor Numerical Methods in Quantum Chemistry

DE GRUYTER

Mathematics Subject Classification 2010
65F30, 65F50, 65N35, 65F10

Authors
Dr. Venera Khoromskaia
Max-Planck Institute for
Mathematics in the Sciences
Inselstr. 22-26
04103 Leipzig
Germany
vekh@mis.mpg.de

DrSci. Boris N. Khoromskij
Max-Planck Institute for
Mathematics in the Sciences
Inselstr. 22-26
04103 Leipzig
Germany
bokh@mis.mpg.de

ISBN 978-3-11-037015-7
e-ISBN (PDF) 978-3-11-036583-2
e-ISBN (EPUB) 978-3-11-039137-4

Library of Congress Control Number: 2018941005

Bibliographic information published by the Deutsche Nationalbibliothek
The Deutsche Nationalbibliothek lists this publication in the Deutsche Nationalbibliografie; detailed bibliographic data are available on the Internet at http://dnb.dnb.de.

Typesetting: VTeX UAB, Lithuania
Printing and binding: CPI books GmbH, Leck
Cover image: Venera Khoromskaia and Boris N. Khoromskij, Leipzig, Germany

www.degruyter.com

Contents

1 Introduction

All truths are easy to understand once they are discovered;
the point is to discover them.
Galileo Galilei

This research monograph describes novel tensor-structured numerical methods in application to problems of computational quantum chemistry. Numerical modeling of the electronic structure of molecules and molecular clusters poses a variety of computationally challenging problems caused by the multi-dimensionality of the governing physical equations. Traditional computer algorithms for the numerical solution of integral-differential equations are usually operating with the discretized representation of the multivariate functions, yielding n^d amount of data for functions in $\mathbb{R}^d$. This problem of the exponential increase in amount of data associated with adding extra dimensions to a mathematical space was described by Richard Bellman [18] as the "curse of dimensionality" (1961). This phenomenon can be hardly eliminated by data-sparse techniques and grid refinement or by using parallelization and high performance computing.

The tensor numerical methods that reduce, and even break the curse of dimensionality, are based on a "smart" rank-structured tensor representation of the multivariate functions and operators using $n \times \cdots \times n$ Cartesian grids. In this book, we discuss how the basic algebraic tensor decompositions originating from chemometrics, and signal processing, recently led to revolutionizing the numerical analysis. However, the multi-linear algebra of tensors is not a single cornerstone for tensor numerical methods. The prior results [93, 94] on theory of the low-rank tensor-product approximation of multivariate functions and operators, based, in particular, on *sinc* quadrature techniques provided a significant background for starting the advanced approaches in scientific computing. Thus the tensor-structured numerical methods resulted from bridging the modern multi-linear algebra and the nonlinear approximation theory for multivariate functions and operators.

Tensors are simply multidimensional arrays of real (complex) numbers. For example, vectors are one-dimensional tensors with the number of entries n, while an $n \times n$-matrix is a two-dimensional tensor of size n^2. A third-order tensor can be generated, for example, by sampling a function of three spatial variables on $n \times n \times n$ 3D Cartesian grid in a volume box, then the number of entries is n^3. Storage size for a tensor of order d grows exponentially in d as n^d, provoking the curse of dimensionality. The work required for computations with this tensor is also of the order of $O(n^d)$. Therefore, it is preferable to find a more efficient way to represent the multidimensional arrays.

A rank-structured representation of tensors reduces the multidimensional array to a sum of tensor products of vectors in d dimensions. In the case of equal size n, for example, the storage and computations would be scaled as $O(dnR)$, where R is

https://doi.org/10.1515/9783110365832-001

the number of summands. This idea is well known since it was proposed by Frank L. Hitchcock in 1927 [131] in the form of so-called "canonical tensors". Thus, the canonical tensor format allows us to avoid the curse of dimensionality. This can be seen as a discrete analogue of the representation of a multivariate function by a sum of separable functions. The main problem of the canonical tensor format is in the absence of stable algorithms for its representation from a full size tensor.

The Tucker tensor decomposition was invented in 1966 by Ledyard R. Tucker and was used in principal component analysis problems in psychometrics, chemometrics, and signal processing for calculating the amount of correlations in experimental data. Usually these data contain rather moderate number of dimensions; the data sizes in every dimension are not large, and the accuracy issues are not significant. The main advantage of the Tucker tensor format is in the existence of stable algorithms for the tensor decomposition based on the higher-order singular value decomposition (HOSVD) introduced by Lieven De Lathauwer et al. in [61, 60]. However, this Tucker algorithm from multilinear algebra requires an available storage for a full format tensor, n^d, and exhibits the complexity of the order of $O(n^{d+1})$ for the HOSVD. Rather low compression rate by the Tucker tensor decomposition in problems of principal component analysis could hardly promote this method for accurate calculations in scientific computing.

The fascinating story with the grid-based tensor numerical methods in scientific computing started in 2006, when it was proven that the error of the Tucker tensor approximation applied to several classes of function-related tensors decays exponentially fast in the Tucker rank [161]. That is, instead of a three-dimensional (3D) tensor having n^3 entries in a full format, one obtains its Tucker tensor approximation given by only $O(3n + \log^3 n)$ numbers, thus gaining an enormous compression rate. The related analytical results on the rank bounds for canonical tensors based on the sinc approximation method had been proven earlier by Ivan Gavrilyuk, Wolfgang Hackbusch, and Boris Khoromskij in [94, 91, 111].

In numerical tests for several classical multivariate functions discretized on $n \times n \times n$ 3D Cartesian grids, it was shown that the Tucker decomposition provides an easily computable low-rank separable representation in a problem-adapted basis [173]. Such beneficial separable representation enables efficient numerical treatment of the integral transforms, and other computationally extensive operations with the multivariate functions. However, the HOSVD in the Tucker decomposition requires full format tensors, which is often not applicable for numerical modeling in physics and quantum chemistry. Thus, the HOSVD does not break the curse of dimensionality, and has, indeed, a limited significance in computational practice.

In this regard, an essential advancement was brought forth by the so-called reduced higher order singular value decomposition (RHOSVD), introduced by Boris Khoromskij and Venera Khoromskaia as part of the canonical-to-Tucker (C2T) transform [173, 174]. The latter works efficiently in cases where the standard Tucker decomposition is unfeasible. It was demonstrated that for the Tucker decomposition of function-related tensors given in the canonical form, for example, resulting from analytic approximation and certain algebraic transforms, there is no need to build a full-size tensor. It is enough to find the orthogonal Tucker basis only by using the directional matrices of the canonical tensor, consisting of skeleton vectors in every single dimension. The C2T decomposition proved to be an efficient tool for reducing the redundant rank parameter in the large canonical tensors. Since the RHOSVD does not require the full size tensor, it promoted further development of tensor methods also to higher dimensions, because it applies to canonical tensors, which are free from the curse of dimensionality.

Furthermore, the orthogonal Tucker vectors, being an adaptive basis of the Tucker tensor representation, exhibited smooth oscillating shapes, which can be viewed as "fingerprints" for a given multivariate function. This property facilitated the multigrid Tucker decomposition proposed in [174, 146], which enables fast 3D tensor calculus in electronic structure calculations using incredibly large grids. Further gainful properties of the multigrid approach for the tensor numerical methods are not yet comprehensively investigated. Since the rank-structured tensor decompositions are basically working on Cartesian grids, the methodology developed for the finite difference methods, including the Richardson extrapolation techniques, yielding $O(h^3)$ accuracy in the mesh size h can be applied.

The traditional methods for numerical solution of the Hartree–Fock equation have been developed in computational quantum chemistry. They are based on the analytical computation of the arising two-electron integrals,[1] convolution-type integrals in $\mathbb{R}^3$, in the problem-adapted naturally separable Gaussian-type basis sets [3, 277, 128], by using *erf*-functions. This rigorous approach resulted in a number of efficient program packages, which required years of development by large scientific groups and which are nowadays widely used in scientific community. See, for example, [299, 292, 88], and other packages listed in Wikipedia. Other models in quantum chemistry, like the density functional theory [251, 107, 268], usually apply a combination of rigorously constructed pseudopotentials, and the grid-based wavefunctions, as well as, the experimentally justified coefficients. In general, for solution of the multidimensional problems in physics and chemistry, it is often best to approximate the multivariate functions by sums of separable functions. However, the initial separable representation of functions may be deteriorated by the integral transforms and other operations, leading to cumbersome computational schemes.

In such a way, the success of the analytical integration methods for the ab-initio electronic structure calculations stems from the big amount of precomputed information based on the physical insight, including the construction of problem-adapted atomic orbitals basis sets, and elaborate nonlinear optimization for calculation of the density-fitting basis. The known limitations of this approach appear due to a strong

1 Also called electron repulsion integrals.

dependence of the numerical efficiency on the size and quality of the chosen Gaussian basis sets. These restrictions might be essential in calculations for larger molecules and heavier atoms. Now, it is a common practice to reduce these difficulties by switching partially or completely to grid-based calculations. The conventional numerical methods quickly encounter tractability limitations even for small molecules, and when using moderate grid sizes. The real space multi-resolution approaches suggest to reduce the grid size by local mesh refinements [122, 305], which may encounter problems with computation of three-dimensional convolution integrals for functions with multiple singularities.

The grid-based tensor-structured numerical methods were first developed for solving challenging problems in electronic structure calculations. The main ingredients include the low-rank grid representation of multivariate functions and operators, and tensor calculation of the multidimensional integral transforms, introduced by the authors in 2007–2010, [166, 187, 145, 146, 147, 168]. An important issue was the possibility for comparison of the results of tensor-based computations with the outputs of benchmark quantum chemical packages, which use the analytical methods for calculating the three-dimensional convolution integrals [300]. It was shown that tensor calculation of the multidimensional convolution operators is reduced to a sequence of one-dimensional convolutions and one-dimensional Hadamard and scalar products [145, 146]. Such reduction to one-dimensional operations enables computations on exceptionally fine tensor grids. The initial multilevel tensor-structured solver for the Hartree–Fock equation was based on the calculation of the Coulomb and exchange integral operators "on-the-fly", using a sequence of refined uniform grids, thus avoiding precomputation and storage of the two-electron integrals tensor [146, 187]. The disadvantage of this version is rather substantial time consumption. This solver is discussed in Chapter 8.

Further progress in tensor methods in electronic structure calculations was promoted by a fast algorithm for the grid-based computation of the two-electron integrals (TEI) [157, 150] in $O(N_b^3)$ storage in the number of basis functions N_b. The fourth-order TEI tensor is calculated in a form of low-rank Cholesky factorization by using an algebraic black-box-type "1D density fitting"scheme, which applies to the products of discretized basis functions. Using the low-rank tensor representation of the Newton convolving kernel and that of the products of basis functions, all represented on $n \times n \times n$ Cartesian grid, the 3D integral transforms are calculated in $O(n \log n)$ complexity. The corresponding algorithms are described in Chapter 10.

The elaborated tensor-based Hartree–Fock solver [147], described in Chapter 11, employs factorized representation of the two-electron integrals and tensor calculation of the core Hamiltonian, including the three-dimensional Laplace and nuclear potential operators [156]. In the course of self-consistent iteration for solving the Hartree–Fock eigenvalue problem, due to factorized representation of TEI, the update of the Coulomb, and exchange parts in the Fock matrix, is reduced to cheap algebraic operations. Owing to grid representation of basis functions, the basis sets are not restricted to Gaussian-type orbitals and are allowed to consist of any well-separable functions defined on a grid. High accuracy is attained because of easy calculations on large 3D grids up to $n^3 = 10^{18}$, so that the high resolution with a mesh size of the order of atomic radii $h \approx 10^{-4}$ Å is possible.

This Hartree–Fock solver is competitive in computational time and accuracy [147] with the solvers in standard packages based on analytical calculations of the Fock operator. It may also have weaker limitations on the size of a molecular system. It works in MATLAB on a laptop for moderate-size molecules, and its efficiency is not yet investigated for larger computing facilities. It is a black-box type scheme; the input needs only the charges and coordinates of nuclei, the number of electron pairs, and the basis set defined on the grid. The tensor approach shows a good potential for the ab-initio calculations for finite lattices [154], and may be used in numerical simulations for small nanostructures.

The progress of tensor methods for three-dimensional problems in quantum chemistry motivated, in particular, development of novel tensor formats. Though the matrix-product states (MPS) format was well known for modeling spin-type systems in many dimensions [302, 294, 293, 264], a considerable impact on further developments of tensor numerical methods in scientific computing was due to the tensor train (TT) format [229, 226], introduced by Ivan Oseledets and Eugene Tyrtyshnikov in 2009. The advanced TT Toolbox developed in the group of Ivan Oseledets [232] provides powerful tools for the function-related multilinear algebra in higher dimensions. A closely related hierarchical Tucker tensor representation was introduced in 2009 by Wolfgang Hackbusch and Stefan Kühn, [115]. Both the tensor train and hierarchical Tucker tensor formats were established on the basis of earlier hierarchical dimension splitting concept in [161].

The quantics tensor train (QTT) approximation, introduced by Boris Khoromskij in 2009,[2] reduces the computational work on discretized multivariate functions to logarithmic complexity, $O(\log n)$ [165, 167]. It was initiated by the idea to test the TT-ranks of long function-related vectors of size 2^d (or q^d), reshaped to multidimensional hypercubes (the quantized image). Then, it was proven that the reshaped n-vectors resulting from discretization of classical functions like exponentials—plane waves or polynomials—have surprisingly small or even constant QTT ranks [167]. Thus, one comes to a paradoxical almost "mesh-free" grid-based calculations, when the size of fine grids used in solution of the multidimensional problems remains practically unrestricted. Numerical study on TT decomposition of $2^d \times 2^d$ matrices, was presented in [225]. A short description of the TT and QTT tensor formats is given in Chapter 4.

Notice that the tensor numerical methods are now recognized as a powerful tool for solving the multidimensional partial differential equations (PDEs) discretized by

2 The paper on QTT approximation was first published in September 2009 as the Preprint 55/2009 of the Max-Planck Institute for Mathematics in the Sciences in Leipzig.

traditional grid-based schemes. The tensor approach established a new branch of numerical analysis, providing efficient algorithms for solving the multidimensional integral-differential equations in $\mathbb{R}^d$, with linear or even logarithmic complexity scaling in the dimension [169, 170, 152, 171]. We also refer to literature surveys on tensor spaces and multilinear algebra [102, 110, 119]. It is still a challenging issue how to make tensors work for problems on complicated geometries.

The rank-structured Cholesky factorization of the two-electron integrals gave rise to a new approach for calculating the excitation energies of molecules introduced in [23] in the framework of the Bethe–Salpeter equation (BSE), [252]. The BSE is a widely used model for ab-initio estimation of the absorption spectra for molecules or surfaces of solids. This eigenvalue problem is a complicated task due to the size of the corresponding matrix that scales as $O(N_b^2)$, where N_b is the number of the basis functions. The new approach based on the diagonal plus low-rank representation of the generating matrix combines the iterative solution of a large rank-structured eigenvalue problem with the reduced basis approach for finding certain number of the smallest in modulo eigenvalues. An interesting solution was found by representing the exchange part of the BSE matrix by a small-size block, which has led to reduction of the overall ranks, and improved accuracy of excitation energies calculations. As a result, the complexity scaling for the numerical solution of the BSE eigenvalue problem is reduced from $O(N_b^6)$ to $O(N_b^2)$.

An efficient interpolation scheme to approximate the spectral density or density of states (DOS) for BSE problem was introduced in [27]. It is based on calculating the traces of parametric matrix resolvents at interpolation points by taking advantage of the block-diagonal plus low-rank matrix structure in the BSE Hamiltonian. It is also shown that a regularized or smoothed DOS discretized on a fine grid of size N can be accurately interpolated using its low rank adaptive QTT tensor representation. The QTT tensor format provides good interpolation properties for strictly oscillating functions with multiple gaps like DOS, and requires asymptotically much fewer (i. e., $O(\log N)$) functional calls compared with the full grid size N. This approach eliminates the computational difficulties of the traditional schemes by avoiding both the need of stochastic sampling, and interpolation by problem-independent functions such as polynomials.

In summary, the tensor-structured approach leads the way to the numerical solution of the Hartree–Fock equation, and subsequent MP2 corrections, based on the efficient grid-based calculation of the two-electron integrals, and the core Hamiltonian. The tensor-based ab-initio calculations also provide good prerequisites for post Hartree–Fock computations of the excitation energies, and of the optical spectra, by using moderate computing requirements. These new techniques in electronic structure calculations can be considered as a starting point for a thorough scientific development and investigation. We notice that the grid-based tensor methodology allows the efficient implementation of the most numerical schemes in a black-box way.

Another challenging problem in computational chemistry is the summation of a large number of the long-range (electrostatic) potentials distributed on finite 3D lattices with vacancies. The recent tensor-based method for summation of the long-range potentials on a finite $L \times L \times L$ lattice, introduced by the authors in [149, 148], provides the computational cost $O(L)$ contrary to $O(L^3 \log L)$ by the traditional Ewald-type methods [79]. It employs the low-rank canonical/Tucker tensor representations of a single Newton potential[3] by using the Laplace transform and sinc-approximation. The required precision is guaranteed by employing large 3D Cartesian grids for the representation of a single reference potential, $\frac{1}{\|x\|}$, $x \in \mathbb{R}^3$. The resulting rank of a tensor representing a sum of a large number, say millions of potentials on a finite three-dimensional lattice, remains the same as for the single reference potential. Indeed, the summation in the volume is reduced to a simple addition of entries in the skeleton vectors of the canonical tensor, thus producing the so-called "assembled" vectors of the collective potential on a lattice. The method remains efficient for multidimensional lattices with step-type geometries and in the presence of multiple defects [149]. The interaction energy of the electrostatic potentials on a lattice is then computed in sub-linear cost, $O(L^2)$ [152].

For multiparticle systems of general type, a novel range separated (RS) tensor format was proposed and analyzed by Peter Benner, Venera Khoromskaia, and Boris Khoromskij in [24]. It provides the rank-structured tensor approximation of highly non-regular functions with multiple singularities in $\mathbb{R}^3$, sampled on the fine $n \times n \times n$ grid. These can be the electrostatic potentials of a large atomic system like a biomolecule or the multidimensional scattered data modeled by radial basis functions, etc. The main advantage of the RS tensor format is that the partition into the long and short range parts is performed just by sorting of a small number of skeleton vectors in the low-rank canonical tensor representation of the generating kernel. It was proven [24] that the sum of long range contributions from all particles can be represented in a form of a low-rank canonical or Tucker tensor at the $O(n)$ storage cost, with a rank parameter only weakly (logarithmically), depending on the number of particles N. The basic tool here is again the RHOSVD algorithm. The representation complexity of the short range part is $O(N)$ with a small prefactor independent on the number of particles.

Note that the RS tensor format differs from the traditional tensor representations in multilinear algebra since it intrinsically applies to function related tensors. The RS format originates from the short-long range splitting within the low-rank representation of a singular potential, for example, the Newton potential $\frac{1}{\|x\|}$ or other radial basis functions in $\mathbb{R}^d$, $d \geq 3$. It essentially extends the applicability of tensor numerical methods in scientific computing.

3 The method works also for other types of multivariate radial basis functions $p(\|x\|)$.

In recent years, due to progress in computer science, the grid-based approaches and real-space numerical methods have attracted more attention in computational quantum chemistry since they allow, in principle, the efficient approximation to the physical entities of interest with a controllable precision [16, 122, 305, 35] and also offer the new techniques for calculation of the molecular excitation energies [23, 27]. On the one side, modern supercomputing facilities enable usage of the computational algorithms [86, 76], which could be considered in former times as unfeasible, and on the other side, there are completely new approaches like the rank-structured tensor numerical methods [187, 157, 147, 152], which suggest a new interpretation to usage of uniform grids in many dimensions. These topics have been recently addressed in a special issue of the PCCP journal devoted to real-space methods in quantum chemistry [85].

Finally, we would like to thank Wolfgang Hackbusch, our former director at the Max-Planck Institute (MPI) in Leipzig, for his continuous attention to our research and for interesting discussions. We are appreciative to Peter Benner, the director at the MPI in Magdeburg, for the fruitful collaboration in the recent years. We thank Felix Otto, the director at the MPI in Leipzig for his encouraging support of this book project. We thank Heinz-Juergen Flad and Reinhold Schneider for our fruitful collaboration which was a significant step in development of tensor methods in computational quantum chemistry. We would like to thank Andreas Savin and Dirk Andrae for valuable collaboration and discussions.

This research monograph on the tensor-structured numerical methods is an introduction to the modern field of numerical analysis with applications in computational quantum chemistry. Many of the presented new topics are based on the papers published by the authors in the recent decade during their research work at the Max-Planck Institute for Mathematics in the Sciences in Leipzig.

This book may be interesting for a wide audience of students and researchers working in computational chemistry and material science, as well as in numerical analysis, scientific computing, and multi-linear algebra. There is already a number of promising results in tensor numerical methods, and there is even more work to be done. We present some algorithms in MATLAB for a quick start in this new field. Numerous pictures can be helpful in explaining the main topics.

Leipzig, 2017

Venera Khoromskaia
Boris Khoromskij

2 Rank-structured formats for multidimensional tensors

2.1 Some notions from linear algebra

From the wide-ranging realm of linear algebra, we discuss here only a set of notions which are essential in describing the main topics of the tensor-structured numerical methods. We refer to a number of standard textbooks on linear algebra, for example, [98, 282, 202, 272].

2.1.1 Vectors and matrices

An ordered set of numbers is called a (column) vector,

$$\mathbf{u} = \begin{bmatrix} u_1 \\ u_2 \\ \vdots \\ u_n \end{bmatrix} \in \mathbb{R}^n.$$

To show that it is a column vector, one can write it explicitly, $\mathbf{u} \in \mathbb{R}^{n\times 1}$. Transpose of a column vector $\mathbf{u}^T$ is a row vector, $\mathbf{u}^T \in \mathbb{R}^{1\times n}$.

Products of column and row vectors give different results depending on the order of multiplication. Multiplying a row vector with a column vector, we obtain a scalar product of vectors, and the result is a number. That is, the scalar (or inner) product of two vectors $\mathbf{u}^T \in \mathbb{R}^{1\times n}$ and $\mathbf{v} \in \mathbb{R}^{n\times 1}$ is the real number given by

$$\mathbf{u}^T\mathbf{v} = [u_1 \quad u_2 \quad \cdots \quad u_n] \begin{bmatrix} v_1 \\ v_2 \\ \vdots \\ v_n \end{bmatrix} = u_1v_1 + u_2v_2 + \cdots + u_nv_n.$$

The scalar product of vectors is the main ingredient of matrix–matrix multiplications, and it is the prototype of contraction operation in tensor algebra.

The Euclidean norm of a vector (or its length) is $\|\mathbf{u}\| = \sqrt{\mathbf{u}^T\mathbf{u}}$.

Multiplying a column vector with a row vector is called a tensor product (or outer product), where vectors may be of different size, and the result is a matrix. Thus a tensor product increases the number of dimensions d of the resulting data array: multiplying two vectors (one-dimensional arrays), we obtain a matrix, i. e., a two-dimensional array corresponding to $d = 2$. Indeed, the tensor product of a column vector $\mathbf{u} \in \mathbb{R}^{m\times 1}$

https://doi.org/10.1515/9783110365832-002

and a row vector $\mathbf{v}^T \in \mathbb{R}^{1\times n}$ is a (rank-1) matrix of size $m \times n$,

$$A = \mathbf{u} \otimes \mathbf{v} = \mathbf{u}\mathbf{v}^T = \begin{bmatrix} u_1 \\ u_2 \\ \vdots \\ u_m \end{bmatrix} \cdot [v_1 \quad v_2 \quad \cdots \quad v_n] = \begin{bmatrix} u_1v_1 & u_1v_2 & \cdots & u_1v_n \\ u_2v_1 & u_2v_2 & \cdots & u_2v_n \\ \vdots & \vdots & \ddots & \vdots \\ u_mv_1 & u_mv_2 & \cdots & u_mv_n \end{bmatrix}. \tag{2.1}$$

A sum of two tensor products of vectors of corresponding lengths will be a matrix of rank-2, and so on. A matrix of rank-R is a sum of R terms, each being a tensor product of two vectors.

In general, an $m \times n$ matrix A is a rectangular array of real numbers arranged into m rows and n columns,

$$A = \begin{bmatrix} a_{11} & a_{12} & \cdots & a_{1n} \\ a_{21} & a_{22} & \cdots & a_{2n} \\ \vdots & \vdots & \ddots & \vdots \\ a_{m1} & a_{m2} & \cdots & a_{mn} \end{bmatrix},$$

where the number a_{ij} is called the entry of the matrix. A matrix A is an element of the linear vector space $\mathbb{R}^{m\times n}$ equipped with an Euclidean scalar product

$$\langle A, B \rangle = \sum_{i=1}^{m} \sum_{j=1}^{n} a_{ij} b_{ij} \tag{2.2}$$

and the Euclidean (Frobenius) norm of a matrix

$$\|A\| = \sqrt{\sum_{i=1}^{m} \sum_{j=1}^{n} a_{ij}^2}. \tag{2.3}$$

Computation of the Frobenius norm of a general matrix needs $O(nm)$ operations. But for the rank-1 matrices, $A = \mathbf{u} \otimes \mathbf{v} = \mathbf{u}\mathbf{v}^T$, the norm

$$\|A\| = \sqrt{\sum_{i=1}^{m} \sum_{j=1}^{n} (u_i v_j)^2} = \sqrt{\sum_{i=1}^{m} \sum_{j=1}^{n} u_i^2 v_j^2} = \sqrt{\sum_{i=1}^{m} u_i^2 \sum_{j=1}^{n} v_j^2} = \|\mathbf{u}\| \cdot \|\mathbf{v}\|$$

can be computed only in $O(m + n)$ operations.

Multiplication of vectors and matrices is based on scalar products of vectors and also depends on their positions with respect to each other. A row vector can be multiplied from the left with a matrix of a proper size (the size of the row vector should be equal to the column length of a matrix), and the result is a row vector of the length equal to the size of a matrix row.

A matrix can be multiplied by a column vector of the size coinciding with the size of matrix rows, thus resulting in a column vector of the size corresponding to matrix columns.

When multiplying the row vector and the matrix, the entries of the resulting vector are computed by scalar products of the vector and columns vectors of a matrix (their sizes should coincide),

$$\mathbf{u}^T A = [u_1 \quad u_2 \quad \cdots \quad u_m] \begin{bmatrix} a_{11} & a_{12} & \cdots & a_{1n} \\ a_{21} & a_{22} & \cdots & a_{2n} \\ \vdots & \vdots & \ddots & \vdots \\ a_{m1} & a_{m2} & \cdots & a_{mn} \end{bmatrix}$$
$$= [u_1 a_{11} + u_2 a_{21} + \cdots + u_m a_{m1} \quad \cdots \quad u_1 a_{1n} + u_2 a_{2n} + \cdots + u_m a_{mn}] \in \mathbb{R}^{1\times n}.$$

Multiplication of a matrix with a column vector is performed by the scalar products of every row of a matrix with this column vector, thus producing a column vector with the size equal to the number of matrix rows,

$$A\mathbf{v} = \begin{bmatrix} a_{11} & a_{12} & \cdots & a_{1n} \\ a_{21} & a_{22} & \cdots & a_{2n} \\ \vdots & \vdots & \ddots & \vdots \\ a_{m1} & a_{m2} & \cdots & a_{mn} \end{bmatrix} \begin{bmatrix} v_1 \\ v_2 \\ \vdots \\ v_n \end{bmatrix} = \begin{bmatrix} c_1 \\ c_2 \\ \vdots \\ c_m \end{bmatrix}, \tag{2.4}$$

where

$$\begin{bmatrix} c_1 \\ c_2 \\ \vdots \\ c_m \end{bmatrix} = \begin{bmatrix} a_{11}v_1 + a_{12}v_2 + \cdots + a_{1n}v_n \\ a_{21}v_1 + a_{22}v_2 + \cdots + a_{2n}v_n \\ \vdots \\ a_{m1}v_1 + a_{m2}v_2 + \cdots + a_{mn}v_n \end{bmatrix} \in \mathbb{R}^m. \tag{2.5}$$

In matrix–matrix multiplications, the above scheme applies column-wise to the right factor or row-wise to the left one.

2.1.2 Matrix–matrix multiplication. Change of basis

Matrix–matrix multiplication can be explained from the point of view of the matrix–vector multiplication (2.4). For two matrices $A \in \mathbb{R}^{m\times n}$ and $B \in \mathbb{R}^{n\times p}$, their product is a matrix

$$C = AB, \quad C \in \mathbb{R}^{m\times p}.$$

As we can see, each entry of the resultant matrix C is obtained as the scalar product of two n-vectors. Complexity of multiplication of two square matrices is $O(n^3)$. If one of matrices is given as an R-term sum of tensor products of vectors, then the complexity of multiplication is $O(Rn^2)$.

A general matrix A in a vector space $\mathbb{R}^{m\times n}$ is supposed to be presented in a basis of a vector space spanned by the unit vectors for rows and columns, $\mathbf{e}_i$ and $\mathbf{e}_j$ in the spaces $\mathbb{R}^{1\times n}$ and $\mathbb{R}^{m\times 1}$, respectively. Here $\mathbf{e}_i$ is a vector with all entries equal to zero except the entry with the number i. That is, an entry of the matrix is given by $a_{ij} = \langle \mathbf{e}_i, A\mathbf{e}_j \rangle$.

One can change a representation basis of a matrix, that is, a given matrix can be presented in a new basis, given by the set of column vectors of transformation matrices. This is done by multiplication (mapping) of the target matrix from both sides by the matrices representing the basis sets. If both matrices are invertible (in which case the size of the mapping matrices equals to the size of the original matrix), then this change of basis is reversible. If the mapping matrices are not square, then the mapping is not reversible.

A matrix $A \in \mathbb{R}^{n\times n}$ in a new basis given by columns of matrix $U \in \mathbb{R}^{n\times n}$ is represented in the factorized form

$$A_U = U^T A U, \quad A_U \in \mathbb{R}^{n\times n}. \tag{2.6}$$

The matrix A_U is the Galerkin projection of A to subspace spanned by columns of U, see Figure 2.1. If U is invertible, then one can get the original matrix A using matrices U^{-1} and U^{-T}, $A = U^{-T} A_U U^{-1}$. If $U \in \mathbb{R}^{n\times m}$, $m < n$, then $A_U \in \mathbb{R}^{m\times m}$, and the operation is not reversible.

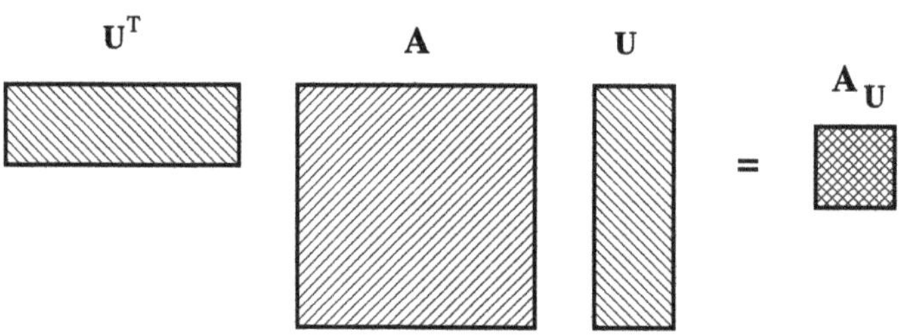

Figure 2.1: A matrix A in the basis of column vectors of the matrix U yields the matrix A_U.

The Kronecker product is an operation on matrices in linear algebra that maps matrices to a matrix. The Kronecker product of matrices $A \in \mathbb{R}^{m\times n}$ and $B \in \mathbb{R}^{p\times q}$ is defined by

$$A \otimes B = \begin{bmatrix} a_{11}B & a_{12}B & \cdots & a_{1n}B \\ a_{21}B & a_{22}B & \cdots & a_{2n}B \\ \vdots & \vdots & \ddots & \vdots \\ a_{m1}B & a_{m2}B & \cdots & a_{mn}B \end{bmatrix} \in \mathbb{R}^{mp\times nq}.$$

In general, this operation is not commutative: $A \otimes B \neq B \otimes A$.

2.1.3 Factorization of matrices

Factorized low-rank representation of matrices reduces the cost of linear algebra operations considerably. There is a number of methods for decomposing a matrix into a sum of tensor products of vectors. In the following, we discuss briefly the singular value decomposition (SVD), the QR-factorization, and the Cholesky decomposition. There is a large number of routines on various platforms that can be applied to calculate these decompositions. For convenience, we refer to the corresponding commands in Matlab.

(1) We start with an eigenvalue decomposition (EVD), which diagonalizes a matrix, that is, finds a basis in which the symmetric matrix becomes diagonal. The eigenvalue decomposition for the symmetric matrix requires the full set of eigenvectors and eigenvalues for the algebraic problem

$$Au = \lambda u.$$

A Matlab command

```
[V,D] = eig(A)
```

produces a diagonal matrix D of eigenvalues and a full orthogonal matrix V whose columns are the corresponding eigenvectors so that $A = VDV^T$ or $AV = VD$.

(2) When A is a rectangular or non-symmetric matrix, the singular value decomposition can be applied. In fact, the SVD suggests to solve the eigenvalue problems for auxiliary symmetric positive definite matrices $AA^T \in \mathbb{R}^{m\times m}$ and $A^TA \in \mathbb{R}^{n\times n}$.

Theorem 2.1. *Let $A \in \mathbb{R}^{m\times n}$, with $m \leq n$, for definiteness. Then there exist $U \in \mathbb{R}^{m\times m}$, $\Sigma \in \mathbb{R}^{m\times n}$, and $V \in \mathbb{R}^{n\times n}$ such that*

$$A = U\Sigma V^T, \tag{2.7}$$

where Σ is a diagonal $m \times n$ matrix whose diagonal entries σ_i, $i = 1, 2, \ldots, m$, are the ordered singular values of A, $\sigma_1 \geq \sigma_2 \geq \cdots \geq \sigma_m \geq 0$, and $U^TU = I_m$ and $V^TV = I_n$, with I_n denoting the $n \times n$ identity matrix.

The algebraic complexity of the SVD transform scales as $O(mn^2)$. We have
- $U \in \mathbb{R}^{m\times m}$ is a matrix composed of orthonormal vectors (columns);
- $V^T \in \mathbb{R}^{n\times n}$ is a matrix composed of orthonormal vectors (rows);
- $\Sigma \in \mathbb{R}^{m\times n}$ is a diagonal matrix of singular values.

Here, matrices U and V include the full set of left and right singular vectors, respectively,

$$U = [\mathbf{u}_1 \quad \mathbf{u}_2 \quad \cdots \quad \mathbf{u}_m], \quad \Sigma = \begin{bmatrix} \sigma_1 & 0 & \cdots & 0 & \cdots & 0 \\ 0 & \sigma_2 & \cdots & 0 & \cdots & 0 \\ \vdots & \vdots & \ddots & \ddots & & \\ 0 & 0 & \cdots & \sigma_m & \cdots & 0 \end{bmatrix}, \quad V^T = \begin{bmatrix} \mathbf{v}_1^T \\ \mathbf{v}_2^T \\ \vdots \\ \mathbf{v}_n^T \end{bmatrix}.$$

The rank of A, $r = \operatorname{rank}(A)$, does not exceed $r \leq m$. If the singular values are rapidly decaying, the SVD gives the possibility to approximate a rectangular matrix A in a factorized form.The best approximation of an arbitrary matrix $A \in \mathbb{R}^{m\times n}$ by a rank-r matrix A_r (say, in Frobenius norm, that is, $\|A\|_F^2 = \sum_{i,j} a_{ij}^2$) can be calculated by the truncated SVD as follows.

Let us consider (2.7) and set $\Sigma_r = \operatorname{diag}\{\sigma_1, \ldots \sigma_r, 0, \ldots, 0\}$. Then the best rank-$r$ approximation is given by

$$A_r := U\Sigma_r V^T = \sum_{i=1}^{r} \sigma_i \mathbf{u}_i \mathbf{v}_i^T,$$

where $\mathbf{u}_i$, $\mathbf{v}_i$ are the respective left and right singular vectors of A. The approximation error in the Frobenius norm is bounded by a sum of squares of discarded singular values:

$$\|A_r - A\|_F \leq \sqrt{\sum_{i=r+1}^{n} \sigma_i^2}. \tag{2.8}$$

The SVD Matlab routine

```
[U,S,V] = svd(A)
```

produces a diagonal matrix S of singular values and the orthogonal matrices U and V whose columns are the corresponding singular vectors so that $A = USV^T$.

(3) LU decomposition represents a matrix as a product of lower and upper triangular matrices. This decomposition is commonly used in the solution of linear systems of equations. For the LU decomposition

$$A = L^T U,$$

the corresponding Matlab routine reads

```
[L,U] = lu(A)
```

so that $A = L^T U$.

(4) The orthogonal-triangular decomposition is called QR factorization that is

$$A = QR.$$

The corresponding MATLAB routine of an m-by-n matrix A reads

```
[Q,R] = qr(A)
```

and produces an m-by-n upper triangular matrix R and an m-by-m unitary matrix Q so that $A = QR$, $Q^T Q = I$.

(5) Cholesky decomposition of a symmetric non-negative definite matrix A,

$$A = R^T R$$

produces an upper triangular matrix R satisfying the equation $R^T R = A$. The *chol* function in MATLAB,

```
R = chol(A)
```

assumes that A is symmetric and positive definite.

In most of applications one deals not with the exact rank of a matrix, but with the so-called ε-rank. This concerns with the rank optimization procedure based on the truncated SVD such that in (2.8) we estimate the ε-rank from the condition

$$\sum_{i=r+1}^{n} \sigma_i^2 \le \varepsilon^2.$$

In the following section we present some numerical experiments illustrating the adaptive low-rank approximation of a matrix.

2.1.4 Examples of rank decomposition for function related matrices

In Example 1 below, we present a simple MATLAB script for testing the decay of the singular values of several matrices. First, a two-dimensional Slater function, $e^{-\alpha\|x\|}$, is discretized in a square box $[-b/2, b/2]^2$ using the $n \times n$ 2D Cartesian grids with n = 65, 257, and n = 513, and the SVD is computed for the resulting matrices. Figure 2.2 (left) shows exponentially fast decay of singular values for all three matrices nearly independently on the matrix size.

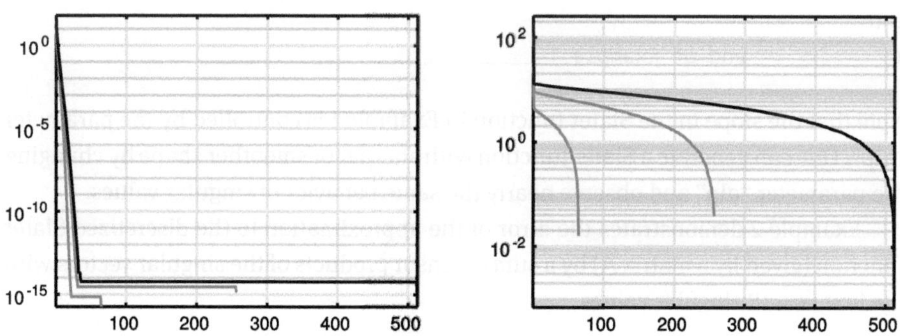

Figure 2.2: Decay of singular values for a matrix generated by a Slater function (left) and for a matrix containing random valued entries (right).

Next, we compose matrices of the same size, but using the generator of random numbers in the interval $[0, 1]$. The singular values of these matrices are shown in Figure 2.2 (right). They are not decaying fast, as was the case for the function related matrix.

```
%____Example 1____________________
clear; b=10; alp=1;
figure(1);
[Fun,sigmas,x,y] = Gener_Slat(65,b,alp); semilogy(sigmas);
hold on; grid on;
[~,sigmas,~,~]= Gener_Slat(257,b,alp); semilogy(sigmas,'r');
[~,sigmas,~,~]= Gener_Slat(513,b,alp); semilogy(sigmas,'black');
grid on; axis tight; set(gca,'fontsize',16);
hold off;
figure(2); mesh(x,y,Fun);
figure(3);
A1 = rand(65,65); [~,S1,~]= svd(A1); semilogy(diag(S1));
hold on; grid on;
A = rand(257,257); [~,S1,~]= svd(A); semilogy(diag(S1),'r');
A = rand(513,513); [~,S1,~]= svd(A); semilogy(diag(S1),'black');
grid on; axis tight; set(gca,'fontsize',16);
hold off;
%______________________
function [Fun1,sigmas,x,y]=Gener_Slat(n1,b,alpha1)
h1=b/(n1-1); x=-b/2:h1:b/2; y=-b/2:h1:b/2;
Fun1=zeros(n1,n1);
for i=1:n1
Fun1(i,:)= exp(-alpha1*sqrt(x(1,i)^2 +y(1,:).^2));
end
[~,S1,~]=svd(Fun1); sigmas=diag(S1);
end
%___________end of Example 1___________________________
```

Note that the slope in the Slater function in Example 1 is controlled by the parameter "alp". One can generate a Slater function with sharper or smoother shape by changing the parameter "alp" and observe nearly the same behavior of singular values.

Example 2 demonstrates the error of the approximation to the discretized Slater function (given by a matrix A) by a sum of tensor products of the singular vectors with the first $m = 18$ singular values.

$$A = \mathbf{u}_1\sigma_1\mathbf{v}_1^T + \cdots + \mathbf{u}_m\sigma_m\mathbf{v}_m^T = \sum_{i=1}^{m} \sigma_i\mathbf{u}_i\mathbf{v}^T.$$

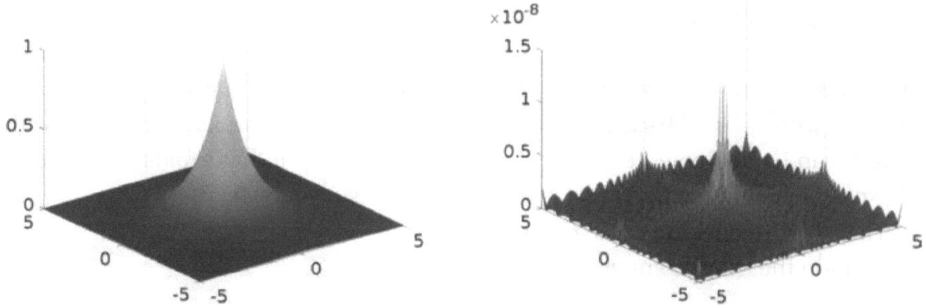

Figure 2.3: A matrix representing a discretized two-dimensional Slater function (left) and the error of its rank-18 factorized representation (right).

When running this program, figure (3) works as an "animation" picture, where one can distinctly notice diminishing of the error of the approximation within the loop while adding to approximation more summands with smaller singular values. Figure 2.3 (left) shows the original discretized function with cusp at zero, and Figure 2.3 (right) shows the final approximation error for rank $r = m = 18$.

```
%____Example 2__________________________________________
b=10; n=412;  h1=b/n;
x=-b/2:h1:b/2; [~,n1]=size(x);
y=-b/2:h1:b/2;
A1=zeros(n1,n1); alpha1=1;
for i=1:n1
A1(i,:)= exp(-alpha1*sqrt(x(1,i)^2 +y(1,:).^2));
end
figure(1); mesh(x,y,A1);
[U1 S1 V1]=svd(A1); sigmas=diag(S1);
figure(5); semilogy(sigmas);

r1=18; Ar1 = zeros(n1,n1);
for i=1:r1
   Ar = sigmas(i,1) *U1(:,i)*V1(:,i)';
   Ar1 =Ar1+Ar;
   figure(2); mesh(x,y,Ar1); drawnow;
   ER_A=abs(A1 - Ar1);
   figure(3); mesh(x,y,ER_A); drawnow;
end
%_______________end of Example 2_______________________
```

2.1.5 Reduced SVD of a rank-R matrix

Let us consider a rank-R matrix $M = AB^T \in \mathbb{R}^{n\times n}$, with the factor matrices $A \in \mathbb{R}^{n\times R}$ and $B \in \mathbb{R}^{n\times R}$, where $R \leq n$. We are interested in the best rank r approximation of M, with $r < R$. It can be implemented using the following algorithm that avoids the singular value decomposition of the target matrix M with possibly large n.

This algorithm includes the following steps:

(1) Perform the QR-decomposition of the side matrices,

$$A = Q_A R_A, \quad B = Q_B R_B,$$

with the unitary matrices $Q_A, Q_B \in \mathbb{R}^{n\times R}$ and the upper triangular matrices $R_A, R_B \in \mathbb{R}^{R\times R}$.

(2) Compute the SVD of the core matrix, $R_A R_B^T \in \mathbb{R}^{R\times R}$

$$R_A R_B^T = U\Sigma V^T,$$

with the diagonal matrix $\Sigma = \operatorname{diag}\{\sigma_1, \dots, \sigma_R\}$ and unitary matrices $U, V \in \mathbb{R}^{R\times R}$.

(3) Compute the best rank-r approximation of the core matrix, $U\Sigma V^T \approx U_r\Sigma_r V_r^T$, by extracting the submatrix $\Sigma_r = \operatorname{diag}\{\sigma_1, \dots, \sigma_r\}$ in Σ, and the first r columns $U_r, V_r \in \mathbb{R}^{R\times r}$ in the unitary matrices U and V, respectively.

(4) Finally, set the rank-r approximation $M_r = Q_A U_r \Sigma_r V_r^T Q_B^T$, where $Q_A U_r$ and $Q_B V_r$ are $n \times r$ unitary matrices.

The approximation error is bounded by $\sqrt{\sum_{i=r+1}^{R} \sigma_i^2}$. The complexity of the above algorithm scales linear in n, $O(nR^2) + O(R^3)$. In the case $R \ll n$, this reduces dramatically the cost $O(n^3)$ of the truncated SVD applied to the full-format $n \times n$ matrix M.

Low-rank approximation of matrices by using only partial information can be computed by heuristic adaptive cross approximation (ACA) methods developed in [286, 99, 287, 288, 15, 289, 14, 223], see also literature therein. Dynamical low-rank approximation of matrices has been analyzed in [191].

2.2 Introduction to multilinear algebra

The ideas and algorithms for the low-rank tensor approximation of multi-dimensional data by using the canonical (CP/CANDECOMP/PARAFAC) and Tucker tensor decompositions have been originally developed in chemometrics, psychometrics [49, 198, 50, 197], and then in signal processing and experimental data analysis [310, 192, 270, 59, 254, 62, 140]. The early papers on the polyadic (canonical) decomposition by F. L. Hitchcock in 1927 [131, 132] and the orthogonal Tucker tensor decomposition, introduced by L. R. Tucker in 1966 [284], gave rise to multilinear algebra of rank-structured tensors. Comprehensive surveys on multi-linear algebra with applications in principal component analysis and image, and signal processing, are presented in [54, 270, 1, 193]. Nowadays, there is an extensive research on tensor decomposition methods in computer science towards big data analysis, see for example [2, 55].

Notice that the tensor decompositions have been used in computer science mostly for quantitative analysis of correlations in the multidimensional data arrays obtained from experiments, without special requirements on the accuracy of decompositions. Usually these data arrays have been considered for a small number of dimensions (modes) and moderate mode sizes.

A mathematical approval and analysis of the Tucker tensor decomposition algorithm was presented in 2000 in the seminal works of L. De Lathauwer, B. De Moor, and J. Vandewalle on the higher-order singular value decomposition [61] and on the best rank-$(r_1, \ldots, r_d)$ orthogonal Tucker approximation of higher-order tensors [60]. The higher-order singular value decomposition (HOSVD) provides a generalization of the matrix singular value decomposition [98]. The main limitation of the Tucker algorithm from computer science [61, 193, 78] is the requirement to have a storage for full size tensor n^d, as well as the complexity of HOSVD, $O(n^{d+1})$, which includes the singular value decomposition of directional unfolding matrices. This makes HOSVD and the corresponding Tucker decomposition algorithm practically unfeasible for the problems in electronic structure calculations, and for solving multidimensional PDEs.

However, multilinear algebra with the Tucker tensor decomposition via HOSVD was one of the starting points for the tensor numerical methods. In what follows, we recall the tensor formats and main algorithms [60, 9] from multilinear algebra, where the techniques are being developed in view of the *arbitrary content* of the multidimensional arrays.

In forthcoming chapters, we shall see that the content of a tensor matters and that for function-related multidimensional arrays, even the standard multilinear algebra algorithms provide amazing results. One can further enhance the schemes by taking into account the predictions from approximation theory [111, 161] on the exponentially fast convergence of the Tucker/CP decompositions in tensor rank applied to the grid-based representation of the multidimensional functions and operators.

Let us start with the multilinear algebra approach to rank-structured tensor approximation, taking into account a general tensor content.

2.2.1 Full format dth order tensors

A tensor of order d is a multidimensional array over a d-tuple index set,

$$\mathbf{A} = [a_{i_1 \ldots i_d}] \in \mathbb{R}^{n_1 \times n_2 \times \cdots \times n_d}, \tag{2.9}$$

where $i_\ell \in I_\ell = \{1, \ldots, n_\ell\}$ is a set of indexes for each mode ℓ, $\ell = 1, \ldots, d$. A tensor $\mathbf{A}$ is an element of the linear vector space

$$\mathbb{V}_{\mathbf{n}} = \bigotimes_{\ell=1}^{d} \mathbb{R}^{n_\ell},$$

where $\mathbf{n} = (n_1, \dots, n_d)$, with the entry-wise addition

$$(\mathbf{A} + \mathbf{B})_{\mathbf{i}} = a_{\mathbf{i}} + b_{\mathbf{i}}$$

and the multiplication by a constant

$$(c\mathbf{A})_{\mathbf{i}} = ca_{\mathbf{i}} \quad (c \in \mathbb{R}).$$

The linear vector space $\mathbb{V}_{\mathbf{n}}$ of tensors is equipped with the Euclidean *scalar product* $\langle\cdot,\cdot\rangle : \mathbb{V}_{\mathbf{n}} \times \mathbb{V}_{\mathbf{n}} \to \mathbb{R}$ defined as

$$\langle \mathbf{A}, \mathbf{B} \rangle := \sum_{(i_1 \dots i_d) \in \mathcal{I}} a_{i_1 \dots i_d} b_{i_1 \dots i_d} \quad \text{for } \mathbf{A}, \mathbf{B} \in \mathbb{V}_{\mathbf{n}}, \tag{2.10}$$

where $\mathbf{i}$ is the d-tuple index set $\mathbf{i} = (i_1, \dots, i_d)$. The related norm

$$\|\mathbf{A}\|_F := \sqrt{\langle \mathbf{A}, \mathbf{A} \rangle}$$

is called the Frobenius norm, as for matrices.

Notice that a vector is an order-1 tensor, whereas a matrix is an order-2 tensor, so the Frobenius tensor norm coincides with the Euclidean norm of vectors and the Frobenius norm of matrices, respectively.

The number of entries in a tensor scales exponentially in the dimension,

$$N = \prod_{\ell=1}^{d} n_\ell, \quad \text{that is, for } n_\ell = n,\ N = n^d.$$

This phenomenon is often called the "curse of dimensionality". As a result, any multilinear operations with tensors given in full format (2.9), for example, computation of a scalar product, have an exponential complexity scaling $O(n^d)$.

Some multilinear algebraic operations with tensors of order d ($d \geq 3$), can be reduced to the standard linear algebra by *unfolding of a tensor* into a matrix.

Unfolding of a tensor $\mathbf{A} \in \mathbb{R}^{I_1 \times \dots \times I_d}$ along the ℓ-mode[1] arranges the ℓ-mode columns of a tensor to be the columns of the resulting unfolding matrix. Figure 2.4 shows unfolding of a 3D tensor. The unfolding of a tensor is a matrix whose columns are the respective fibers[2] along ℓ-mode, $\ell = 1, \dots, d$.

---

1 Note that in multilinear algebra the notion "mode" is often used for designating the particular dimension. ℓ-mode means the dimension number ℓ. Also, tensors of order d are called d-dimensional tensors.

2 Fibers along mode ℓ are generalization of notions of rows and columns for matrices.

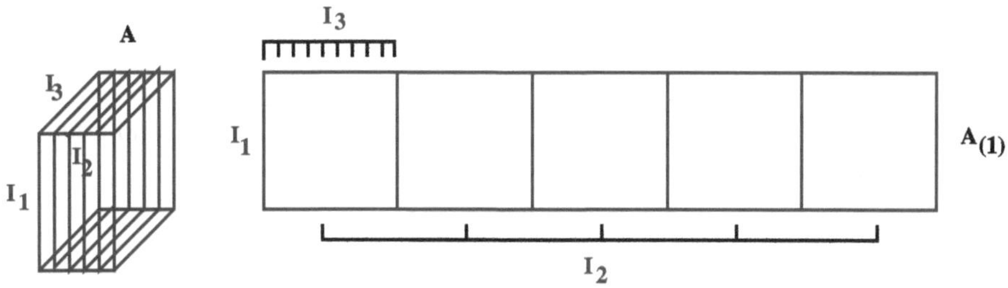

Figure 2.4: Unfolding of a 3D tensor for mode $\ell = 1$.

Specifically, the unfolding of a tensor along mode ℓ is a matrix of size $n_\ell \times (n_{\ell+1} \cdots n_d n_1 \cdots n_{\ell-1})$, further denoted by

$$A_{(\ell)} = [a_{ij}] \in \mathbb{R}^{n_\ell \times (n_{\ell+1} \cdots n_d n_1 \cdots n_{\ell-1})}, \tag{2.11}$$

whose columns are the respective fibers [193] of **A** along the ℓth mode such that the tensor entry $a_{i_1 i_2 \ldots i_d}$ is mapped into the matrix element $a_{i_\ell j}$ where the long index is given by

$$j = 1 + \sum_{k=1, k \neq \ell}^{d} (i_k - 1) J_k, \quad \text{with } J_k = \prod_{m=1, m \neq \ell}^{k-1} n_m.$$

In Matlab, the unfolding operation is performed by a simple `reshape` command. Then unfolding is done with respect to the given (first) variable. For unfolding along other variables, it is necessary to make a corresponding permutation (reordering of the dimensions) by using `permute` command as shown below.

The following script demonstrates the Matlab implementation for unfolding of a 3D tensor of size $5 \times 7 \times 10$:

```
n1=5; n2=7; n3=10;
A=rand(n1,n2,n3)  % generate a 3D tensor with random coefficients
% unfolding along mode 1:
A1= reshape(A,n1,n2*n3);
% or for unfolding along mode 3,
B=permute(A,[3,2,1]); A3= reshape(B,n3,n1*n2);
```

The size of the unfolding matrix $A_{(1)}$ in the above Matlab example is 5×70, whereas the size of the unfolding matrix $A_{(3)}$ is 10×35.

Another important tensor operation is the so-called *contracted product* of two tensors. This operation is similar to matrix–matrix multiplication with the difference that for matrices it is important that matrices are positioned in a proper way for multiplication over the compatible size; in the case of tensors, one explicitly determines the contraction mode ℓ. In the following, we frequently use the *tensor–matrix multiplication* along mode ℓ.

Definition 2.2 ([59]). *Contracted product*: Given a tensor $\mathbf{A} \in \mathbb{R}^{I_1\times\cdots\times I_d}$ and a matrix $M \in \mathbb{R}^{J_\ell\times I_\ell}$, we define the respective mode-ℓ tensor–matrix product by

$$\mathbf{B} = \mathbf{A} \times_\ell M \in \mathbb{R}^{I_1\times\cdots\times I_{\ell-1}\times J_\ell\times I_{\ell+1}\times\cdots\times I_d}, \tag{2.12}$$

where[3]

$$b_{i_1\dots i_{\ell-1}j_\ell i_{\ell+1}\dots i_d} = \sum_{i_\ell=1}^{n_\ell} a_{i_1\dots i_{\ell-1}i_\ell i_{\ell+1}\dots i_d} m_{j_\ell i_\ell}, \quad j_\ell \in J_\ell.$$

Contraction can be easily performed by using the following sequence of operations,
- matrix unfolding (reshaping) of the tensor;
- matrix–matrix multiplication over the corresponding dimension;
- reshaping of the resulting matrix back to a tensor.

The examples of contractions of a tensor with a matrix are shown in the subroutine for the Tucker decomposition algorithm presented in Section 3.

The tensor–matrix contracted product can be applied successively along several modes, and it can be shown to be commutative:

$$(\mathbf{A} \times_\ell M) \times_m P = (\mathbf{A} \times_m P) \times_\ell M = \mathbf{A} \times_\ell M \times_m P, \quad \ell \neq m.$$

We notice the convenience of notation of type $\times_\ell$ since it gives explicitly the mode number, which is subjected to contraction.

Figure 2.5 illustrates a sequence of contracted products of a tensor $\mathbf{A} \in \mathbb{R}^{n_1\times n_2\times n_3}$ with matrices $M_3 \in \mathbb{R}^{r_3\times n_3}$, $M_2 \in \mathbb{R}^{r_2\times n_2}$, and $M_1 \in \mathbb{R}^{r_1\times n_1}$ as follows:
- Contraction of tensor $\mathbf{A}$ in mode $\ell = 3$ with the matrix $M_3 \in \mathbb{R}^{r_3\times n_3}$ yields a tensor A_3 of size $n_1 \times n_2 \times r_3$,
$$\mathbf{A} \times_3 M_3 = \mathbf{A}_3 \in \mathbb{R}^{n_1\times n_2\times r_3}.$$
- Contraction of tensor $\mathbf{A}_3$ in mode $\ell = 2$ with the matrix $M_2 \in \mathbb{R}^{r_2\times n_2}$ yields a tensor $\mathbf{A}_2$ of size $n_1 \times r_2 \times r_3$,
$$\mathbf{A}_3 \times_2 M_2 = \mathbf{A}_2 \in \mathbb{R}^{n_1\times r_2\times r_3}.$$
- Contraction in the mode 1 with the matrix $M_1 \in \mathbb{R}^{r_1\times n_1}$ yields the tensor $\mathbf{A}_1 \in \mathbb{R}^{r_1\times r_2\times r_3}$,
$$\mathbf{A}_2 \times_1 M_1 = \mathbf{A}_1 \in \mathbb{R}^{r_1\times r_2\times r_3}.$$

As a result of all contractions, the original tensor $\mathbf{A}$ is represented in the basis given by matrices M_1, M_2, and M_3.

3 Here the sign "$\times_\ell$" denotes contraction over the mode number ℓ.

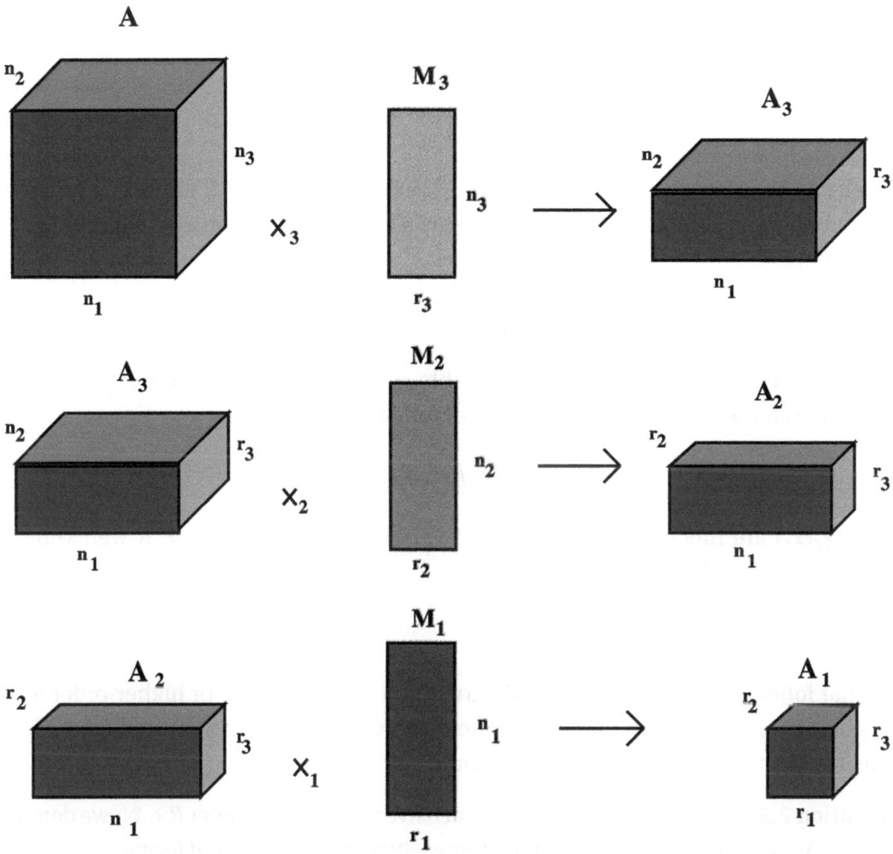

Figure 2.5: A sequence of contracted products in all three modes of a tensor **A** with the corresponding matrices M_3, M_2, and M_1.

2.2.2 Canonical and Tucker tensor formats

As we mentioned in the previous section, the number of entries in a full format tensor grows exponentially in dimension *d*.

To get rid of exponential scaling in the dimension, we are interested in the rank-structured representations of tensors. The simplest rank-structured tensor is constructed by tensor product of vectors $\mathbf{u}^{(\ell)} = \{u_{i_\ell}^{(\ell)}\}_{i_\ell=1}^{n} \in \mathbb{R}^{n_\ell}$, which forms the canonical rank-1 tensor

$$\mathbf{A} \equiv [u_{\mathbf{i}}]_{\mathbf{i}\in\mathcal{I}} = \mathbf{u}^{(1)} \otimes \cdots \otimes \mathbf{u}^{(d)} \in \mathbb{V}_{\mathbf{n}},$$

with entries given by $u_{\mathbf{i}} = u_{i_1}^{(1)} \cdots u_{i_d}^{(d)}$. Notice that a rank-1 tensor requires only *dn* numbers to store it (now linear scaling in the dimension). Moreover, the scalar product of two rank-1 tensors $\mathbf{U}, \mathbf{V} \in \mathbb{V}_{\mathbf{n}}$ is a product of *d* componentwise univariate scalar products

$$\langle \mathbf{U}, \mathbf{V} \rangle := \prod_{\ell=1}^{d} \langle \mathbf{u}^{(\ell)}, \mathbf{v}^{(\ell)} \rangle,$$

which can be calculated in $O(dn)$ operations. Recall that for $d = 2$, the tensor product of two vectors, $\mathbf{u} \in \mathbb{R}^I$ and $\mathbf{v} \in \mathbb{R}^J$, represents a rank-1 matrix (see also equation (2.1) in Section 2.1),

$$\mathbf{u} \otimes \mathbf{v} = \mathbf{u}\mathbf{v}^T \in \mathbb{R}^{I\times J}.$$

An analogue of a rank-1 tensor is a separable multivariate function $f(x_1, x_2, \ldots, x_d) \in \mathbb{R}^d$, which can be presented as a product of univariate functions,

$$f(x_1, x_2, \ldots, x_d) = f_1(x_1) f_2(x_2) \cdots f_d(x_d),$$

where $f_\ell(x_\ell)$ are functions over the single variable x_ℓ, $\ell = 1, 2, \ldots, d$. A well-known example is the multivariate Gaussian,

$$f(x_1, x_2, \ldots, x_d) = e^{-(\alpha_1 x_1^2 + \cdots + \alpha_d x_d^2)} = e^{-\alpha_1 x_1^2} \cdots e^{-\alpha_d x_d^2}.$$

In what follows, we consider the rank-structured representation of higher-order tensors based on sums of rank-1 tensors. There are two basic rank-structured tensor formats frequently used in multilinear algebra.

Definition 2.3. The canonical tensor format: Given a rank parameter $R \in \mathbb{N}$, we denote by $\mathcal{C}_R \subset \mathbb{V}_\mathbf{n}$ a set of tensors that can be represented in *the canonical format*,

$$\mathbf{U} = \sum_{\nu=1}^{R} \xi_\nu \mathbf{u}_\nu^{(1)} \otimes \cdots \otimes \mathbf{u}_\nu^{(d)}, \quad \xi_\nu \in \mathbb{R}, \tag{2.13}$$

with normalized vectors $\mathbf{u}_\nu^{(\ell)} \in \mathbb{V}_\ell$ ($\ell = 1, \ldots, d$). The minimal parameter R in representation (2.13) is called the rank (or canonical rank) of a tensor.

The storage for a tensor in the canonical format is $dRn \ll n^d$. Figure 2.6 visualizes a canonical tensor in 3D.

Note that an analogue of the canonical tensor is the representation of a multivariate function $f(x_1, x_2, \ldots, x_d) \in \mathbb{R}^d$ by a sum of R separable functions:

$$f(x_1, x_2, \ldots, x_d) = \sum_{k=1}^{R} f_{1,k}(x_1) f_{2,k}(x_2) \cdots f_{d,k}(x_d),$$

where $f_{\ell,k}(x_\ell)$ are functions over the single variable x_ℓ, $\ell = 1, 2, \ldots, d$.

Introducing the side matrices corresponding to representation (2.13),

$$U^{(\ell)} = [\mathbf{u}_1^{(\ell)} \quad \cdots \quad \mathbf{u}_R^{(\ell)}] \in \mathbb{R}^{n_\ell \times R},$$

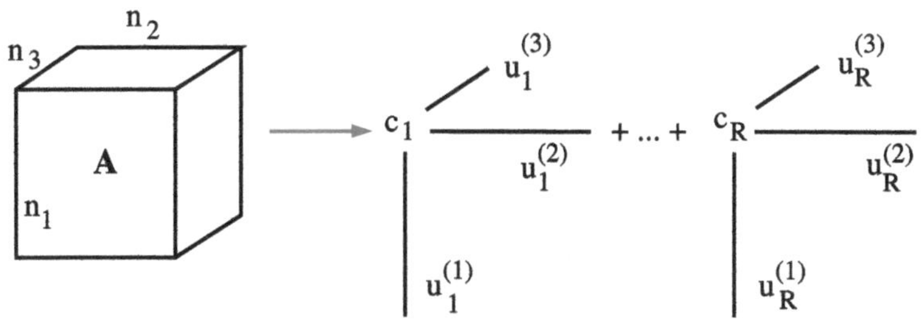

Figure 2.6: Visualizing canonical tensor decomposition of a third-order tensor.

and the diagonal tensor $\boldsymbol{\xi} := \operatorname{diag}\{\xi_1, \ldots, \xi_R\}$ such that $\xi_{\nu_1,\ldots,\nu_d} = 0$ except when $\nu_1 = \cdots = \nu_d$ with $\xi_{\nu,\ldots,\nu} = \xi_\nu$ ($\nu = 1, \ldots, R$), we obtain the equivalent contracted product representation of the rank-R canonical tensor,

$$\mathbf{U} = \boldsymbol{\xi} \times_1 U^{(1)} \times_2 U^{(2)} \cdots \times_d U^{(d)}. \tag{2.14}$$

The canonical tensor representation is helpful for the multilinear tensor operations. In Section 2.2.4 it is shown that the bilinear tensor operations with tensors in the rank-R canonical format have linear complexity

$$O\left(\sum_{\ell=1}^{d} n_\ell\right), \quad \text{or } O(dRn) \quad \text{if } n_\ell = n,$$

with respect to both the univariate grid size n of a tensor and the dimension parameter d. The *disadvantage* of this representation is lack of fast and stable algorithms for the best approximation of arbitrary tensors in the fixed-rank canonical format.

The other commonly used tensor format, introduced by Tucker [284], is the *rank-$(r_1, \ldots, r_d)$ Tucker tensor format.* It is based on a representation in subspaces

$$\mathbb{T}_\mathbf{r} := \bigotimes_{\ell=1}^{d} \mathbb{T}_\ell \quad \text{of } \mathbb{V}_\mathbf{n} \quad \text{for certain } \mathbb{T}_\ell \subset \mathbb{V}_\ell$$

with fixed dimension parameters $r_\ell := \dim \mathbb{T}_\ell \le n$.

Definition 2.4. The Tucker tensor format: For given rank parameter $\mathbf{r} = (r_1, \ldots, r_d)$, we denote by $\mathcal{T}_\mathbf{r}$ the subset of tensors in $\mathbb{V}_\mathbf{n}$ represented in the Tucker format

$$\mathbf{A} = \sum_{\nu_1=1}^{r_1} \cdots \sum_{\nu_d=1}^{r_d} \beta_{\nu_1,\ldots,\nu_d} \mathbf{v}_{\nu_1}^{(1)} \otimes \cdots \otimes \mathbf{v}_{\nu_d}^{(d)} \in \mathbb{V}_\mathbf{n}, \tag{2.15}$$

with some vectors $\mathbf{v}_{\nu_\ell}^{(\ell)} \in \mathbb{V}_\ell = \mathbb{R}^{I_\ell}$ ($1 \le \nu_\ell \le r_\ell$), which form an orthonormal basis of r_ℓ-dimensional subspaces $\mathbb{T}_\ell = \operatorname{span}\{\mathbf{v}_\nu^{(\ell)}\}_{\nu=1}^{r_\ell}$ ($\ell = 1, \ldots, d$).

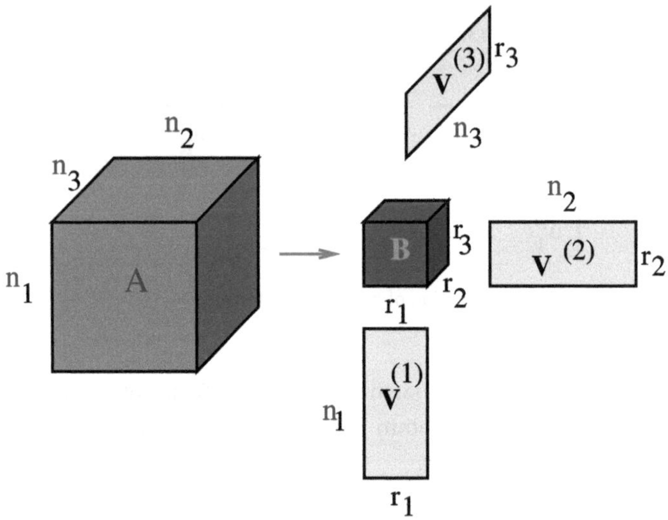

Figure 2.7: Visualizing the Tucker decomposition for a 3D tensor.

The coefficients tensor $\boldsymbol{\beta} = [\beta_{\nu_1,\dots,\nu_d}]$, which is an element of a tensor space

$$\mathbb{B}_{\mathbf{r}} = \mathbb{R}^{r_1\times\cdots\times r_d}, \tag{2.16}$$

is called the *core tensor*. We call the parameter $r = \min_\ell\{r_\ell\}$ the minimal Tucker rank. Figure 2.7 visualizes a Tucker tensor decomposition of a tensor $\mathbf{A}$ in $\mathbb{R}^{n_1\times n_2\times n_3}$.

Note that for problems in signal processing or principal component analysis, some of the mode sizes of the core tensor, i. e., the Tucker rank r_ℓ, may be close to the original tensor size n_ℓ in the corresponding mode.

Introducing the (orthogonal) side matrices $V^{(\ell)} = [\mathbf{v}_1^{(\ell)} \cdots \mathbf{v}_{r_\ell}^{(\ell)}]$ such that ${V^{(\ell)}}^T V^{(\ell)} = I_{r_\ell\times r_\ell}$, we then use a tensor-by-matrix contracted product notation to represent the Tucker decomposition of $\mathbf{A}_{(\mathbf{r})} \in \mathcal{T}_{\mathbf{r}}$ in a compact form,

$$\mathbf{A}_{(\mathbf{r})} = \boldsymbol{\beta} \times_1 V^{(1)} \times_2 V^{(2)} \cdots \times_d V^{(d)}. \tag{2.17}$$

Remark 2.5. Notice that the representation (2.17) is not unique, since the tensor $A_{(\mathbf{r})}$ is invariant under directional rotations. In fact, for any set of orthogonal $r_\ell \times r_\ell$ matrices Y_ℓ ($\ell = 1,\dots,d$), we have the equivalent representation

$$\mathbf{A}_{(\mathbf{r})} = \widehat{\boldsymbol{\beta}} \times_1 \widehat{V}^{(1)} \times_2 \widehat{V}^{(2)} \cdots \times_d \widehat{V}^{(d)},$$

with

$$\widehat{\boldsymbol{\beta}} = \boldsymbol{\beta} \times_1 Y_1 \times_2 Y_2 \cdots \times_d Y_d, \quad \widehat{V}^{(\ell)} = V^{(\ell)} Y_\ell^T, \quad \ell = 1,\dots,d.$$

Remark 2.6. If the subspaces $\mathbb{T}_\ell = \operatorname{span}\{v_\nu^{(\ell)}\}_{\nu=1}^{r_\ell} \subset \mathbb{V}_\ell$ are fixed, then the approximation $\mathbf{A}_{(\mathbf{r})} \in \mathcal{T}_\mathbf{r}$ of a given tensor $\mathbf{A} \in \mathbb{V}_\mathbf{n}$ is reduced to the orthogonal projection of $\mathbf{A}$ onto the particular linear space $\mathbb{T}_\mathbf{r} = \bigotimes_{\ell=1}^d \mathbb{T}_\ell \subset \mathcal{T}_{\mathbf{r},\mathbf{n}}$, that is,

$$\begin{aligned}\mathbf{A}_{(\mathbf{r})} &= \sum_{\nu_1,\dots,\nu_d=1}^{\mathbf{r}} \langle v_{\nu_1}^{(1)} \otimes \cdots \otimes v_{\nu_d}^{(d)}, \mathbf{A}\rangle v_{\nu_1}^{(1)} \otimes \cdots \otimes v_{\nu_d}^{(d)} \\ &= (\mathbf{A} \times_1 V^{(1)^T} \times_2 \cdots \times_d V^{(d)^T}) \times_1 V^{(1)} \times_2 \dots \times_d V^{(d)}.\end{aligned}$$

This property plays an important role in the computation of the best orthogonal Tucker approximation, where the "optimal" subspaces $\mathbb{T}_\ell$ are recalculated within a nonlinear iteration process.

In the following, to simplify the discussion of complexity issues, we assume that $r_\ell = r$ ($\ell = 1,\dots,d$). The storage requirements for the Tucker decomposition are estimated by $r^d + drn$, where usually r is noticeably smaller than n. In turn, the maximal canonical rank of the Tucker representation is bounded by r^{d-1} (see Remark 3.17).

2.2.3 Tucker tensor decomposition for full format tensors

The Tucker approximation of dth-order tensors is the higher-order extensions of the best rank-r matrix approximation in linear algebra, based on the truncated SVD.

Since the subset of Tucker tensors $\mathcal{T}_{\mathbf{r},\mathbf{n}}$ is not linear space the best Tucker approximation problem leads to the challenging *nonlinear minimization* problem

$$\mathbf{A}_0 \in \mathcal{S}_0 \subset \mathbb{V}_\mathbf{n}: \quad f(\mathbf{A}) := \|\mathbf{A}_0 - \mathbf{A}\|^2 \to \min \tag{2.18}$$

over all tensors $\mathbf{A} \in \mathcal{S} = \{\mathcal{T}_{\mathbf{r},\mathbf{n}}\}$. Here, $\mathcal{S}_0$ might be the set of Tucker or CP tensors with the rank parameter substantially larger than r.

As the basic nonlinear approximation scheme, we consider the best orthogonal rank-$(r_1,\dots,r_d)$ Tucker approximation for the full format input, corresponding to the choice $\mathcal{S}_0 = \mathcal{T}_{\mathbf{r},\mathbf{n}}$. Tensors $\mathbf{A} \in \mathcal{T}_\mathbf{r}$, are parameterized as in (2.17), with the orthogonality constraints

$$V^{(\ell)} \in \mathcal{V}_{n_\ell,r_\ell} \quad (\ell = 1,\dots,d),$$

where

$$\mathcal{V}_{n,r} := \{Y \in \mathbb{R}^{n\times r} : Y^T Y = I_{r\times r} \in \mathbb{R}^{r\times r}\} \tag{2.19}$$

is the so-called Stiefel manifold of $n \times r$ orthogonal matrices. This minimization problem on the product of Stiefel manifolds was first addressed in [197].

In the following, we denote by $\mathcal{G}_\ell$ the Grassman manifold, that is, the factor space of the Stiefel manifold $\mathcal{V}_{n_\ell,r_\ell}$ ($\ell = 1,\dots,d$) in (2.19), with respect to all possible rotations. See Remark 2.5.

The key point for the efficient solution of the minimization problem (2.18) over tensor manifold $\mathcal{S} = \mathcal{T}_{\mathbf{r},\mathbf{n}}$ is its equivalent reformulation as the *dual maximization problem* [60],

$$[Z^{(1)},\dots,Z^{(d)}] = \operatorname{argmax}\|[\langle v_{\nu_1}^{(1)} \otimes \cdots \otimes v_{\nu_d}^{(d)}, \mathbf{A}\rangle]_{\mathbf{v}=\mathbf{1}}^{\mathbf{r}}\|_{\mathbb{B}_\mathbf{r}}^2 \tag{2.20}$$

over the set of side-matrices $V^{(\ell)} = [v_1^{(\ell)} \cdots v_{r_\ell}^{(\ell)}]$ in the Stiefel manifold $\mathcal{V}_{n_\ell,r_\ell}$, as in (2.19).

The following lemma by De Lathauwer, De Moor, and Vandewalle [60] shows that the minimization of the original quadratic functional is reduced to the dual maximization problem, thus eliminating the core tensor $\boldsymbol{\beta}$ from the optimization process.

Lemma 2.7 ([60]). *For given $\mathbf{A}_0 \in \mathbb{R}^{I_1\times\cdots\times I_d}$, the minimization problem* (2.18) *on $\mathcal{T}_\mathbf{r}$ is equivalent to the dual maximization problem*

$$g(V^{(1)},\dots,V^{(d)}) := \|\mathbf{A}_0 \times_1 V^{(1)^T} \times_2 \cdots \times_d V^{(d)^T}\|^2 \to \max \tag{2.21}$$

over a set $V^{(\ell)} \in \mathbb{R}^{n_\ell\times r_\ell}$ from the Grassman manifold, i. e., $V^{(\ell)} \in \mathcal{G}_\ell$ ($\ell = 1,\dots,d$). For given maximizing matrices $Z^{(m)}$ ($m = 1,\dots,d$), the core tensor $\boldsymbol{\beta}$ minimizing (2.18) *is represented by*

$$\boldsymbol{\beta} = \mathbf{A}_0 \times_1 Z^{(1)^T} \times_2 \cdots \times_d Z^{(d)^T} \in \mathbb{R}^{r_1\times\cdots\times r_d}. \tag{2.22}$$

In view of Remark 2.5, the rotational non-uniqueness of the maximizer in (2.20) can be avoided if one solves this maximization problem in the Grassmann manifold. The dual maximization problem (2.21) posed on the compact manifold can be proven to have at least one global maximum (see [161, 78]). For the size consistency of the arising tensors, we require the natural compatibility conditions

$$r_\ell \le \bar{r}_\ell := r_1\cdots r_{\ell-1}r_{\ell+1}\cdots r_d, \quad \ell = 1,\dots,d. \tag{2.23}$$

The best (nonlinear) Tucker approximation by solving the dual maximization problem (2.20) is usually solved numerically by the ALS iteration, combined with the so-called higher order SVD (HOSVD), introduced by De Lathauwer et al. in [61] and [60], respectively. We recall the theorem from [61].

Theorem 2.8 (*d*th-order SVD, HOSVD, [61]). *Every real (complex) $n_1\times n_2\times\cdots\times n_d$-tensor $\mathbf{A}$ can be written as the product*

$$\mathbf{A} = \mathcal{S} \times_1 V^{(1)} \times_2 V^{(2)} \cdots \times_d V^{(d)}, \tag{2.24}$$

in which
1. *$V^{(\ell)} = [V_1^{(\ell)}\, V_2^{(\ell)} \cdots V_{n_\ell}^{(\ell)}]$ is a unitary $n_\ell \times n_\ell$-matrix;*
2. *$\mathcal{S}$ is a complex $n_1 \times n_2 \times \cdots \times n_d$-tensor of which the subtensors $\mathcal{S}_{i_\ell=\alpha}$, obtained by fixing the ℓth index to α, have the following properties:*

(i) *all-orthogonality: two subtensors $\mathcal{S}_{i_\ell=\alpha}$ and $\mathcal{S}_{i_\ell=\beta}$ are orthogonal for all possible values of ℓ, α, and β subject to $\alpha \neq \beta$:*

$$\langle \mathcal{S}_{i_\ell=\alpha}, \mathcal{S}_{i_\ell=\beta} \rangle = 0 \quad \text{when } \alpha \neq \beta$$

(ii) *ordering: $\|\mathcal{S}_{i_\ell=1}\| \geq \|\mathcal{S}_{i_\ell=2}\| \geq \cdots \geq \|\mathcal{S}_{i_\ell=n_\ell}\| \geq 0$ for all positive values of ℓ.*

The Frobenius norms $\|\mathcal{S}_{i_\ell=i}\|$, symbolized by $\sigma_i^{(\ell)}$, are ℓ-mode singular values of $A_{(\ell)}$ and the vector $U_i^{(\ell)}$ is the ith ℓ-mode left singular vector of $A_{(\ell)}$.

Another theorem from [61] proves the error bound for the truncated HOSVD. It states that for the HOSVD of **A**, as given in Theorem 2.8 with the ℓ-mode rank of **A**, $\operatorname{rank}(A_{(\ell)}) = R_\ell$ ($\ell = 1, \ldots, d$), the tensor $\tilde{\mathbf{A}}$ obtained by discarding the smallest ℓ-mode singular values $\sigma_{r_\ell+1}^{(\ell)}, \sigma_{r_\ell+2}^{(\ell)}, \ldots, \sigma_{R_\ell}^{(\ell)}$ for given values of r_ℓ ($\ell = 1, \ldots, d$), (i. e., setting the corresponding parts of $\mathcal{S}$ equal to zero) provides the following approximation error

$$\|\mathbf{A} - \tilde{\mathbf{A}}\|^2 \leq \sum_{i_1=r_1+1}^{R_1} \sigma_{i_1}^{(1)^2} + \sum_{i_2=r_2+1}^{R_2} \sigma_{i_2}^{(2)^2} + \cdots + \sum_{i_d=r_d+1}^{R_d} \sigma_{i_d}^{(d)^2}.$$

We refer to the original papers [60, 61] on the detailed discussions of above theory which was an important step for applying tensor decompositions in scientific computing.

Figure 2.8 illustrates the statements of above theorems by an example of a cubic third-order tensor **A**. It shows the core tensor $\mathcal{S}$ and the matrices $V^{(1)}$, $V^{(2)}$, and $V^{(3)}$ from (2.24). The size of the core tensor $\mathcal{S}$ is the same as the size of the original tensor **A**, except that it is now represented in the orthogonal basis, given by matrices V_1, V_2, and, V_3. The core tensor of the truncated HOSVD is colored by yellow.

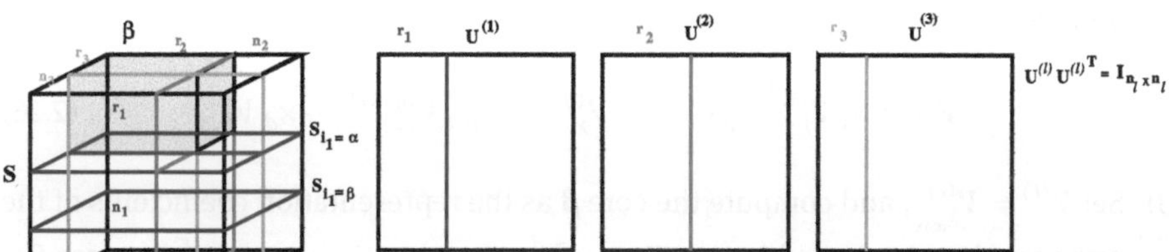

Figure 2.8: Illustration to Theorem 2.8.

The orthogonality of subtensors $\mathcal{S}_{i_\ell=\alpha}$ and $\mathcal{S}_{i_\ell=\beta}$ follows from the fact that these matrices originate from reshaping of the orthogonal vectors in the matrix $(W^{(\ell)})^T$ of the SVD of the respective matrix unfolding of **A** for modes ℓ, $\ell = 1, 2, 3$,

$$A_{(\ell)} = V^{(\ell)} \Sigma^{(\ell)} \left(W^{(\ell)}\right)^T.$$

Note that matrices $V^{(\ell)}$, $\ell = 1,2,3$, obtained as a result of the singular value decomposition of the corresponding matrix unfolding of $\mathbf{A}$ for modes ℓ, $\ell = 1,2,3$, initially have the same size as the original tensor. Based on the truncated HOSVD, their size reduction can be performed taking into account the decay of singular values in $\Sigma^{(\ell)}$ and then discarding the smallest singular values subject to some threshold $\varepsilon > 0$. It corresponds to the choice of first $r^{(\ell)}$ vectors in $V^{(\ell)}$, as shown in Figure 2.8. The sizes $r^{(\ell)}$ may be different, depending on the chosen threshold and the structure of the initial tensor $\mathbf{A}$.

Next, we recall the Tucker decomposition algorithm for full format tensors, introduced by De Lathauwer et al. in [60]. It is based on the initial guess by HOSVD and the alternating least square (ALS) iteration.

Tucker decomposition algorithm for full format tensor ($\mathbb{V}_{\mathbf{n}} \to \mathcal{T}_{\mathbf{r},\mathbf{n}}$)

Given the input tensor $\mathbf{A} \in \mathbb{V}_{\mathbf{n}}$, the Tucker rank $\mathbf{r}$, and the maximum number of ALS iterations $k_{\max} \geq 1$.

(1) Compute the truncated HOSVD of $\mathbf{A}$ to obtain an initial guess $V_0^{(\ell)} \in \mathbb{R}^{n_\ell \times r_\ell}$ for the ℓ-mode side-matrices $V^{(\ell)}$ ($\ell = 1,\dots,d$) ("truncated" SVD applied to each matrix unfolding $A_{(\ell)}$). Figure 2.9 illustrates this step of the algorithm for a 3D tensor ($\ell = 1,2,3$).

(2) For $k = 1 : k_{\max}$ perform ALS iteration:
for each $q = 1,\dots,d$, and with fixed side-matrices $V_{k-1}^{(\ell)} \in \mathbb{R}^{n_\ell \times r_\ell}$, $\ell \neq q$, the ALS iteration optimizes the q-mode matrix $V_k^{(q)}$ via computing the dominating r_q-dimensional subspace (truncated SVD) for the respective matrix unfolding

$$B_{(q)} \in \mathbb{R}^{n_q \times \bar{r}_q}, \quad \bar{r}_q = r_1 \cdots r_{q-1} r_{q+1} \cdots r_d = O(r^{d-1}), \tag{2.25}$$

corresponding to the tensor obtained by the "single-hole" contracted product in q-mode:

$$\mathbf{B} = \mathbf{A} \times_1 V_k^{(1)^T} \times_2 \cdots \times_{q-1} V_k^{(q-1)^T} \times_{q+1} V_{k-1}^{(q+1)^T} \cdots \times_d V_{k-1}^{(d)^T}. \tag{2.26}$$

(3) Set $V^{(\ell)} = V_{k_{\max}}^{(\ell)}$, and compute the core $\boldsymbol{\beta}$ as the representation coefficients of the orthogonal projection of $\mathbf{A}$ onto $\mathbb{T}_{\mathbf{n}} = \bigotimes_{\ell=1}^{d} \mathbb{T}_\ell$, with $\mathbb{T}_\ell = \operatorname{span}\{\mathbf{v}_\nu^{(\ell)}\}_{\nu=1}^{r_\ell}$ (see Remark 2.6),

$$\boldsymbol{\beta} = \mathbf{A} \times_1 V^{(1)^T} \times_2 \cdots \times_d V^{(d)^T} \in \mathbb{B}_{\mathbf{r}}.$$

The computational costs are the following: (1) the HOSVD cost is $W = O(dn^{d+1})$; (2) the costs of ALS procedure: each iteration has the cost $O(dr^{d-1} n \min\{r^{d-1}, n\} + dn^d r)$, which represents the expense of SVDs and the computation of matrix unfoldings $B_{(q)}$. The last step, i. e., computation of the core tensor, has the cost $O(r^d n)$.

Let us comment the Tucker decomposition algorithm from [61] for a third-order tensor $\mathbf{A} \in \mathbb{R}^{n_1 \times n_2 \times n_3}$ by using Figures 2.9, 2.10, and 2.11. Set the Tucker ranks as r_1, r_2, and r_3, respectively.

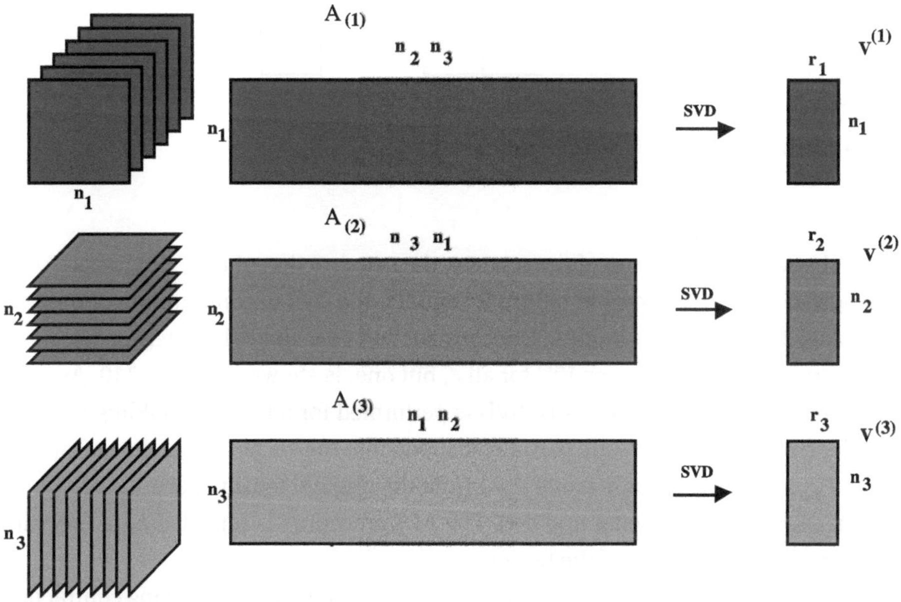

Figure 2.9: The initial guess for the Tucker decomposition is computed by HOSVD via SVD of the ℓ-mode unfolding matrices, $\ell = 1, 2, 3$.

(1) At the first step, the truncated SVD is computed for three unfolding matrices $A_{(1)} \in \mathbb{R}^{n1 \times n2\, n3}$, $A_{(2)} \in \mathbb{R}^{n2 \times n3\, n1}$, and $A_{(3)} \in \mathbb{R}^{n3 \times n1\, n2}$, as shown in Figure 2.9. Every SVD needs $O(n^4)$ computer operations if we simplify $n_\ell = n$. Thus, it is the most storage/time consuming part of the algorithm.

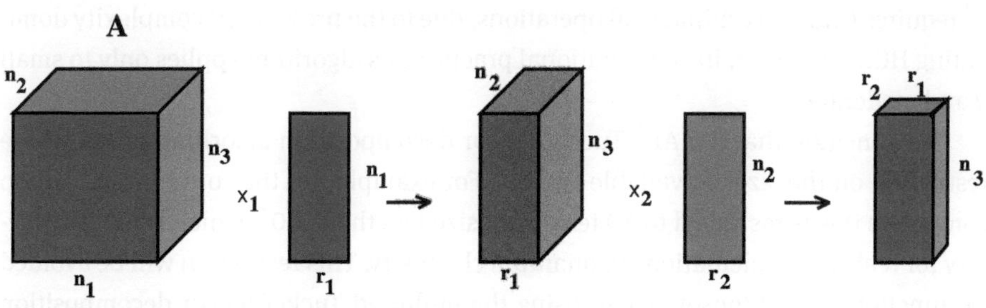

Figure 2.10: Construction of a "single-hole" tensor by contractions.

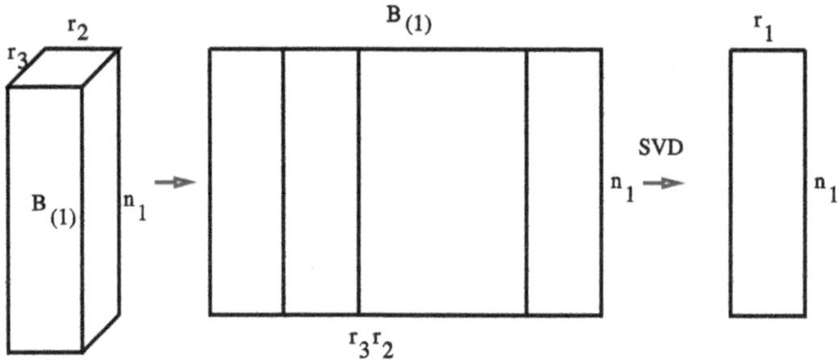

Figure 2.11: Unfolding of a "single-hole" tensor.

(2) At the ALS iteration step of the scheme, the construction of the "single-hole" tensors, given by (2.26), allows to reduce essentially the cost of computation of the best mappings for the Tucker modes. The construction of a single-hole tensor for $\ell = 3$ by contractions with matrices $V^{(\ell)}$ for all ℓ, but one, is shown in Figure 2.10. As illustrated in Figure 2.11, the truncated SVD is performed for a tensor unfolding of much smaller size, since it is already partially mapped into the Tucker projection subspaces $\mathbb{T}_\ell \in \mathcal{T}_{\mathbf{r},\mathbf{n}}$, except the single mode $\ell = 1$ from the original tensor space $\mathbb{V}_\mathbf{n}$, for which the mapping matrix is being updated. The ALS procedure is repeated $k_{\max}$ times for every mode ℓ, $\ell = 1, \ldots, d$, of the tensor.

(3) At the last step of the algorithm, the core tensor is computed using contraction of the original tensor with updated side matrices $V_{r_\ell}^{(\ell)}$, $\ell = 1, \ldots, d$.

With fixed $k_{\max}$, the overall complexity of the algorithm for $d = 3$, $n_\ell = n$, and $r_\ell = r$, $\ell = 1, 2, 3$, is estimated by

$$W_{F\to T} = O(n^4 + n^3 r + n^2 r^2 + n^3 r) = O(n^4),$$

where different summands denote the cost of initial HOSVD of $\mathbf{A}$, computation of unfolding matrices $B_{(q)}$, related SVDs, and computation of the core tensor, respectively.

Notice that the Tucker model applied to the general fully populated tensor of size n^d requires $O(dn^{d+1})$ arithmetical operations, due to the presence of complexity dominating HOSVD. Hence, in computational practice this algorithm applies only to small d and moderate n.

We conclude that the ALS Tucker tensor decomposition algorithm poses severe restriction on the size of available tensors. For example, for the conventional laptop computers this is restricted to 3D tensors of size less than 200^3, which is not satisfactory for real space calculations in quantum chemistry. This restriction will be avoided for function-related tensors when using the multigrid Tucker tensor decomposition discussed in Section 3.1.

2.2.4 Basic bilinear operations with rank-structured tensors

We have observed that the canonical and Tucker tensor formats provide representations by using sums of tensor product of vectors. Hence, the standard operations with tensors are reduced to one-dimensional operations in corresponding dimensions, exactly in the same way, as it is done for rank-structured matrices (see Section 2.1).

The main point here is the rank of the tensor, that is, the number of tensor product summands. However, the separation rank parameter is hard to be controlled for tensors containing unstructured or experimental data. Due to addition/multiplication of ranks in every rank-structured operation, after several steps we may have a "curse of ranks" instead of curse of dimensions. However, it will be shown in Chapter 3 that for *function related tensors* things become different, due to their intrinsically low ε-ranks. Moreover, for tensors approximating functions and operators, it is possible to provide means for reducing their ranks after a sequence of tensor operations.

For the sake of clarity (and without loss of generality), in this section we assume that $r = r_\ell$, $n = n_\ell$ ($\ell = 1, \ldots, d$). If there is no confusion, the index $\mathbf{n}$ can be skipped. We denote by W the complexity of various tensor operations (say, $W_{\langle\cdot,\cdot\rangle}$) or the related storage requirements (say, $W_{\mathrm{st}(\boldsymbol{\beta})}$). We estimate the storage demands W_{st} and complexity of the following standard tensor-product operations: the scalar product, the Hadamard (component-wise) product, and the convolution transform. We consider the multilinear operations in $\mathcal{T}_{\mathbf{r},\mathbf{n}}$ and $\mathcal{C}_{R,\mathbf{n}}$ tensor classes.

The Tucker model requires

$$W_{\mathrm{st},T} = drn + r^d \tag{2.27}$$

storage to represent a tensor. The storage for the rank-R canonical tensor scales linearly in d,

$$W_{\mathrm{st},C} = dRn. \tag{2.28}$$

Setting $R = \alpha r$ with $\alpha \geq 1$, we can specify the range of parameters where the Tucker model is less storage consuming compared with the canonical one

$$r^{d-1} \leq d(\alpha - 1)n \quad (\text{for } d = 3 : \; r^2 \leq 3(\alpha - 1)n).$$

In general, the numerical Tucker decomposition leads to a fully populated core tensor that is represented by r^d nonzero elements. However, the special data structure of the Tucker core can be imposed, which reduces the complexity of the corresponding tensor operations (cf. [161]). In particular, for the mixed (two-level) Tucker-canonical decomposition, the core tensor is represented in the rank-R CP format (see Definition 3.15), so that storage demands scale linearly in d,

$$W_{\mathrm{st,TC}} = dr(n + R).$$

Bilinear operations in the Tucker format
For given tensors $\mathbf{A}_1 \in \mathcal{T}_{\mathbf{r}_1}$, $\mathbf{A}_2 \in \mathcal{T}_{\mathbf{r}_2}$ represented in the form (2.15), i. e.,

$$\begin{aligned} \mathbf{A}_1 &= \sum_{\nu_1=1}^{r_1} \cdots \sum_{\nu_d=1}^{r_d} \beta_{\nu_1,\ldots,\nu_d} \mathbf{u}_{\nu_1}^{(1)} \otimes \cdots \otimes \mathbf{u}_{\nu_d}^{(d)} \in \mathbb{V}_{\mathbf{n}}, \\ \mathbf{A}_2 &= \sum_{\mu_1=1}^{r_1} \cdots \sum_{\mu_d=1}^{r_d} \zeta_{\mu_1,\ldots,\nu_d} \mathbf{v}_{\mu_1}^{(1)} \otimes \cdots \otimes \mathbf{v}_{\mu_d}^{(d)} \in \mathbb{V}_{\mathbf{n}}, \end{aligned} \tag{2.29}$$

the *scalar product* (2.10) is computed by

$$\langle \mathbf{A}_1, \mathbf{A}_2 \rangle := \sum_{\mathbf{k}=1}^{\mathbf{r}_1} \sum_{\mathbf{m}=1}^{\mathbf{r}_2} \beta_{k_1 \ldots k_d} \zeta_{m_1 \ldots m_d} \prod_{\ell=1}^{d} \langle \mathbf{u}_{k_\ell}^{(\ell)}, \mathbf{v}_{m_\ell}^{(\ell)} \rangle. \tag{2.30}$$

In fact, applying the definition of the scalar product in (2.10) to the rank-1 tensors (with $R = \mathbf{r} = 1$), we have

$$\begin{aligned} \langle \mathbf{A}_1, \mathbf{A}_2 \rangle &:= \sum_{\mathbf{i} \in \mathcal{I}} u_{i_1}^{(1)} \cdots u_{i_d}^{(d)} v_{i_1}^{(1)} \cdots v_{i_d}^{(d)} \\ &= \sum_{i_1=1}^{n_1} u_{i_1}^{(1)} v_{i_1}^{(1)} \cdots \sum_{i_d=1}^{n_d} u_{i_d}^{(d)} v_{i_d}^{(d)} = \prod_{\ell=1}^{d} \langle \mathbf{u}^{(\ell)}, \mathbf{v}^{(\ell)} \rangle. \end{aligned} \tag{2.31}$$

Then, the above representation follows by combining all rank-1 terms in the left-hand side in (2.30).

We further simplify and suppose that $\mathbf{r} = \mathbf{r}_1 = \mathbf{r}_2 = (r, \ldots, r)$. The calculation in (2.30) then includes dr^2 scalar products of vectors of size n plus r^{2d} multiplications, leading to the overall complexity

$$W_{\langle \cdot, \cdot \rangle} = O(dnr^2 + r^{2d}),$$

whereas for calculation of the respective tensor norm, the second term reduces to $O(r^d)$.

Note that in the case of mixed Tucker-canonical decomposition (see Definition 3.15), the scalar product can be computed in $O(R^2 + dr^2 n + dR^2 r)$ operations (cf. [161], Lemma 2.8).

For given tensors $\mathbf{A}, \mathbf{B} \in \mathbb{R}^{\mathcal{I}}$, the *Hadamard product* $\mathbf{A} \odot \mathbf{B} \in \mathbb{R}^{\mathcal{I}}$ of two tensors of the same size $\mathcal{I}$ is defined by the componentwise product,

$$(\mathbf{A} \odot \mathbf{B})_{\mathbf{i}} = a_{\mathbf{i}} \cdot b_{\mathbf{i}}, \quad \mathbf{i} \in \mathcal{I}.$$

Hence, for $\mathbf{A}_1, \mathbf{A}_2 \in \mathcal{T}_{\mathbf{r}}$, as in (2.29), we tensorize the Hadamard product by

$$\mathbf{A}_1 \odot \mathbf{A}_2 := \sum_{k_1, m_1=1}^{r} \cdots \sum_{k_d, m_d=1}^{r} \beta_{k_1 \ldots k_d} \zeta_{m_1 \ldots m_d} (\mathbf{u}_{k_1}^{(1)} \odot \mathbf{v}_{m_1}^{(1)}) \otimes \cdots \otimes (\mathbf{u}_{k_d}^{(d)} \odot \mathbf{v}_{m_d}^{(d)}). \tag{2.32}$$

Again, applying definition (2.10) to the rank-1 tensors (with $\boldsymbol{\beta} = \boldsymbol{\zeta} = 1$), we obtain

$$\begin{aligned}(\mathbf{A}_1 \odot \mathbf{A}_2)_{\mathbf{i}} &= (u_{i_1}^{(1)} v_{i_1}^{(1)}) \cdots (u_{i_d}^{(d)} v_{i_d}^{(d)}), \quad \mathbf{i} \in \mathcal{I}, \\ \mathbf{A}_1 \odot \mathbf{A}_2 &= (\mathbf{u}^{(1)} \odot \mathbf{v}^{(1)}) \otimes \cdots \otimes (\mathbf{u}^{(d)} \odot \mathbf{v}^{(d)}). \end{aligned} \tag{2.33}$$

Then, (2.32) follows by summation over all rank-1 terms in $\mathbf{A}_1 \odot \mathbf{A}_2$. Relation (2.32) leads to the storage requirement

$$W_{\mathrm{st}(\odot)} = O(dr^2 n + r^{2d}),$$

which includes the memory size for d modes $n \times r \times r$ Tucker vectors, and for the new Tucker core of size $(r^2)^d$.

Summation of two tensors is performed by concatenation of the side matrices, their orthogonalization and recomputation of the Tucker core.

Summary on tensor operations in rank-*R* canonical format

We consider tensors $\mathbf{A}_1$, $\mathbf{A}_2$, represented in the rank-R canonical format (2.13):

$$\mathbf{A}_1 = \sum_{k=1}^{R_1} c_k \mathbf{u}_k^{(1)} \otimes \cdots \otimes \mathbf{u}_k^{(d)}, \quad \mathbf{A}_2 = \sum_{m=1}^{R_2} b_m \mathbf{v}_m^{(1)} \otimes \cdots \otimes \mathbf{v}_m^{(d)}, \tag{2.34}$$

with normalized vectors $\mathbf{u}_k^{(\ell)}, \mathbf{v}_m^{(\ell)} \in \mathbb{R}^{n_\ell}$. For simplicity of discussion, we assume that $n_\ell = n$, $\ell = 1, \dots, d$. We have

(1) A sum of two canonical tensors, given by (2.34), can be written as

$$\mathbf{A}_1 + \mathbf{A}_2 = \sum_{k=1}^{R_1} c_k \mathbf{u}_k^{(1)} \otimes \cdots \otimes \mathbf{u}_k^{(d)} + \sum_{m=1}^{R_2} b_m \mathbf{v}_m^{(1)} \otimes \cdots \otimes \mathbf{v}_m^{(d)}, \tag{2.35}$$

resulting in the canonical tensor with the rank, at most, $R_S = R_1 + R_2$. This operation has no cost since it is simply a concatenation of side matrices.

(2) For given canonical tensors $\mathbf{A}_1$, $\mathbf{A}_2$, the *scalar product* (2.10) is computed by (see (2.31))

$$\langle \mathbf{A}_1, \mathbf{A}_2 \rangle := \sum_{k=1}^{R_1} \sum_{m=1}^{R_2} c_k b_m \prod_{\ell=1}^{d} \langle \mathbf{u}_k^{(\ell)}, \mathbf{v}_m^{(\ell)} \rangle. \tag{2.36}$$

Calculation of (2.36) includes $R_1 R_2$ scalar products of vectors in $\mathbb{R}^n$, leading to the overall complexity

$$W_{\langle \cdot, \cdot \rangle} = O(dnR_1 R_2).$$

(3) For $\mathbf{A}_1$, $\mathbf{A}_2$ given by (2.34), we tensorize the Hadamard product by (see (2.33))

$$\mathbf{A}_1 \odot \mathbf{A}_2 := \sum_{k=1}^{R_1} \sum_{m=1}^{R_2} c_k b_m (\mathbf{u}_k^{(1)} \odot \mathbf{v}_m^{(1)}) \otimes \cdots \otimes (\mathbf{u}_k^{(d)} \odot \mathbf{v}_m^{(d)}). \tag{2.37}$$

The complexity of this operation is estimated by $O(dnR_1 R_2)$.

Convolution of tensors will be considered in Section 5.

3 Rank-structured grid-based representations of functions in $\mathbb{R}^d$

3.1 Super-compression of function-related tensors

In the numerical solution of multidimensional problems, it is advantageous to have separable representations of multivariate functions, since then all calculations are reduced to operations with univariate functions. A well-known example is a Gaussian function, which can be represented as a product of one-dimensional Gaussians. However, the multivariate functions can loose their initial separability after undergoing nonlinear (integral) transformations involved in multidimensional PDEs, leading to cumbersome error-prone calculation schemes.

When solving the d-dimensional PDEs by using the standard finite difference or finite element methods via the grid-based representation of the multidimensional functions and operators, the storage and, consequently, the number of operations scale exponentially with respect to the dimension size d. This redundancy of the grid representation in conventional numerical methods can be only mildly released for low-dimensional problems when using mesh refinement in finite element approaches or sparse grids methods.

In the previous section, it was shown that there are algebraically separable representations of the multidimensional arrays by using the Tucker or canonical tensor decomposition. The question is how to compute such representations in numerical analysis of multidimensional PDEs? Another question is concerned with the numerical efficiency of the tensor approach and the possibilities to reduce the separation rank parameters, adapting them to the approximation threshold.

The sinc-quadrature based canonical approximation to analytic functions and certain operator-valued functions have been analyzed in [94, 91, 111, 112, 162, 161, 166]. The related results in approximation theory can be found in [5, 89, 90, 114, 117, 118]. However, this kind of canonical-type approximation in the explicit analytic form is limited to spherically symmetric functions while algebraic canonical approximation algorithms suffer from slow and unstable convergence. In 2006, it was proven by Boris Khoromskij that for some classes of function related tensors, the Tucker approximation error via the minimization (2.18), decays exponentially in the Tucker rank [161]:

$$\|\mathbf{A}_{(\mathbf{r})} - \mathbf{A}_0\| \le Ce^{-\alpha\hat{r}} \quad \text{with } \hat{r} = \min_{\ell} r_{\ell}, \tag{3.1}$$

where $\mathbf{A}_{(\mathbf{r})}$ is a minimizer in (2.18). As a consequence, the approximation error $\varepsilon > 0$ can be achieved with the moderate rank parameter

$$\hat{r} = O(|\log \varepsilon|).$$

https://doi.org/10.1515/9783110365832-003

In this section, we demonstrate that, for a wide class of function-related tensors, the Tucker decomposition provides a separable approximation with exponentially fast decay of the error with respect to the rank parameter.

The particular properties of the Tucker decomposition for function-related tensors led to invention of the multigrid Tucker decomposition method, which allows one to reduce numerical complexity dramatically. Moreover, for function-related tensors, the novel canonical-to-Tucker (C2T) transform and the reduced higher-order singular-value decomposition (RHOSVD) were developed in [174], which made a tremendous impact on the evolution of tensor numerical methods. C2T transform provides a stable algorithm for reducing a large canonical tensor rank arising in the course of bilinear matrix–tensor and tensor–tensor operations.

The C2T algorithm has pushed forward the grid-based numerical methods for calculating the 3D integral convolution operators for functions with multiple singularities [174], providing the accuracy level comparable with the analytical evaluation of the same integrals. In turn, the RHOSVD applies to the tensor in the canonical form (say, resulting from certain algebraic transforms or analytic approximations), and it does not need building a full tensor for the Tucker decomposition. Indeed, it is enough to find the orthogonal Tucker basis only for directional matrices of the canonical tensor, which consists of skeleton vectors in every single dimension [174]. Presumably, the invention of the RHOSVD and the C2T algorithm anticipated development of the tensor formats avoiding the "curse of dimensionality".

We conclude that coupling the multi-linear algebra of tensors and the nonlinear approximation theory resulted in the tensor-structured numerical methods for multidimensional PDEs. The prior results on theory of tensor-product approximation of multivariate functions and operators [94, 91, 111, 161] were significant prerequisites for understanding and developing the tensor numerical methods. First, we sketch some of the basic results in approximation theory.

3.1.1 Prediction of approximation theory: $O(\log n)$ ranks

In the following, we choose the set $\mathcal{S}$ of rank-structured (formatted) tensors within the above defined tensor classes and call the elements in $\mathcal{S}$ as $\mathcal{S}$-tensors.

To perform computation in the low-parametric tensor formats (say, in the course of rank-truncated iteration), we need to perform a nonlinear "projection" of the current iterand onto $\mathcal{S}$. This action is fulfilled by using the tensor truncation operator $T_{\mathcal{S}} : \mathbb{V}_{\mathbf{n},d} \to \mathcal{S}$ defined by

$$\mathbf{A}_0 \in \mathcal{S}_0 \subset \mathbb{V}_{\mathbf{n},d} : \quad T_{\mathcal{S}}\mathbf{A}_0 = \operatorname*{argmin}_{\mathbf{T}\in\mathcal{S}} \|\mathbf{A}_0 - \mathbf{T}\|_{\mathbb{V}_{\mathbf{n}}}, \tag{3.2}$$

which is a challenging nonlinear approximation problem. In practice, the computation of the minimizer $T_{\mathcal{S}}\mathbf{A}_0$ can be performed only approximately. The replacement of

$\mathbf{A}_0$ by its approximation in the tensor class $\mathcal{S}$ is called the *tensor truncation* to $\mathcal{S}$ and is denoted by $T_{\mathcal{S}}\mathbf{A}_0$.

There are analytic and algebraic methods of approximate solution to the problem (3.2) applicable to different classes of rank-structured tensors $\mathcal{S}$. The target tensor may arise, in particular, as the grid-based representation of regular enough functions, say, solutions of PDEs or some classical Green's kernels. The storage and numerical complexity for the elements in $\mathcal{S}$ are strictly determined by the rank parameters involved in the parametrization within given tensor format. In view of the relation $\hat{r} = O(|\log \varepsilon|)$ between the Tucker rank and the corresponding approximation error $\|\mathbf{A}_0 - \mathbf{T}\|_{\mathbb{V}_\mathbf{n}}$, which is the case for the wide class of function-related tensors [161], one may expect in the PDE-related applications the $O(\log n)$ ranks asymptotic in the univariate mode size of the $n^{\otimes d}$ tensors living on $n \times \cdots \times n$ tensor grid. Our experience in the numerical simulations in electronic structure calculations confirms this hypothesis.

Such optimistic effective rank bounds justify the benefits of tensor numerical methods in large-scale scientific computing, indicating that these methods are not just heuristic tools but rigorously approved techniques.

3.1.2 Analytic methods of separable approximation of multivariate functions and operators

In what follows, we discuss the low rank approximation of a special class of higher-order tensors, also called function-related tensors (FRTs), obtained by sampling the multi-variate function over $n \times \cdots \times n$ tensor grid in $\mathbb{R}^d$. These data directly arise from:

(a) A separable approximation of multi-variate functions;
(b) Nyström/collocation/Galerkin discretization of integral operators with the Green's kernels;
(c) The tensor-product approximation of some analytic matrix-valued functions.

The constructive analytic approximation methods are based on sinc-quadrature representations for analytic functions [271, 215]. These techniques apply, in particular, to the class of Green's kernels (the Poisson, Yukawa, Helmholtz potentials), cf. [220, 113], to certain kernel functions arising in the Boltzmann equation [162], in electronic structure calculations [122, 30, 174, 187], and to correlation functions in the construction of the Karhünen–Loéve expansion [266, 182, 185], as well as in multidimensional data analysis [71, 206].

Error estimate for tensor approximation of analytic generating function

In the following, we define FRTs corresponding to collocation-type discretization, see [113]. The Nyström and Galerkin approximations to function-related tensors have been discussed in [111, 161].

Given the function $g : \Omega := \Pi_1 \times \cdots \times \Pi_d \to \mathbb{R}$ with $\Pi_\ell = \Pi = [a,b]^p$ and $p = 1,2,3$, for $\ell = 1,\ldots,d$. Define the univariate grid-size $n \in \mathbb{N}$ and the mesh-size $h = (b-a)/n$. We denote by $\{x^{(1)}_{\mathbf{i}_1},\ldots,x^{(d)}_{\mathbf{i}_d}\}$ a set of collocation points living in the midpoints of the tensor grid with mesh-size h, where $\mathbf{i}_\ell = (i_{\ell,1},\ldots,i_{\ell,p}) \in \mathcal{I}_\ell := I_{\ell,1}\times\cdots\times I_{\ell,p}$ $(\ell = 1,\ldots,d)$. Here we have $i_{\ell,m} \in I_n := \{1,\ldots,n\}$ $(m = 1,\ldots,p)$.

We consider the case $d \geq 2$ with some fixed $p \in \{1,2,3\}$. In particular, the case of functions in $\mathbb{R}^d$ is treated with $p = 1$, whereas the matrix (operator) decompositions correspond to the choice $p = 2$. In the latter case, we introduce the reordered index set of pairs

$$\mathcal{M}_\ell := \{\mathbf{m}_\ell : \mathbf{m}_\ell = (i_\ell, j_\ell),\ i_\ell, j_\ell \in I_n\} \quad (\ell = 1,\ldots,d)$$

such that $\mathcal{I} = \mathcal{M}_1 \times \cdots \times \mathcal{M}_d$ with $\mathcal{M}_\ell = I_n \times I_n$.

Here we follow [113] and focus on the collocation-type schemes that are based on tensor-product ansatz functions

$$\psi^{\mathbf{i}}(y_1,\ldots,y_d) = \prod_{\ell=1}^{d} \psi_\ell^{i_\ell}(y_\ell), \quad \mathbf{i} = (i_1,\ldots,i_d) \in \mathcal{I}_1 \times \cdots \times \mathcal{I}_d,\ y_\ell \in \Pi_\ell. \tag{3.3}$$

Definition 3.1 (FRT by collocation). Let $p = 2$. Given the function $g :\in \mathbb{R}^{pd} \to \mathbb{R}$ and tensor-product basis set (3.3), we introduce the coupled variable $\zeta^{(\ell)}_{i_\ell} := (x^{(\ell)}_{i_\ell}, y_\ell)$ including the collocation point $x^{(\ell)}_{i_\ell}$ and $y_\ell \in \Pi$, the pair $\mathbf{m}_\ell := (i_\ell, j_\ell) \in \mathcal{M}_\ell$ and define the collocation-type dth-order FRT by $\mathbf{A} \equiv \mathbf{A}(g) := [a_{\mathbf{m}_1\ldots\mathbf{m}_d}] \in \mathbb{R}^{\mathcal{M}_1\times\cdots\times\mathcal{M}_d}$ with the tensor entries

$$a_{\mathbf{m}_1\ldots\mathbf{m}_d} := \int_\Omega g(\zeta^{(1)}_{i_1},\ldots,\zeta^{(d)}_{i_d})\psi^{\mathbf{j}}(y_1,\ldots,y_d)dy, \quad \mathbf{m}_\ell \in \mathcal{M}_\ell. \tag{3.4}$$

In the case $p = 1$, we simplify to $\zeta^{(\ell)}_{i_d} = y_\ell$, $\mathbf{m}_\ell := j_\ell$.

The key observation is that there is a natural duality between separable approximation of the multivariate generating function g and the tensor-product decomposition of the related multidimensional array $\mathbf{A}(g)$. As result, the canonical decompositions of $\mathbf{A}(g)$ can be derived by using a corresponding separable expansion of the generating function g (see [111, 116] for more details).

Lemma 3.2 ([113]). *Suppose that a multivariate function $g : \Omega \subset \mathbb{R}^{pd} \to \mathbb{R}$ can be accurately approximated by a separable expansion*

$$g_R(\boldsymbol{\zeta}) := \sum_{k=1}^{R} \mu_k \Phi_k^{(1)}(\zeta^{(1)})\cdots\Phi_k^{(d)}(\zeta^{(d)}) \approx g(\boldsymbol{\zeta}), \quad \boldsymbol{\zeta} = (\zeta^{(1)},\ldots,\zeta^{(d)}) \in \mathbb{R}^{pd}, \tag{3.5}$$

where $\mu_k \in \mathbb{R}$ and $\Phi_k^\ell : \Pi \subset \mathbb{R}^2 \to \mathbb{R}$. Introduce the canonical decomposition of $\mathbf{A}(g)$ via $\mathbf{A}_{(R)} := \mathbf{A}(g_R)$ (cf. Definition 3.1) where the canonical skeleton vectors are defined by

$$V_k^{(\ell)} = \left\{\int \Phi_k^{(\ell)}(\zeta_i^{(\ell)})\psi_\ell^j(y_\ell)dy_\ell\right\}_{(i,j)\in\mathcal{M}_\ell} \in \mathbb{R}^{\mathcal{I}_\ell\times\mathcal{J}_\ell}, \quad \ell = 1,\ldots,d,\ k = 1,\ldots,R. \tag{3.6}$$

Then the FRT $\mathbf{A}_{(R)}$ *approximates* $\mathbf{A}(g)$ *with the error estimated by*

$$\|\mathbf{A}(g) - \mathbf{A}_{(R)}(g_R)\|_\infty \le C\|g - g_R\|_{L^\infty(\Omega)}.$$

Though in general a decomposition (3.5) with small separation rank R is a complicated numerical task, in many interesting applications efficient approximation methods are available. In particular, for a class of multivariate functions (say, for radial basis functions in $\mathbb{R}^d$), it is possible to obtain a dimension independent bound on the separation rank $R = \mathcal{O}(\log n|\log\varepsilon|)$, e. g., based on sinc-quadrature methods or a direct approximation by exponential sums (see examples in [39, 40, 111, 161, 206]).

Next, we discuss the constructive canonical and Tucker tensor decomposition of FRTs applied to a general class of *analytic generating functions* represented in terms of their generalized Laplace transform.

sinc-quadrature approximation in the Hardy space

We use constructive approximation, based on the classical sinc-quadrature methods. For the readers convenience, we recall the well-known approximation results by the sinc-methods (cf. [271, 92, 96, 95]). Recall that the Hardy space $H^1(D_\delta)$ is defined as the set of all complex-valued functions f that are analytic in the strip

$$D_\delta := \{z \in \mathbb{C} : |\Im m\, z| < \delta\}$$

and such that

$$N(f, D_\delta) := \int_{\partial D_\delta} |f(z)||dz| = \int_{\mathbb{R}} (|f(x + i\delta)| + |f(x - i\delta)|)dx < \infty.$$

Given $f \in H^1(D_\delta)$, the step size of the quadrature $\mathfrak{h} > 0$, and $M \in \mathbb{N}_0$, the corresponding $(2M + 1)$-point sinc-quadrature approximating the integral $\int_{\mathbb{R}} f(\xi)d\xi$ reads

$$T_M(f, \mathfrak{h}) := \mathfrak{h} \sum_{k=-M}^{M} f(k\mathfrak{h}) \approx \int_{\mathbb{R}} f(\xi)d\xi. \tag{3.7}$$

Proposition 3.3. *Let* $f \in H^1(D_\delta)$, $\mathfrak{h} > 0$, *and* $M \in \mathbb{N}_0$ *be given. If*

$$|f(\xi)| \le C\exp(-b|\xi|) \qquad \text{for all } \xi \in \mathbb{R} \text{ with } b, C > 0, \tag{3.8}$$

then the quadrature error satisfies

$$\left|\int_{\mathbb{R}} f(\xi)d\xi - T_M(f, \mathfrak{h})\right| \le Ce^{-\sqrt{2\pi\delta bM}} \quad \text{with } \mathfrak{h} = \sqrt{2\pi\delta/bM}$$

with a positive constant C *depending only on* f, δ, b *(cf.* [271]*).*

If f possesses the hyper-exponential decay

$$|f(\xi)| \leq C\exp(-be^{a|\xi|}) \qquad \textit{for all } \xi \in \mathbb{R} \quad \textit{with } a, b, C > 0, \tag{3.9}$$

then the choice $\mathfrak{h} = \log(\frac{2\pi aM}{b})/(aM)$ leads to the improved error bound (cf. [95])

$$\left|\int_{\mathbb{R}} f(\xi)d\xi - T_M(f,\mathfrak{h})\right| \leq CN(f, D_\delta)e^{-2\pi\delta aM/\log(2\pi aM/b)}.$$

Note that $2M+1$ is the number of quadrature/interpolation points. If f is an even function, then this number reduces to $M+1$.

Error bounds for canonical and Tucker decomposition

Following [113], we consider a class of multivariate functions $g : \mathbb{R}^d \to \mathbb{R}$ parameterized by $g(\boldsymbol{\zeta}) = G(\rho(\boldsymbol{\zeta})) \equiv G(\rho)$ with

$$\rho \equiv \rho(\boldsymbol{\zeta}) = \rho_1(\zeta^{(1)}) + \cdots + \rho_d(\zeta^{(d)}) > 0, \quad \rho_\ell : \mathbb{R}^2 \to \mathbb{R}_+,$$

where the univariate function $G : \mathbb{R}_+ \to \mathbb{R}$ is supposed to be represented via the Laplace transform

$$G(\rho) = \int_{\mathbb{R}_+} \mathcal{G}(\tau)e^{-\rho\tau}d\tau. \tag{3.10}$$

Consider the FRT approximation corresponding to $p = 2$, $\zeta^{(\ell)} = (x_\ell, y_\ell)$ (cf. Definition 3.1). Without loss of generality, we introduce one and the same scaling function

$$\psi^i(\cdot) = \psi(\cdot + (i-1)h), \quad i \in I_n, \tag{3.11}$$

for all spatial dimensions $\ell = 1, \ldots, d$, where $h > 0$ is the mesh parameter of the spacial grid. For the ease of exposition, we simplify further and set $\rho \equiv \rho(\boldsymbol{\zeta}) = \sum_{\ell=1}^{d} \rho_0(\zeta^{(\ell)})$, i. e., $\rho_\ell = \rho_0(x_\ell, y_\ell)$ ($\ell = 1, \ldots, d$) with $\rho_0 : [a,b]^2 \to \mathbb{R}_+$. For $i \in I_n$, let $\{\bar{x}_i\}$ be the set of cell-centered collocation points on the univariate grid of step size h in $[a,b]$. For each $i, j \in I_n$, we introduce the parameter dependent integral

$$\Psi_{i,j}(\tau) := \int_{\mathbb{R}^2} e^{-\rho_0(\bar{x}_i, y)\tau}\psi(y + (j-1)h)dy, \quad \tau \geq 0, \tag{3.12}$$

where τ is the integration variable in (3.10).

Theorem 3.4 (FRT approximation [113]). *Assume that:*

(a) *The Laplace transform $\mathcal{G}(\tau)$ in (3.10) has an analytic extension $\mathcal{G}(w)$, $w \in \Omega_{\mathcal{G}}$, into a certain domain $\Omega_{\mathcal{G}} \subset \mathbb{C}$ that can be transformed by a conformal map onto the strip D_δ such that $w = \varphi(z)$, $z \in D_\delta$, and $\varphi^{-1} : \Omega_{\mathcal{G}} \to D_\delta$;*

(b) *for all* $(\mathbf{i}, \mathbf{j}) \in \mathcal{I} \times \mathcal{J}$, *the transformed integrand*

$$f(z) := \varphi'(z)\mathcal{G}(\varphi(z)) \prod_{\ell=1}^{d} \Psi_{i_\ell j_\ell}(\varphi(z)) \tag{3.13}$$

belongs to the Hardy space $H^1(D_\delta)$ *with* $N(f, D_\delta) < \infty$ *uniformly in* $(\mathbf{i}, \mathbf{j})$;

(c) *the function* $f(t)$, $t \in \mathbb{R}$, *in* (3.13) *has either exponential* (c1) *or hyper-exponential* (c2) *decay as* $t \to \pm\infty$ *(see Proposition* 3.3*).*

Under the assumptions (a)–(c), *we have that, for each* $M \in \mathbb{N}$, *the FRT* $\mathbf{A}(g)$, *defined on* $[a, b]^d$, *allows an exponentially convergent symmetric*[1] *canonical approximation* $\mathbf{A}_{(R)} \in \mathcal{C}_R$ *with* $V_k^{(\ell)}$ *as in* (3.6), *where the expansion* (3.5) *is obtained by substitution of* f *from* (3.13) *into the* sinc-*quadrature* (3.7) *such that we have*

$$\|\mathbf{A}(g) - \mathbf{A}_{(R)}\|_\infty \leq Ce^{-\alpha M^\nu} \quad \text{with } R = 2M + 1, \tag{3.14}$$

where $\nu = \frac{1}{2}$ *and* $\alpha = \sqrt{2\pi\delta b}$ *in case* (c1), *and* $\nu = 1$ *and* $\alpha = \frac{2\pi\delta b}{\log(2\pi a M/b)}$ *in case* (c2).

Theorem 3.4 proves the existence of the canonical decomposition of the FRT $\mathbf{A}(g)$ with the Kronecker rank $r = \mathcal{O}(|\log\varepsilon| \log 1/h)$ (in case (c2)) or $r = \mathcal{O}(\log^2\varepsilon)$ (in case (c1)), which provide an approximation of order $\mathcal{O}(\varepsilon)$. In our applications, we usually have $1/h = \mathcal{O}(n)$, where n is the number of grid-points in one spacial direction. Theorem 5.12 applies to translation invariant or spherically symmetric (radial) functions, in particular, to the classical Newton, Yukawa, Helmholtz, and Slater-type kernels

$$\frac{1}{\|x-y\|}, \quad \frac{e^{-\lambda\|x-y\|}}{\|x-y\|}, \quad \frac{\cos(\lambda\|x-y\|)}{\|x-y\|}, \quad \text{and} \quad e^{-\lambda\|x-y\|},$$

where $x, y \in \mathbb{R}^3$, $\lambda > 0$; see [111] for the case of Newton kernel. We refer to [163, 164], where the sinc-based CP approximations to the Yukawa and Helmholtz kernels have been analyzed. In particular, the low-rank Tucker approximations to the Slater and Yukawa kernels have been proven in [161] and in [164].

3.1.3 Tucker decomposition of function-related tensors

In what follows, we apply the Tucker decomposition algorithm to tensors generated by a number of commonly used radial basis functions, including classical Green kernels in 3D, and study numerically their properties. Recall that for a given continuous

1 A dth-order tensor is called symmetric if it is invariant under arbitrary permutations of indices in $\{1, \ldots, d\}$.

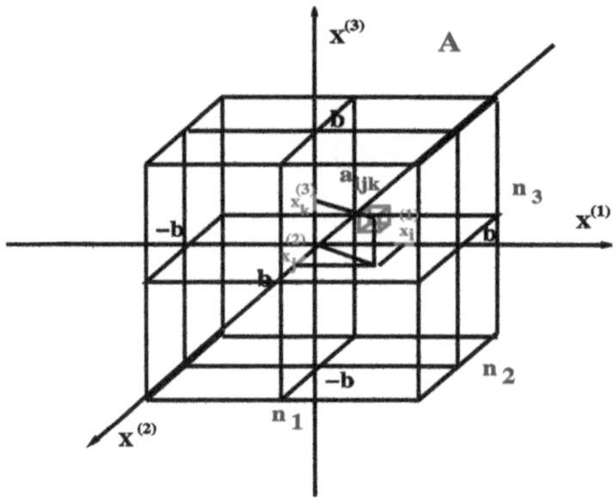

Figure 3.1: Generation of a function-related tensor in 3D on $n_1 \times n_2 \times n_3$ Cartesian grid. Tensor value in a voxel is computed by using sample of a function at the corresponding centers of the grid intervals.

function $g : \Omega \to \mathbb{R}$, $\Omega := \prod_{\ell=1}^{d}[-b_\ell, b_\ell] \subset \mathbb{R}^d$, $0 < b_\ell < \infty$, the collocation-type function-related tensor of order d is defined by

$$\mathbf{A}_0 \equiv \mathbf{A}_0(g) := [a_{i_1 \dots i_d}] \in \mathbb{R}^{I_1 \times \dots \times I_d} \quad \text{with } a_{i_1 \dots i_d} := g(x_{i_1}^{(1)}, \dots, x_{i_d}^{(d)}), \tag{3.15}$$

where $(x_{i_1}^{(1)}, \dots, x_{i_d}^{(d)}) \in \mathbb{R}^d$ are grid collocation points, indexed by $\mathcal{I} = I_1 \times \dots \times I_d$,

$$x_{i_\ell}^{(\ell)} = -b_\ell + (i_\ell - 1)h_\ell, \quad i_\ell = 1, 2, \dots, n_\ell, \ \ell = 1, \dots, d, \tag{3.16}$$

which are the nodes of equally spaced subintervals, with the mesh size $h_\ell = 2b_\ell/(n_\ell - 1)$; see Figure 3.1. When using an odd discretization parameter, the function is samples in the nodes of the grid, for example, for $d = 3$,

$$x_{i_\ell}^{(\ell)} = -b_\ell + (i_\ell - 1/2)(2b_\ell/n_\ell), \quad i_\ell = 1, 2, \dots, n_\ell, \ \ell = 1, 2, 3. \tag{3.17}$$

For functions in $\mathbb{R}^3$, we generate a tensor $\mathbf{A} \in \mathbb{R}^{n_1 \times n_2 \times n_3}$ with entries $a_{ijk} = g(x_i^{(1)}, x_j^{(2)}, x_k^{(3)})$. We test the rank-dependence of the Tucker approximation to the function-related tensors $\mathbf{A}$. Based on the examples of some classical Green's kernels, one can figure out if it is possible to rely on the Tucker tensor approximation to obtain algebraically their low-rank separable tensor representations. We consider the Slater-type, Newton, and Helmholtz kernels in $\mathbb{R}^3$, which have the typical singularity at the origin.

The initial tensor $\mathbf{A}_0$ is approximated by a rank $\mathbf{r} = (r, \dots, r)$ Tucker representation $\mathbf{A}_{(r)}$, where the rank-parameter r increases from $r = 1, 2, \dots$ to some predefined value, $r_{\max}$. Then the orthogonal Tucker vectors and the core tensor of size $r \times r \times r$ are used for the construction back to full size tensor corresponding to $\mathbf{A}_{(r)}$, for estimating the error of tensor decomposition, $\|\mathbf{A}_0 - \mathbf{A}_{(r)}\|$ for the given rank. For every Tucker rank $\mathbf{r}$ in the respective range, we compute the relative error in the Frobenius norm as in (2.10)

$$E_{\mathrm{FN}} = \frac{\|\mathbf{A}_0 - \mathbf{A}_{(r)}\|}{\|\mathbf{A}_0\|} \tag{3.18}$$

and the relative difference of norms (ℓ^2-energy norm)

$$E_{\mathrm{FE}} = \frac{\|\mathbf{A}_0\| - \|\mathbf{A}_{(r)}\|}{\|\mathbf{A}_0\|}. \tag{3.19}$$

Notice that the projection property of the Tucker decomposition we have reads $\|\mathbf{A}_{(r)}\| \le \|\mathbf{A}_0\|$.

(1) Slater function. The Slater-type functions play a significant role in electronic structure calculations. For example, the Slater function given by

$$g(x) = \exp(-\alpha\|x\|) \quad \text{with } x = (x_1, x_2, x_3)^T \in \mathbb{R}^3,$$

represents the electron "orbital" ($\alpha = 1$) and the electron density function ($\alpha = 2$) corresponding to the Hydrogen atom. Here and in the following, $\|x\| = \sqrt{\sum_{\ell=1}^{d} x_\ell^2}$ denotes the Euclidean norm of $x \in \mathbb{R}^d$.

We compute the rank-(r, r, r) Tucker approximation to the function-related tensor defined in the nodes of the $n_1 \times n_2 \times n_3$ 3D Cartesian grid with $n_1 = 65$, $n_2 = 67$, and $n_3 = 69$ in the interval $2b = 10$. The slice of the discretized Slater function at the middle of the z-axis is shown in Figure 3.2, top-left. Figure 3.2, top-right, shows the fast exponential convergence of the approximation errors E_{FN}, (3.18), and E_{FE}, (3.19), with respect to the Tucker rank. Thus, the Slater function can be efficiently approximated by low-rank Tucker tensors. In fact, Tucker rank $r = 10$ provides a maximum absolute error of the approximation of order 10^{-5}, and $r = 18$ provides approximation with accuracy $\sim 10^{-10}$. Note that the error of the Tucker tensor approximation only slightly depends on the discretization parameter n. The corresponding numerical tests will be demonstrated further in the section on multigrid tensor decomposition, since the standard Tucker algorithm is practically restricted to univariate grid size of the order of $n_\ell \approx 200$.

Figure 3.2, bottom-left, shows the example of the orthogonal vectors of the dominating subspaces of the Tucker tensor decomposition. Note that the vectors corresponding to the largest entries in the Tucker core exhibit essentially smooth shapes. Figure 3.2, bottom-right, presents the entries of the Tucker core tensor $\boldsymbol{\beta} \in \mathbb{R}^{7\times 7\times 7}$ by displaying its first four matrix slices $M_{\boldsymbol{\beta},\nu_r} \in \mathbb{R}^{7\times 7\times 1}$, $\nu_r = 1, \ldots, 4$. Numbers inside the figure indicate the maximum values of the core entries at a given slice $M_{\boldsymbol{\beta},\nu_r} \in \mathbb{R}^{7\times 7\times 1}$ of $\boldsymbol{\beta}$. Figure 3.2 shows that the "energy" of the decomposed function is concentrated in several upper slices of the core tensor, and the entries of the core are also decreasing fast from slice to slice.

(2) Newton kernel. The best rank-$\mathbf{r}$ Tucker decomposition algorithm with $\mathbf{r} = (r, \ldots, r)$ is applied for approximating the Newton kernel [173]

$$g(x) = \frac{1}{\|x\|}, \quad x \in \mathbb{R}^3,$$

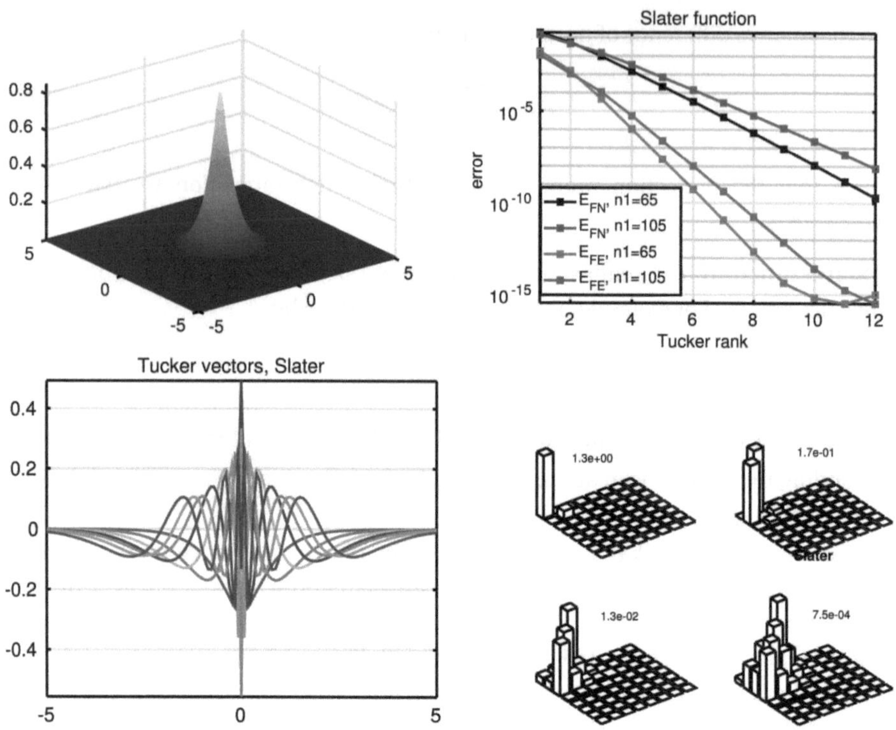

Figure 3.2: Top: discretized Slater function (left) and the error of its Tucker tensor approximation versus the Tucker rank (right). Bottom: orthogonal vectors of the Tucker decomposition (left) and entries of the Tucker core.

in the cube $[-b,b]^3$ with $b = 5$, on the cell-centered uniform grid with discretization parameter $n = 64$.

We consider the sampling points $x_i^{(\ell)} = h/2 + (i-1)h$, $\ell = 1,2,3$, for three space variables $x^{(\ell)}$. Figure 3.3, top-left, shows the potential at the plane close to zero point (at $z = h/2$), and the top-right figure displays the absolute error of its approximation with the Tucker rank $r = 18$, demonstrating the accuracy about 10^{-10}. Figure 3.3, left-bottom, shows stable exponential convergence of the errors (3.18) and (3.19) with respect to the Tucker rank. In particular, it follows that accuracy of the order of 10^{-5}, is achieved with the Tucker rank $r = 10$, and for 10^{-3}, one can chose the rank $r = 7$. The right hand side of Figure 3.3, bottom, shows the orthogonal vectors $\mathbf{v}_k^{(1)}$, $k = 1,\dots,6$, for the mode $\ell = 1$ ($x^{(1)}$-axis).

(3) Helmholtz potential. In the next example, we consider a Tucker approximation of the third-order FRT generated by the Helmholtz functions given by

$$g_1(x) = \frac{\sin(\kappa\|x\|)}{\|x\|} \quad \text{with } x = (x_1, x_2, x_3)^T \in \mathbb{R}^3,$$

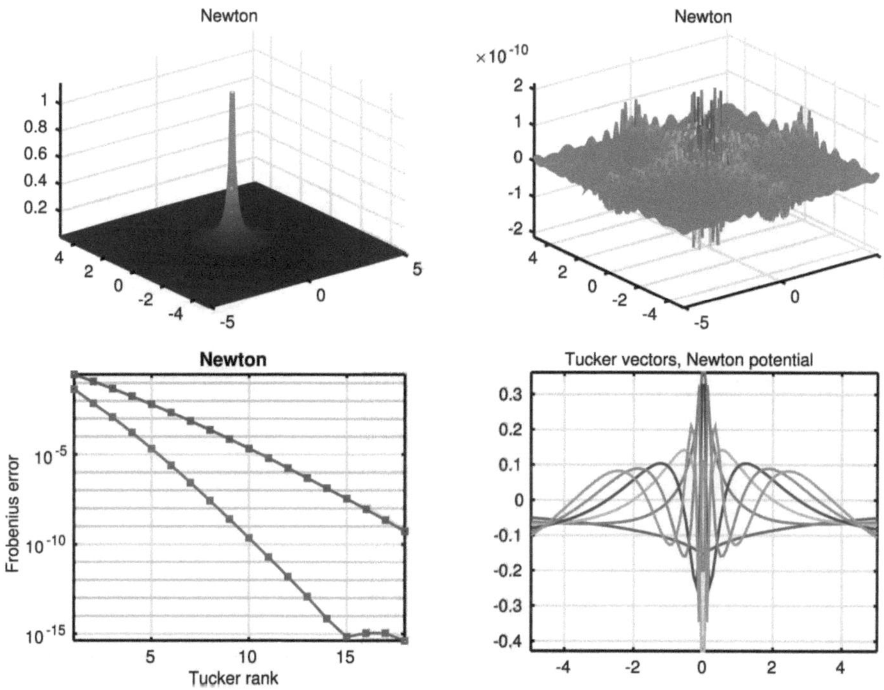

Figure 3.3: Top: the plane of the 3D Newton potential (left), and the error of its Tucker tensor approximation with the rank $r = 18$. Bottom: decay of the Tucker approximation error versus the Tucker rank (left) and the orthogonal Tucker vectors $\mathbf{v}_k^{(1)}$, $k = 1, \ldots, 6$.

with $\kappa = 1$ and $\kappa = 3$ and

$$g_2(x) = \frac{\cos(\kappa \|x\|)}{\|x\|} \quad \text{with } x = (x_1, x_2, x_3)^T \in \mathbb{R}^3.$$

We consider the FRT with the same "voxel-centered" collocation points with respect to the $n \times n \times n$ grid over $[-b, b]^3$, $b = 5$, as in the previous examples.

Figure 3.4 shows the potential for $\kappa = 1$ (top-left) and the error (top-right) of its Tucker tensor approximation with the rank 7, which is of the order of 10^{-15}. Figure 3.4, bottom, indicates the exponential convergence of the Tucker tensor approximation in the rank parameter (left) and shows the examples of the orthogonal vectors of the Tucker tensor decomposition. Figures in 3.5 show a similar decay for the Helmholz potential with $\kappa = 3$. The approximation with the Tucker rank $r = 10$ provides the error of the order of 10^{-10}. Figure 3.6 illustrates the results for the singular kernel $\frac{\cos \|x\|}{\|x\|}$.

Other numerical results on the Tucker tensor decomposition can be found in [173, 146]. Recent numerics on the Tucker tensor decomposition for the Matérn radial basis functions is presented in [206].

The following lemma proves (see [173], Lemma 2.4 and [146] for more details) that the relative difference of norms of the best rank-$(r_1, \ldots, r_d)$ Tucker approximation $\mathbf{A}_{(\mathbf{r})}$

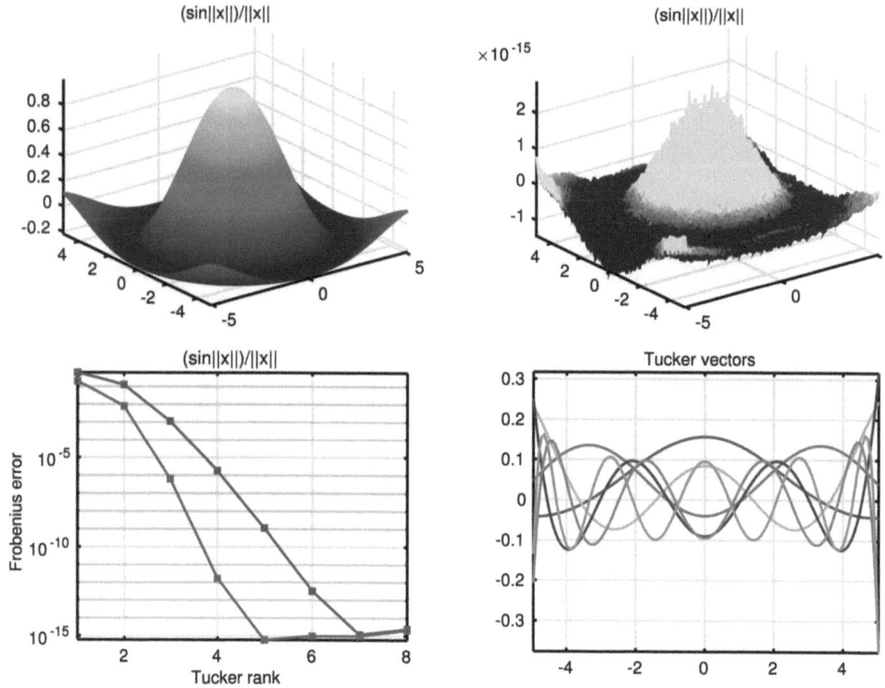

Figure 3.4: Top: the plane of the 3D Helmholz potential $\frac{\sin\|x\|}{\|x\|}$ over cross-section (left) and the absolute error for its Tucker approximation with the rank $r = 6$ (right). Bottom: decay of the Tucker approximation error in the Frobenius norm with respect to the Tucker rank (left) and the orthogonal Tucker vectors $v_k^{(1)}$, $k = 1, \ldots, 6$, for the Helmholz potential $\frac{\sin\|x\|}{\|x\|}$ (right).

and the target $\mathbf{A}_0$ is estimated by the square of the relative Frobenius norm of $\mathbf{A}_{(\mathbf{r})} - \mathbf{A}_0$, which was confirmed by the numerics above.

Lemma 3.5 (Quadratic convergence in norms). *Let* $\mathbf{A}_{(\mathbf{r})} \in \mathbb{R}^{I_1\times\cdots\times I_d}$ *solve the minimization problem* (2.18) *over* $\mathbf{A} \in \mathcal{T}_{\mathbf{r}}$. *Then we have the "quadratic" relative error bound*

$$\frac{\|\mathbf{A}_0\| - \|\mathbf{A}_{(\mathbf{r})}\|}{\|\mathbf{A}_0\|} \le \frac{\|\mathbf{A}_{(\mathbf{r})} - \mathbf{A}_0\|^2}{\|\mathbf{A}_0\|^2}. \tag{3.20}$$

Moreover, $\|\boldsymbol{\beta}\| = \|\mathbf{A}_{(\mathbf{r})}\| \le \|\mathbf{A}_0\|$.

The presented numerical experiments may be reproduced by the readers by using the Matlab code for the Tucker decomposition algorithm for 3D tensors presented below. The Example 3 contains the main program (the reader can name it as "Test_Tucker.m") and the subroutines 1, 2, and 3 contain all necessary functions:

Tucker_full_3D_ini(A3,NR,kmax,ir,Ini),
Tuck_2_F(LAM3F,U1,U2,U3),
TnormF(A).

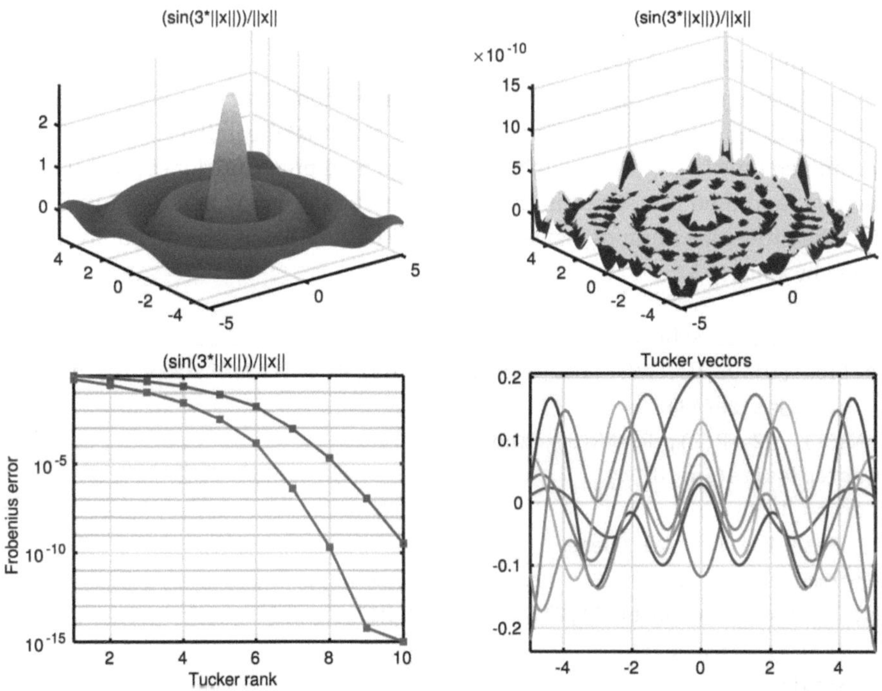

Figure 3.5: Top: the slice of the 3D Helmholz potential $\frac{\sin\|3x\|}{\|x\|}$ over cross-section (left) and the absolute error for its Tucker approximation with rank $r = 10$ (right). Bottom: decay of the Tucker approximation error in the Frobenius norm with respect to the Tucker rank (left) and the orthogonal Tucker vectors $\mathbf{v}_k^{(1)}$, $k = 1, \dots, 6$, corresponding to $\frac{\sin\|3x\|}{\|x\|}$ (right).

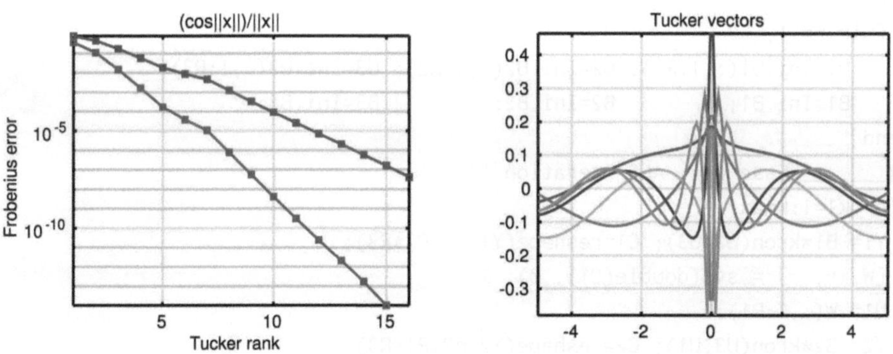

Figure 3.6: Decay of the Tucker approximation error in the Frobenius norm with respect to the Tucker rank (left) and the orthogonal Tucker vectors $\mathbf{v}_k^{(1)}$, $k = 1, \dots, 6$, for the Helmholz potential $\frac{\cos\|x\|}{\|x\|}$ (right).

We recommend to copy and paste first the main program from Example 3, and then add all subroutines to the end of this file.

The Tucker_full_3D_ini function in subroutine 1 computes the Tucker decomposition of a 3D tensor A3 for given Tucker ranks *ir*. Note that the initial guess by HOSVD is computed only in the first call (since it is a repeated procedure for every call of the function), and then it is stored in the auxiliary structure Ini. Function TnormF computes the Frobenius norm of a given tensor A. The number of ALS iterations here is chosen by $kmax = 3$.

```
%_______________subroutine 1_____________________________________________
function [U1,U2,U3,LAM3F,Ini] = Tucker_full_3D_ini(A3,NR,kmax,ir,Ini)
[n1,n2,n3]=size(A3);
R1=NR(1); R2=NR(2); R3=NR(3); %nd=3;
[~,nstep]=size(Ini.U1);
if ir == 1
%___ Fase I - Initial Guess
D= permute(A3,[1,3,2]); B1= reshape(D,n1,n2*n3);
[Us, ~, ~]= svd(double(B1),0); U1=Us(:,1:R1);
Ini.U1=Us(:,1:nstep);
Ini.B1=B1;
D= permute(A3,[2,1,3]); B2= reshape(D,n2,n1*n3);
[Us, ~, ~]= svd(double(B2),0); U2=Us(:,1:R2);
Ini.U2=Us(:,1:nstep);
Ini.B2=B2;
D= permute(A3,[3,2,1]); B3= reshape(D,n3,n1*n2);
[Us, ~, ~]= svd(double(B3),0); U3=Us(:,1:R3);
Ini.U3=Us(:,1:nstep);
Ini.B3=B3;
end
if ir ~= 1;
    U1=Ini.U1(:,1:R1); U2=Ini.U2(:,1:R2); U3=Ini.U3(:,1:R3);
    B1=Ini.B1;          B2=Ini.B2;          B3=Ini.B3;
end
%_______ Fase II - ALS Iteration
for k1=1:kmax
 Y1= B1*kron(U2,U3); C1=reshape(Y1,n1,R2*R3);
 [W, ~, ~] = svd(double(C1), 0);
 U1= W(:,1:R1);
 Y2= B2*kron(U3,U1); C2=reshape(Y2,n2,R1*R3);
 [W, ~, ~] = svd(double(C2), 0);
 U2= W(:,1:R2);
 Y3= B3*kron(U1,U2); C3=reshape(Y3,n3,R2*R1);
```

```
 [W, ~, ~] = svd(double(C3), 0);
 U3= W(:,1:R3);
end
Y1= B1*kron(U2,U3); LAM3 = U1'*Y1 ; LLL=reshape(LAM3,R1,R3,R2);
LAM3F=permute(LLL,[1,3,2]);
end

%_________subroutine 2 _____________________________________
function f = Tnorm(A)
NS = size(A); nd =length(NS); nsa =1;
for i = 1:nd
nsa = nsa*NS(i);
end
B = reshape(A,1,nsa); f = norm(B);
end
%__________________________________________________________

%_________subroutine 3_____________________________________
function A3F=Tuck_2_F(LAM3F,U1,U2,U3)
[R1,R2,R3]=size(LAM3F);
[n1,~]=size(U1); [n2,~]=size(U2); [n3,~]=size(U3);
  LAM31=reshape(LAM3F,R1,R2*R3);
  CNT1=LAM31'*U1';
  CNT2=reshape(CNT1,R2,R3*n1);
  CNT3=CNT2'*U2';
  CNT4=reshape(CNT3,R3,n1*n2);
  CNT5=CNT4'*U3';
  A3F=reshape(CNT5,n1,n2,n3);
end
%__________________________________________________________
```

In the main program, given in Example 3, first, the 3D tensor related to a Slater function is generated in the rectangular computational box with the grid sizes $n_1 = 65$, $n_2 = 67$, and $n_3 = 69$ along x-, y-, and z-axis, respectively.

The Tucker tensor decomposition is performed with equal ranks for all three variables x, y, and z, in a loop starting from a rank equal to one up to a maximum rank given by the parameter “max_Tr”. Number of ALS iterations is given by the parameter “kmax”.

The main program prints out the same data as shown in Figure 3.5: the error of the Tucker decomposition with respect to the Tucker rank in figure (1), displays the generated tensor in figure (2), and examples of the Tucker vectors for one of the modes are provided in figure (3).

```
%________________Example 3___ main program ________
% Test_Tucker.m
clear; close all; max_Tr=10; % maximum Tucker rank
b=10.0; % size of the interval
T_error=zeros(1,max_Tr); nd=3; kmax =3;
T_en_error=zeros(1,max_Tr);
coef=1; al=2.0;
n= 64;  b2=b/2; h1 = b/n;
x=-b2:h1:b2; [~,n1]=size(x); % interval in x: [-b,b]
y=-b2-h1:h1:b2+h1; [~,n2]=size(y);  % interval in y: [-b-h1,b+h1]
z=-b2-2*h1:h1:b2+2*h1; [~,n3]=size(z);  % interval in y: [-b-2h1,b+2h1]
A3=zeros(n1,n2,n3); A3F=zeros(n1,n2,n3);
%__________ generate a 3D tensor__________________
for i=1:n1
    for j=1:n2
       A3(i,j,:)=coef*exp(-al*sqrt(x(1,i)^2 + y(1,j)^2 + z(1,:).^2));
    end
end
Ini.U1=zeros(n1,max_Tr); Ini.U2=zeros(n2,max_Tr);
Ini.U3=zeros(n3,max_Tr);
for ir=1:max_Tr
   NR=[ir ir ir];
   [U1,U2,U3,LAM3F,Ini] = Tucker_full_3D_ini(A3,NR,kmax,ir,Ini);
   A3F=Tuck_2_F(LAM3F,U1,U2,U3);
   err=Tnorm(A3F - A3)/Tnorm(A3);
   T_error(1,ir)=abs(err);
   enr=(Tnorm(A3F) -Tnorm(A3))/Tnorm(A3);
   T_en_error(1,ir)=abs(enr);
   fprintf(1, '\n iter = %d , err_Fro =  %5.4e \n', ir, err);
end
figure(1); %____ draw convergence of the error____
semilogy(T_error(1,1:max_Tr),'Linewidth',2,'Marker','square');
hold on;
semilogy(T_en_error(1,1:max_Tr),'r','Linewidth',2,'Marker','square');
set(gca,'fontsize',16);
xlabel('Tucker rank','fontsize',16);
ylabel('error','fontsize',16);
grid on; axis tight;
%______________draw the function________________________________
A2=A3F(:,:,(n3-1)/2); % take a 2D plane of the 3D tensor
figure(2); mesh(y,x,A2); axis tight;
%________________draw__Tucker vectors_________________________
```

```
figure(3)
plot(x,U1(:,1),'Linewidth',2);
hold on;
for j=2:max_Tr-2
    plot(x,U1(:,j),'Linewidth',2);
end
set(gca,'fontsize',16);
str3='Tucker vectors';
str2=[str3,'  al= ',num2str(al) ', rank = ',num2str(max_Tr)];
title(str2,'fontsize',16); axis tight; grid on; hold off;
%_____________________________________________________________
```

When changing the grid papameter “n”, please note that due to the restrictions for the HOSVD, the size of the tensor should be not larger that $n1\, n2\, n3 \leq 128^3$.

The following conclusions are the consequences from above numerics [146].

Remark 3.6. The Tucker approximation error for the considered class of function-related tensors decays exponentially with respect to the Tucker rank.

Remark 3.7. The shape of the orthogonal vectors in the unitary matrices of the Tucker decomposition for the class of function-related tensors is almost independent on n.

Remark 3.8. The entries of the core tensor of the Tucker decomposition for the considered function-related tensors decay fast vs. index $k_\ell = 1, \dots, r$, $\ell = 1, 2, 3$.

Properties of the Tucker decomposition for the function-related tensors described in Remarks 3.7 and 3.8 will be used further in the development of the multigrid Tucker algorithms.

3.2 Multigrid Tucker tensor decomposition

In the previous section, we have discussed the Tucker decomposition algorithm that provides complexity of the Tucker tensor approximation of the order of

$$W_{\mathrm{F2T}} = O(n^{d+1}) \tag{3.21}$$

for full format target tensors of size $n^{\otimes d}$. This bound restricts application of this standard Tucker scheme from multilinear algebra to small dimensions d and moderate grid sizes n. Thus, we have the computational work for the Tucker decomposition of the full format tensors in 3D,

$$W_{\mathrm{F2T}} = O(n^4), \tag{3.22}$$

which practically restricts the maximum size of the input tensors to $n \approx 200^3$ for conventional computers. Our goal is to reach linear in a volume complexity $O(n^3)$ by avoiding the HOSVD transform, thus allowing the maximum size of the input tensors corresponding to available computer storage.

The *multigrid Tucker tensor decomposition*, which gives the way to avoid storage limitations of the standard Tucker algorithm, was introduced by V. Khoromskaia and B. Khoromskij in 2008 [174, 146]. The idea of the multilevel Tucker approximation originates from investigating the numerical examples of the orthogonal Tucker decomposition for the function-related tensors, in particular, the regularity of the orthogonal Tucker vectors and the weak dependence of their shapes on the univariate grid parameter n.

The nonlinear multigrid Tucker tensor approximation problem for minimizing the functional

$$\mathbf{A}_{0,M} \in \mathcal{S}_0 \subset \mathbb{V}_{\mathbf{n}_M}: \quad f(\mathbf{A}) := \|\mathbf{A}_{0,m} - \mathbf{A}_m\|^2 \to \min \tag{3.23}$$

is solved over a sequence of nested subspaces

$$\mathbb{T}_{\mathbf{r},0} \subset \cdots \subset \mathbb{T}_{\mathbf{r},m} \subset \cdots \subset \mathbb{T}_{\mathbf{r},M},$$

using the sequence of dyadically refined grids of size $n = n_0 2^{m-1}$ with $m = 1, \ldots, M$. Thus, the Tucker decomposition problem for a tensor $\mathbf{A}_{0,M} \in \mathbb{V}_{\mathbf{n}_M}$ obtained as discretization of a function over the fine grid of size n_M is based on the successive reiteration of the ALS Tucker approximation on a sequence of refined grids. In this case, the initial guess is computed by HOSVD only at the coarsest grid with $n_0 \ll n_M$, at the moderate cost $O(n_0^{d+1})$. Then, on finer grids, for the first run of ALS iterations, the initial guess for $\{V_m^{(\ell)}\}_{\ell=1}^d$ is computed by interpolation of the orthogonal Tucker vectors $\{V_{m-1}^{(\ell)}\}_{\ell=1}^d$ from the previous grid level. At every current ALS iteration, it is updated by contractions with the full tensor $\mathbf{A}_{0,m} \in \mathbb{V}_{\mathbf{n}_m}$ at the corresponding grid. Thus, the "single-hole" tensors obtained by using the interpolated orthogonal matrices appear to be of sufficient "quality", as if they were obtained by the projections using the HOSVD for the corresponding grid.

The resulting complexity of the Multigrid Tucker decomposition for full format tensors is estimated by

$$w_{\mathrm{F2T}} = O(n^3),$$

which currently makes possible the application of multigrid accelerated full-to-Tucker (F2T) algorithm to the 3D function-related tensors with the maximal possible grid size that is only bounded by the storage size for the input tensor. We refer also to [231] where the adaptive cross approximation to 3D Tucker tensors has been applied.

The concept of the multigrid Tucker approximation applies to the multidimensional data obtained as a discretization of a regular (say, continuous or analytic) multivariate function on a sequence of refined spatial grids. The typical application areas include the tensor approximation of multi-dimensional operators and functionals, the solution of integral-differential equations in $\mathbb{R}^d$, data-structured representation of physically relevant quantities, see for example [36].

For a fixed grid parameter n, let us introduce the equidistant tensor grid

$$\omega_{\mathbf{d},n} := \omega_1 \times \omega_2 \times \cdots \times \omega_d, \tag{3.24}$$

where $\omega_\ell := \{-b + (k-1)h : k = 1, \ldots, n+1\}$ $(\ell = 1, \ldots, d)$ with mesh-size $h = 2b/n$. Define a set of collocation points $\{x_{\mathbf{i}}\}$ in $\Omega := [-b, b]^d \in \mathbb{R}^d$, located at the midpoints of the grid-cells and numbered by $\mathbf{i} \in \mathcal{I} := \{1, \ldots, n\}^d$ (see the explicit definition in (3.16)). For fixed n, the target tensor

$$\mathbf{A}_m = [a_{n_m, \mathbf{i}_m}] \in \mathbb{R}^{\mathcal{I}_m} \tag{3.25}$$

is defined by sampling of the given continuous multivariate function $f : \Omega \to \mathbb{R}$ on the set of collocation points $\{x_{\mathbf{i}}\}$ as follows:

$$a_{n_m, \mathbf{i}} = f(x_{\mathbf{i}_m}), \quad \mathbf{i}_m \in \mathcal{I}_m.$$

The algorithm for the multigrid Tucker tensor approximation for full size tensors is described as follows.

Algorithm Multigrid Tucker ($\mathbb{V}_{\mathbf{n}} \to \mathcal{T}_{\mathbf{r}}$) (Multigrid full-to-Tucker approximation).

(1) Given $\mathbf{A}_m \in \mathbb{V}_{\mathbf{n}}$ in the form (3.25), corresponding to a sequence of grid parameters $n_m := n_0 2^m$, $m = 0, 1, \ldots, M$. Fix the Tucker rank $\mathbf{r}$, and the iteration number $k_{\max}$.

(2) For $m = 0$, solve the approximation problem by Tucker algorithm ($\mathbb{V}_{\mathbf{n}_0} \to \mathcal{T}_{\mathbf{r}}$) by using HOSVD and $k_{\max}$ steps of ALS iteration.

(3) For $m = 1, \ldots, M$, perform the *cascadic multigrid Tucker approximation*:

(3a) Compute the initial guess for the side matrices on level m by interpolation $\mathbf{I}_{(m-1,m)}$ from level $m-1$ (using piecewise linear or cubic splines)

$$V^{(\ell)} = V_m^{(\ell)} = \mathbf{I}_{(m-1,m)}(V_{m-1}^{(\ell)}), \quad \ell = 1, \ldots, d.$$

(3b) Starting with the initial guess $V^{(\ell)}$ $(\ell = 1, \ldots, d)$, perform $k_{\max}$ steps of the ALS iteration as in Step (2) of Basic Tucker algorithm (see Section 2.2.3).

(4) Compute the core $\boldsymbol{\beta}$ by the orthogonal projection of A onto $\mathbb{T}_{\mathbf{n}} = \bigotimes_{\ell=1}^{d} \mathbb{T}_\ell$ with $\mathbb{T}_\ell = \operatorname{span}\{v_\nu^{(\ell)}\}_{\nu=1}^{r_\ell}$ (see Remark 2.6),

$$\boldsymbol{\beta} = A \times_1 V^{(1)^T} \times_2 \cdots \times_d V^{(d)^T} \in \mathbb{B}_{\mathbf{r}},$$

at the cost $O(r^d n)$.

Figure 3.7, left, shows the numerical example of the multigrid Tucker approximation to fully populated tensors given by the 3D Slater function $e^{-\|x\|}$ $(x \in [-b, b]^3$, $b = 5.0)$, sampled over large $n \times n \times n$ uniform grids with $n = 128, 256$, and 512. The corresponding computation times (in MATLAB) (sec) of the multigrid Tucker decomposition algorithm are shown in Figure 3.7, right.

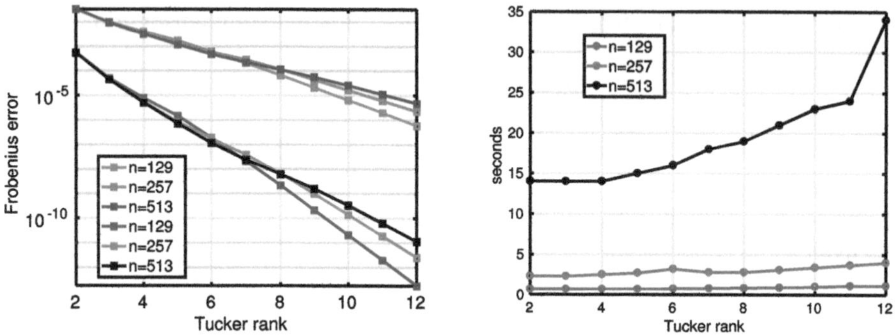

Figure 3.7: Convergence of the multigrid Tucker approximation with respect to the Tucker rank r (left) and times for the multigrid algorithm (right).

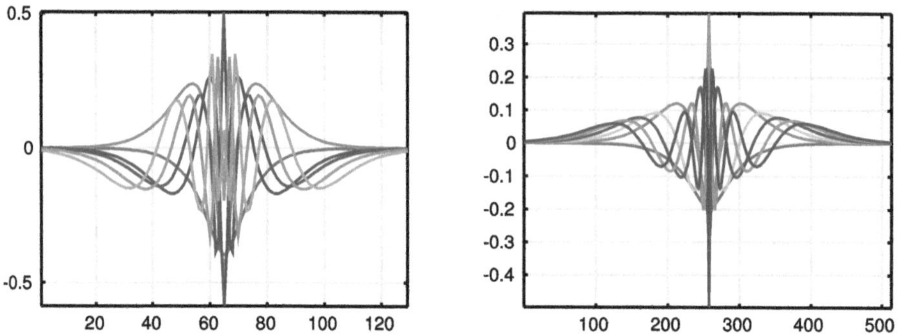

Figure 3.8: Tucker vectors for the Slater potential on the grid with $n = 129$ (left) and $n = 513$ (right).

Figure 3.8 shows the shape of the Tucker vectors for the values of the discretization parameter $n = 129$ and $n = 513$.

For testing the programs for multigrid Tucker tensor decomposition, first generate 3D tensors by the program in Example 4, with $n = 32, 64, 128, 256$, (if storage allows, also 512). Before starting the program in Example 5, add the subroutines MG, and Interpolation, as well as subroutines 2, 3 from the previous example.

Then one can start the program, choosing the parameter MG (MG = 3, 4, or 5 correspond to the largest grid sizes 128, 256, or 512, respectively).

```
%________________Example 4__Main program 4_________________________
clear;  max_Tr=18; % maximum Tucker rank
b=10.0; % size of the interval
T_error=zeros(1,max_Tr); T_energy=zeros(1,max_Tr);
nd=3; kmax =3;
coef=1; al=2.0;
n= 32; b2=b/2; h1 = b/n;
```

```
x=-b2:h1:b2; [~,n1]=size(x); % interval in x: [-b,b]
y=-b2-h1:h1:b2+h1; [~,n2]=size(y);  % interval in y: [-b-h1,b+h1]
z=-b2-2*h1:h1:b2+2*h1; [~,n3]=size(z);  % interval in y: [-b-2h1,b+2h1]
A3=zeros(n1,n2,n3); A3F=zeros(n1,n2,n3);
for i=1:n1
    for j=1:n2
       A3(i,j,:) = coef*exp(-al*sqrt(x(1,i)^2 +  y(1,j)^2 +  z(1,:).^2));
    end
end

filename2 = ['A3_' int2str(n1) '_N_Slat.mat']
save(filename2,'A3');
%_______________end of Example 4________________________________
```

The complexity of the multigrid Tucker approximation by the ALS algorithm applied to full format tensors is given in the following lemma.

Lemma 3.9 ([174]). *Suppose that $r^2 \leq n_m$ for large m. Then the numerical cost of the multigrid Tucker algorithm is estimated by*

$$W_{F\to T} = O(n_M^3 r + n_M^2 r^2 + n_M r^4 + n_0^4) = O(n_M^3 r + n_0^4).$$

Proof. In Step (2), the HOSVD on the coarsest grid level requires $O(n_0^4)$ operations (which for large $n = n_m$ is negligible compared to other costs in the algorithm). Next, for fixed $n = n_m$, the assumption $r^2 \leq n$ implies that at every step of the ALS iterations the costs of the consequent contractions to compute the $n \times r^2$ unfolding matrix $B_{(q)}$ is estimated by $O(n^3 r + n^2 r^2)$, whereas the SVD of $B_{(q)}$ requires $O(nr^4)$ operations. Summing up over the levels completes the proof, taking into account that the Tucker core is computed in $O(n_M^3 r)$ operations. □

```
%_______________Example 5___Main program_5___________________________
clear; nstep=10; MG=3;  b=10.0;
nd=3; kmax =5; ng=32;
T_error=zeros(nstep,MG); T_energy=zeros(nstep,MG);

 for nr=1:nstep
 NR=[nr nr nr];  disp(nr);
   for im=1:MG
   n1=ng*2^(im-1)+1; disp(n1);
   b2=b/2; Hunif = b/(n1-1); xcol1=-b2:Hunif:b2;
   ycol1=-b2-Hunif:Hunif:b2+Hunif;
   zcol1=-b2-2*Hunif:Hunif:b2+2*Hunif;
   filename1 = ['A3_' int2str(n1) '_N_Slat.mat'];
```

```
   load(filename1);
   [n1,n2,n3]=size(A3);
   if im==1; UC1=zeros(n1,nr); UC2=zeros(n2,nr); UC3=zeros(n3,nr);
    save('INTER_COMPS_MG.mat','UC1','UC2','UC3');
    else
    load INTER_COMPS_MG.mat;
    end
    Kopt = 0;  if im >1; Kopt = 1;  end  %MG
   [U1,U2,U3,LAM3F] = TensR_3sub_OPT_MG(A3,NR,kmax,Kopt,UC1,UC2,UC3);
   if im < MG
    n11=2*n1-1;
    Hunif11 = b/(n11-1); xcol11=-b2:Hunif11:b2;
    ycol11=-b2-Hunif11:Hunif11:b2+Hunif11;
    zcol11=-b2-2*Hunif11:Hunif11:b2+2*Hunif11;
    [UC1,UC2,UC3] = Make_Inter_Vect_xyz(xcol1,xcol11,ycol1,ycol11,...
    zcol1,zcol11,n11,U1,U2,U3);
   end
   save INTER_COMPS_MG.mat UC1 UC2 UC3;
   A3F=Tuck_2_F(LAM3F,U1,U2,U3);
   err=Tnorm(A3F - A3)/Tnorm(A3);
   enr=(Tnorm(A3F) -Tnorm(A3))/Tnorm(A3);
   T_error(nr,im)=abs(err);
   T_energy(nr,im)=abs(enr);
   fprintf(1, '\n iter = %d , err_Fro =  %5.4e \n', nr, err);
  end
 end
 figure(20);
      for i=1:MG
       semilogy(T_error(2:nstep,i),'Linewidth',2,'Marker','square');
       hold on;
       semilogy(T_energy(2:nstep,i),':','Linewidth',2,'Marker','square');
       set(gca,'fontsize',16);
       xlabel('Tucker rank','fontsize',16);
       ylabel('error','fontsize',16);
       grid on; axis tight;
      end
%___________________end of main program_________________________________

%_______________subroutine_MG____________________________________
function [U1,U2,U3,LAM3F] = TensR_3sub_OPT_MG(A3,NR,kmax,...
    Kopt,UC1,UC2,UC3)
[n1,n2,n3]=size(A3);
```

```
R1=NR(1); R2=NR(2); R3=NR(3);
if Kopt == 1
U1 = UC1;  U2 = UC2; U3 = UC3;
end
D= permute(A3,[1,3,2]); B1= reshape(D,n1,n2*n3);
D= permute(A3,[2,1,3]); B2= reshape(D,n2,n1*n3);
D= permute(A3,[3,2,1]); B3= reshape(D,n3,n1*n2);
if Kopt == 0
%____Fase I - Initial Guess ________
[Us,~,~]= svd(double(B1),0); U1=Us(:,1:R1);
[Us,~,~]= svd(double(B2),0); U2=Us(:,1:R2);
[Us,~,~]= svd(double(B3),0); U3=Us(:,1:R3);
end
%_______ Fase II - ALS Iteration ____
for k1=1:kmax
Y1= B1*kron(U2,U3);
C1=reshape(Y1,n1,R2*R3);
[W,~,~] = svd(double(C1), 0);
U1= W(:,1:R1); %
Y2= B2*kron(U3,U1);
C2=reshape(Y2,n2,R1*R3);
[W,~,~] = svd(double(C2), 0);
U2= W(:,1:R2); %
Y3= B3*kron(U1,U2);
C3=reshape(Y3,n3,R2*R1);
[W,~,~] = svd(double(C3), 0);
U3= W(:,1:R3); %
end
Y1= B1*kron(U2,U3);
LAM3 = U1'*Y1 ;      %'
LLL=reshape(LAM3,R1,R3,R2);
LAM3F=permute(LLL,[1,3,2]);
end
%_____________________________________________________________________

%______subroutine Interpolation______
function [U10,U20,U30] = Make_Inter_Vect_xyz(xcol,ixcol,ycol,iycol,...
zcol,izcol,n11,UT1,UT2,UT3)
n12=n11+2; n13=n11+4;
[~,R1]=size(UT1); [~,R2]=size(UT2); [~,R3]=size(UT3);
U10=zeros(n11,R1); U20=zeros(n12,R2); U30=zeros(n13,R3);
for i=1:R1
```

```
U10(:,i) = interp1(xcol,UT1(:,i),ixcol,'spline');
end
for i=1:R2
U20(:,i) = interp1(ycol,UT2(:,i),iycol,'spline');
end
for i=1:R3
U30(:,i) = interp1(zcol,UT3(:,i),izcol,'spline');
end
end
%-------------------end of example-----------------
```

3.2.1 Examples of potentials on lattices

(5) Periodic structures of Slater functions. Finally, we analyze the "multi-centered Slater potential" obtained by displacing a single Slater function with respect to the $L \times L \times L$ lattice nodes, with a distance between nodes $H > 0$, specifying centers of Slater functions,

$$g(x) = c \sum_{i=1}^{m} \sum_{j=1}^{m} \sum_{k=1}^{m} e^{-\alpha\sqrt{(x_1 - iH + \frac{m+1}{2}H)^2 + (x_2 - jH + \frac{m+1}{2}H)^2 + (x_3 - kH + \frac{m+1}{2}H)^2}}. \tag{3.26}$$

Figure 3.9 (top-left) recalls a single Slater function. The corresponding convergence of the multigrid Tucker approximation error in Frobenius norm for the grids 65^3, 129^3, and 257^3, respectively, are shown in Figure 3.9 (top-right). Figure 3.9 (bottom-left) shows the cross-section of a multi-centered Slater potential on an $8 \times 8 \times 8$ lattice and the corresponding Tucker tensor approximation error for the same grids is shown in Figure 3.9 (bottom-right).

Inspection of these periodic structures shows that the convergence rate of the rank-(r, r, r) Tucker approximation practically does not depend on the size of the lattice-type structure. And accuracies are nearly the same. For example, for the Tucker rank $r = 10$, it is exactly 10^{-5} for all versions of the single/multicentered Slater function. These properties were first demonstrated on numerical examples of multicentered Slater function in [146] for $L \times L \times L$ lattices with $L = 10$ and $L = 16$. These features can be valuable in the grid-based modeling of periodic (or nearly periodic) structures in the density functional theory. It indicates that the Tucker decomposition can be helpful in constructing of a small number of problem-adapted basis functions for large lattice-type clusters of atoms.

Figure 3.10 shows the Tucker vectors of the multi-Slater function for $L = 10$. Grid size is 129^3. Next remark, see V. Khoromskaia [146], became a prerequisite for the development of powerful methods for summation of the long-range potentials on large finite 3D lattices [148, 149, 153].

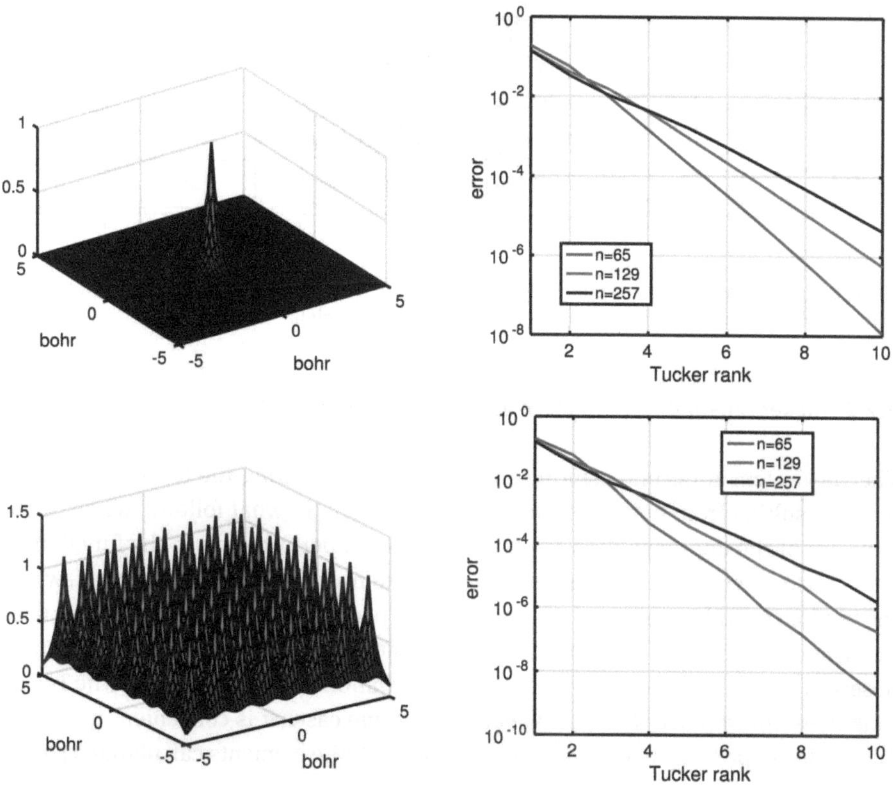

Figure 3.9: Comparison of the decay of the Tucker tensor decomposition error vs. r for a single Slater function and for Slater functions positioned at 3D lattice nodes.

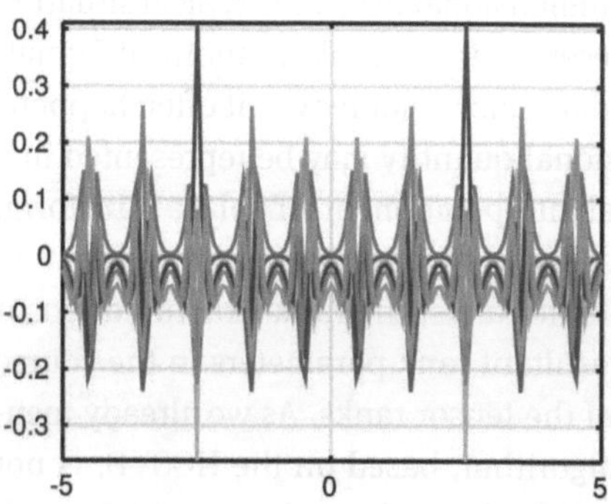

Figure 3.10: Tucker vectors of a 3D multi-centered Slater potential with 10^3 centers.

Remark 3.10. For a fixed approximation error, the Tucker rank of lattice-type structures practically does not depend on the number of cells included in the computational box.

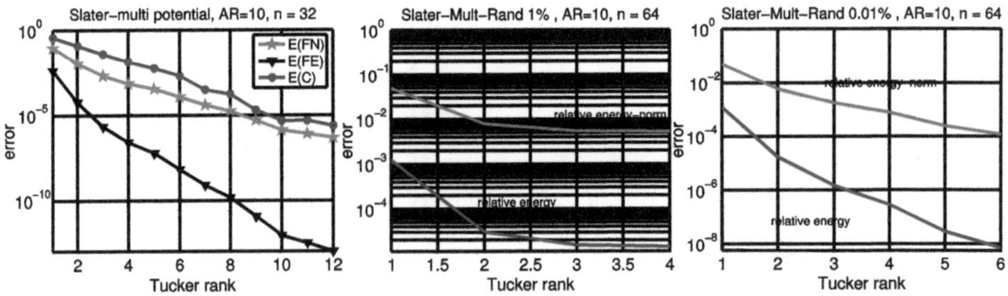

Figure 3.11: Convergence of the approximation error for the multi-centered unperturbed (left panel) and randomly perturbed Slater potential (middle, and right panels).

3.2.2 Tucker tensor decomposition as a measure of randomness

The Tucker tensor decomposition can be used for measuring the level of noise in a tensor resulting from finite element calculations [173]. In what follows, we show the behavior of the approximation error, under random perturbation in the function related tensor. Figure 3.11 demonstrates such an example for the Slater potential, where the random complement equals to 1 percent, 0.1 percent, and 0.01 percent of the maximum amplitude. It can be seen that the exponential convergence in the Tucker rank is observed only till the order of the random perturbation. Further increase in the Tucker rank does not improve the approximation. In some cases it is convenient to use the Tucker decomposition to estimate the accuracy of finite elements calculations [36].

3.3 Reduced higher order SVD and canonical-to-Tucker transform

In applications related to the solution of high dimensional PDEs, the typical situation may arise when the target tensor is already presented in the rank-R canonical format, $\mathbf{A} \in \mathcal{C}_{R,\mathbf{n}}$, but with relatively large R and large mode size n. Moreover, it often happens that a tensor corresponding to a multidimensional quantity may be represented in a discretized separable form, using a polynomial interpolation, or a Laplace transform, resulting in a sum of a huge number of Gaussians, due to the accurate sinc quadrature approximation. Even in case of initially moderate ranks, the rank-structured tensor operations (2.34)–(2.37), lead to increase of resultant rank parameters in the course of multilinear algebra, due to multiplication of the tensor ranks. As we already mentioned, the standard Tucker decomposition algorithm, based on the HOSVD, is not computationally feasible for the case of large mode size n, due to its complexity scaling as $O(n^{d+1})$.

To that end, an essential advance was brought by the so-called reduced higher-order singular value decomposition (RHOSVD) as part of the canonical-to-Tucker (C2T) transform introduced in 2008, [174]. It was demonstrated that for the Tucker decomposition of function-related tensors, given in the canonical form, there is no need to build a full tensor. Instead, the orthogonal basis is computed using only the directional (side) matrices of the canonical tensor, that consist of skeleton vectors in every single dimension. The RHOSVD can be considered as the generalization of the reduced SVD for rank-R matrices (see Section 2.1.5) to higher-order canonical rank-R tensors. Actually, RHOSVD is considered as the SVD in many dimensions that can be performed without tensor–matrix unfolding, and it is free of the so-called “curse of dimensionality”.

3.3.1 Reduced higher order SVD for canonical target

We consider the function-related tensor presented in the rank-R canonical format, $\mathbf{A} \in \mathcal{C}_{R,\mathbf{n}}$ (here we recall (2.13) from section 2.2.2 for convenience of presentation),

$$\mathbf{A} = \sum_{k=1}^{R} \xi_k \mathbf{u}_k^{(1)} \otimes \cdots \otimes \mathbf{u}_k^{(d)}, \quad \xi_k \in \mathbb{R}.$$

In what follows, we use the equivalent contacted product tensor representation of $\mathbf{A}$ (see Figure 3.12),

$$\begin{aligned} &\mathbf{A} = \boldsymbol{\xi} \times_1 U^{(1)} \times_2 U^{(2)} \times_3 \cdots \times_d U^{(d)}, \\ &\boldsymbol{\xi} = \operatorname{diag}\{\xi_1, \ldots, \xi_d\} \in \mathbb{R}^{R^{\otimes d}}, \quad U^{(\ell)} = [\mathbf{u}_1^{(\ell)} \ldots \mathbf{u}_R^{(\ell)}] \in \mathbb{R}^{n_\ell \times R}, \end{aligned} \tag{3.27}$$

where the core tensor $\boldsymbol{\xi}$ is a diagonal of a hypercube.

Canonical tensors with large ranks appear, for example, in electronic structure calculations. The electron density of a molecule is a Hadamard product of two large sums of discrete Gaussians representing the Gaussian-type orbitals, and it may be represented in a rank-R canonical tensor with relatively large rank $R \approx 10^4$. At the same time, the $n \times n \times n$ 3D Cartesian grid for representation of the electron density should be large enough, with n of the order of 10^4, to resolve multiple nuclear cusps, corresponding to locations of nuclei in a molecule. Then the size of the respective tensor in full format is about $n^3 = 10^{12}$, which is far from being feasible for the Tucker tensor decomposition algorithm discussed in the previous sections. The same applies to the multigrid version.

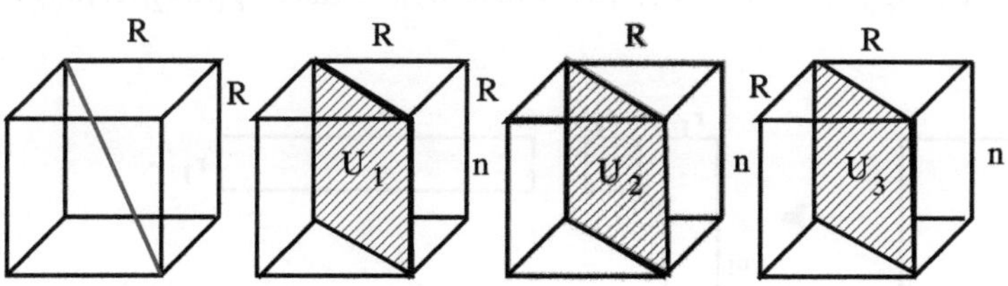

Figure 3.12: Representation of the 3D canonical rank-R tensor as contractions (3.27) with side matrices.

The canonical-to-Tucker (C2T) transform, introduced in [174], proved to be an efficient tool for reducing the redundant rank parameter for the canonical rank-R tensors, in a large scale of the sizes of their grids and ranks. The favorable feature of the C2T algorithm is the capability to produce the Tucker decomposition of a canonical tensor without the need to compute its representation in the full tensor format.

It is worth to note that this transform could hardly appear in multilinear algebra since it is appropriate particularly for function-related rank-R canonical tensors exhibiting exponentially fast decay of singular values with respect to the Tucker rank. In fact, this algorithm, as well as, the theory on low-rank approximation of the multidimensional functions and operators [161, 111], was the starting point for tensor numerical methods for PDEs. This is because it provides an efficient tool for reducing the ranks increasing due to multiplication of tensor ranks in the course of tensor operations.

The main idea of the C2T transform is as follows:

- Reduced HOSVD, which provides the initial guess for the Tucker tensor decomposition, is computed by the SVD of side matrices $U^{(\ell)}$. There is no need to construct the full size tensor and make its HOSVD at enormous cost.
- The Tucker core tensor is computed by simple contractions; ALS iteration for the nearly best Tucker approximation practically requires only few steps.
- C2T transform accomplished with the CP decomposition of the small size Tucker core provides the algorithm for the rank reduction in the canonical tensor (Tucker-to-canonical (T2C) transform will be discussed in the next section).

Instead of the Tucker transform decomposition of full size tensors that requires the HOSVD via SVD of full unfolding matrices of size $n \times n^{d-1}$, it is sufficient to make the *reduced* HOSVD based on the SVD of small side matrices $U^{(\ell)}$ of size $n \times R$, $\ell = 1, \dots, d$,

$$U^{(\ell)} = V^{(\ell)} \Sigma_\ell W^{(\ell)}, \qquad \ell = 1, \dots, d. \tag{3.28}$$

Figure 3.13 shows the SVD step in RHOSVD for the dimension $\ell = 1$.

The RHOSVD transform is defined as follows, [174].

Definition 3.11 (RHOSVD). Given $\mathbf{A} = \boldsymbol{\xi} \times_1 U^{(1)} \times_2 U^{(2)} \cdots \times_d U^{(d)} \in \mathcal{C}_{R,\mathbf{n}}$, $\boldsymbol{\xi} = \operatorname{diag}\{\xi_1, \dots, \xi_R\}$, and the Tucker rank parameter $\mathbf{r} = (r_1, \dots, r_d)$, introduce the truncated SVD of the side-matrices $U^{(\ell)}$, $V_0^{(\ell)} D_{\ell,0} W_0^{(\ell)T}$ $(\ell = 1, \dots, d)$, where $D_{\ell,0} = \operatorname{diag}\{\sigma_{\ell,1}, \sigma_{\ell,2}, \dots, \sigma_{\ell,r_\ell}\}$,

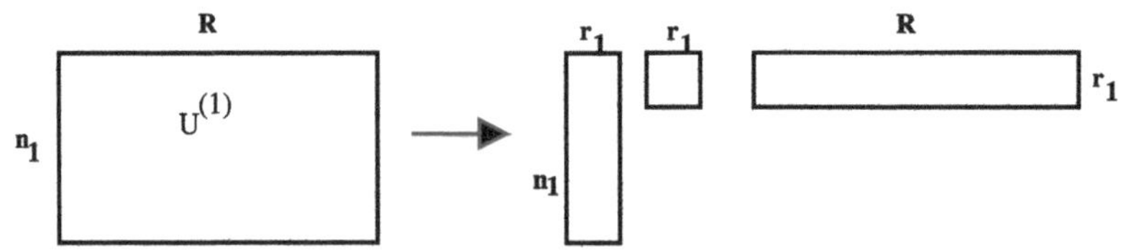

Figure 3.13: RHOSVD: truncated SVD of the side matrix $U^{(1)}$ in the C2T transform.

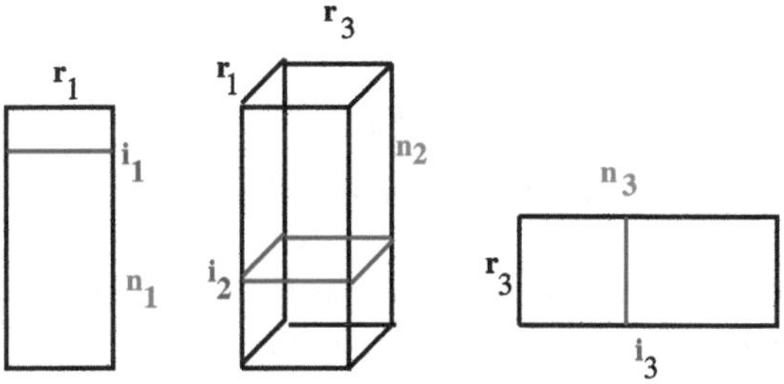

Figure 3.14: C2T transform converts a 3D canonical tensor (3.27) into a contracted product of two orthogonal matrices and a single-hole tensor.

whereas $V_0^{(\ell)} \in \mathbb{R}^{n\times r_\ell}$ and $W_0{}^{(\ell)} \in \mathbb{R}^{R\times r_\ell}$ are the respective submatrices of $V^{(\ell)}$ and $W^{(\ell)}$ in SVD of $U^{(\ell)}$ in (3.28). Then the RHOSVD approximation of **A** is given by

$$\mathbf{A}^0_{(\mathbf{r})} = \boldsymbol{\xi} \times_1 [V_0^{(1)} D_{1,0} {W_0^{(1)}}^T] \times_2 [V_0^{(2)} D_{2,0} {W_0^{(2)}}^T] \cdots \times_d [V_0^{(d)} D_{d,0} {W_0^{(d)}}^T]. \tag{3.29}$$

Notice that $\mathbf{A}^0_{(\mathbf{r})}$ in (3.29) is obtained by the projection of the tensor **A** onto the matrices of left singular vectors $V_0^{(\ell)}$. Using projections of the initial CP tensor **A** onto the orthogonal matrices $V_0^{(\ell)}$, it is possible to construct the single-hole tensor for every mode of **A**. For example, if $d = 3$, the tensor given in (3.27) converts into a contraction of two orthogonal matrices and the single-hole tensor, that is actually the tensor train (TT) representation, see Figure 3.14.

The C2T decomposition with RHOSVD was originally developed for reducing the ranks of the canonical tensor representation of the electron density. Now it is used in many other applications, for example, in summation of many-particle potentials and low-rank representation of the radial basis functions. In fact, the RHOSVD can be used as a first step in multiplicative tensor formats like TT and HT, when the original tensor is given in the canonical tensor format.

In what follows, we recall Theorem 2.5 from [174] describing the algorithm of canonical-to-Tucker approximation and proving the error estimate.

Theorem 3.12 (Canonical to Tucker approximation).

(a) *Let* $\mathbf{A} \in \mathcal{C}_{R,\mathbf{n}}$ *be given by* (2.13). *For given Tucker rank* $\mathbf{r} = (r_1, \ldots, r_d)$, *the minimization problem*

$$\mathbf{A} \in \mathcal{C}_{R,\mathbf{n}} \subset \mathbb{V}_\mathbf{n} : \quad \mathbf{A}_{(\mathbf{r})} = \operatorname{argmin}_{\mathbf{T}\in\mathcal{T}_{\mathbf{r},\mathbf{n}}} \|\mathbf{A} - \mathbf{T}\|_{\mathbb{V}_\mathbf{n}} \tag{3.30}$$

is equivalent to the dual maximization problem

$$[\widehat{V}^{(1)}, \ldots, \widehat{V}^{(d)}] = \underset{Y^{(\ell)}\in\mathcal{G}_\ell}{\operatorname{argmax}} \left\| \sum_{\nu=1}^{R} \xi_\nu ({Y^{(1)}}^T \mathbf{u}_\nu^{(1)}) \otimes \cdots \otimes ({Y^{(d)}}^T \mathbf{u}_\nu^{(d)}) \right\|^2_{\mathbb{B}_\mathbf{r}} \tag{3.31}$$

over the Grassman manifolds $\mathcal{G}_\ell$, $Y^{(\ell)} = [\mathbf{y}_1^{(\ell)} \cdots \mathbf{y}_{r_\ell}^{(\ell)}] \in \mathcal{G}_\ell$ $(\ell = 1, \ldots, d)$, *where* ${Y^{(\ell)}}^T \mathbf{u}_\nu^{(\ell)} \in \mathbb{R}^{r_\ell}$.

(b) *The compatibility condition* (2.23) *is simplified to*

$$r_\ell \leq \operatorname{rank}(U^{(\ell)}) \quad \text{with } U^{(\ell)} = [\mathbf{u}_1^{(\ell)} \quad \cdots \quad \mathbf{u}_R^{(\ell)}] \in \mathbb{R}^{n\times R},$$

and we have the solvability of (3.31), *assuming that the above relation is valid. The maximizer in* (3.31) *is given by orthogonal matrices* $V^{(\ell)} = [\mathbf{v}_1^{(\ell)} \cdots \mathbf{v}_{r_\ell}^{(\ell)}] \in \mathbb{R}^{n\times r_\ell}$, *which can be computed similarly to the Tucker decomposition for full size tensors, where the truncated HOSVD at Step (1) is now substituted by RHOSVD; see* (3.29).

(c) *The minimizer in* (3.30) *is then calculated by the orthogonal projection*

$$\mathbf{A}_{(\mathbf{r})} = \sum_{\mathbf{k}=\mathbf{1}}^{\mathbf{r}} \mu_{\mathbf{k}} \mathbf{v}_{k_1}^{(1)} \otimes \cdots \otimes \mathbf{v}_{k_d}^{(d)}, \quad \mu_{\mathbf{k}} = \langle \mathbf{v}_{k_1}^{(1)} \otimes \cdots \otimes \mathbf{v}_{k_d}^{(d)}, \mathbf{A}\rangle,$$

so that the core tensor $\boldsymbol{\mu} = [\mu_{\mathbf{k}}]$ *can be represented in the rank-R canonical format*

$$\boldsymbol{\mu} = \sum_{\nu=1}^{R} \xi_\nu (V^{(1)^T} \mathbf{u}_\nu^{(1)}) \otimes \cdots \otimes (V^{(d)^T} \mathbf{u}_\nu^{(d)}) \in \mathcal{C}_{R,\mathbf{r}}. \tag{3.32}$$

(d) *Let* $\sigma_{\ell,1} \geq \sigma_{\ell,2} \geq \cdots \geq \sigma_{\ell,\min(n,R)}$ *be the singular values of the* ℓ*-mode side-matrix* $U^{(\ell)} \in \mathbb{R}^{n\times R}$ ($\ell = 1,\ldots,d$). *Then the RHOSVD approximation* $\mathbf{A}^0_{(\mathbf{r})}$, *as in* (3.29), *exhibits the error estimate*

$$\|\mathbf{A} - \mathbf{A}^0_{(\mathbf{r})}\| \leq \|\boldsymbol{\xi}\| \sum_{\ell=1}^{d} \Bigg(\sum_{k=r_\ell+1}^{\min(n,R)} \sigma_{\ell,k}^2 \Bigg)^{1/2}, \quad \text{where } \|\boldsymbol{\xi}\| = \sqrt{\sum_{\nu=1}^{R} \xi_\nu^2}. \tag{3.33}$$

The complexity of the C2T transform for the 3D canonical tensor is estimated by

$$W_{C\to T} = O(nR^2).$$

We notice that the error estimate (3.33) in Theorem 3.12 actually provides the control of the RHOSVD approximation error via the computable ℓ-mode error bounds since, by the construction, we have

$$\|U^{(\ell)} - V_0^{(\ell)} D_{\ell,0} W_0^{(\ell)}\|_F^2 = \sum_{k=r_\ell+1}^{n} \sigma_{\ell,k}^2, \quad \ell = 1,\ldots,d.$$

This result is similar to the well-known error estimate for the HOSVD approximation; see [61].

3.3.2 Canonical-to-Tucker transform via RHOSVD

In the following, we specify the details of the C2T computational scheme for the case $d = 3$. To define the RHOSVD-type rank-$\mathbf{r}$ Tucker approximation to the tensor in (2.13), we set $n_\ell = n$ and suppose for definiteness that $n \leq R$. Now the SVD of the side-matrix $U^{(\ell)}$ is given by

$$U^{(\ell)} = V^{(\ell)} D_\ell W^{(\ell)^T} = \sum_{k=1}^{n} \sigma_{\ell,k} \mathbf{v}_k^{(\ell)} \mathbf{w}_k^{(\ell)^T}, \quad \mathbf{v}_k^{(\ell)} \in \mathbb{R}^n,\ \mathbf{w}_k^{(\ell)} \in \mathbb{R}^R, \tag{3.34}$$

with the orthogonal matrices $V^{(\ell)} = [\mathbf{v}_1^{(\ell)}, \dots, \mathbf{v}_n^{(\ell)}]$, and $W^{(\ell)} = [\mathbf{w}_1^{(\ell)}, \dots, \mathbf{w}_n^{(\ell)}]$, $\ell = 1, 2, 3$. Given the rank parameter $\mathbf{r} = (r_1, r_2, r_3)$ with $r_1, r_2, r_3 < n$, we recall the truncated SVD of the side-matrix

$$U^{(\ell)} \mapsto U_0^{(\ell)} = \sum_{k=1}^{r_\ell} \sigma_{\ell,k} \mathbf{v}_k^{(\ell)} \mathbf{w}_k^{(\ell)^T} = V_0^{(\ell)} D_{\ell,0} W_0^{(\ell)^T}, \quad \ell = 1, 2, 3,$$

where $D_{\ell,0} = \operatorname{diag}\{\sigma_{\ell,1}, \sigma_{\ell,2}, \dots, \sigma_{\ell,r_\ell}\}$ and matrices $V_0^{(\ell)} \in \mathbb{R}^{n\times r_\ell}$, ${W_0}^{(\ell)} \in \mathbb{R}^{R\times r_\ell}$ represent the orthogonal factors being the respective sub-matrices in the SVD factors of $U^{(\ell)}$.

Based on Theorem 3.12, the corresponding algorithm C2T for the rank-R input data can be designed. The algorithm *Canonical-to-Tucker* (for the 3D tensor) includes the following steps [174]:

Input data: Side matrices $U^{(\ell)} = [\mathbf{u}_1^{(\ell)} \dots \mathbf{u}_R^{(\ell)}] \in \mathbb{R}^{n_\ell \times R}$, $\ell = 1, 2, 3$, composed of the vectors $\mathbf{u}_k^{(\ell)} \in \mathbb{R}^{n_\ell}$, $k = 1, \dots, R$, see (2.13); maximal Tucker-rank parameter $\mathbf{r}$; maximal number of the ALS iterations $m_{\max}$ (usually a small number).

(I) Compute the SVD of the side matrices:

$$U^{(\ell)} = V^{(\ell)} D^{(\ell)} W^{(\ell)}, \quad \ell = 1, 2, 3.$$

Discard the singular vectors in $V^{(\ell)}$ and the respective singular values up to given rank threshold, yielding the small orthogonal matrices $V_0^{(\ell)} \in \mathbb{R}^{n_\ell \times r_\ell}$, $W_0^{(\ell)} \in \mathbb{R}^{R\times r_\ell}$, and diagonal matrices $D_{\ell,0} \in \mathbb{R}^{r_\ell \times r_\ell}$, $\ell = 1, 2, 3$.

(II) Project the side matrices $U^{(\ell)}$ onto the orthogonal basis set defined by $V_0^{(\ell)}$

$$U^{(\ell)} \mapsto \widetilde{U}^{(\ell)} = (V_0^{(\ell)})^T U^{(\ell)} = D_{\ell,0} W_0^{(\ell)^T}, \quad \widetilde{U}^{(\ell)} \in \mathbb{R}^{r_\ell \times R}, \quad \ell = 1, 2, 3, \tag{3.35}$$

and compute $\mathbf{A}_{(\mathbf{r})}^0$ as in (3.29).

(III) (Find dominating subspaces). Perform the following ALS iteration for $\ell = 1, 2, 3$, $m_{\max}$ times at most, starting from the RHOSVD initial guess $\mathbf{A}_{(\mathbf{r})}^0$:

- For $\ell = 1$: construct the partially projected image of the full tensor,

$$\mathbf{A} \mapsto \widetilde{\mathbf{B}}_{(1)} = \sum_{k=1}^{R} c_k \mathbf{u}_k^{(1)} \otimes \widetilde{\mathbf{u}}_k^{(2)} \otimes \widetilde{\mathbf{u}}_k^{(3)}, \quad c_k \in \mathbb{R}. \tag{3.36}$$

  Figure 3.15 shows that it is exactly the same construction as for the so-called single-hole tensor $\mathbf{B}_{(q)}$ appearing at the ALS step in the Tucker decomposition algorithm for the full size tensors.[2] Here $\mathbf{u}_k^{(1)} \in \mathbb{R}^{n_1}$ lives in the physical space for mode $\ell = 1$, whereas $\widetilde{\mathbf{u}}_k^{(2)} \in \mathbb{R}^{r_2}$ and $\widetilde{\mathbf{u}}_k^{(3)} \in \mathbb{R}^{r_3}$, the column vectors of $\widetilde{U}^{(2)}$, and $\widetilde{U}^{(3)}$, respectively, live in the index sets of $V^{(\ell)}$-projections.

2 But now we are not restricted by the storage for the full size tensor.

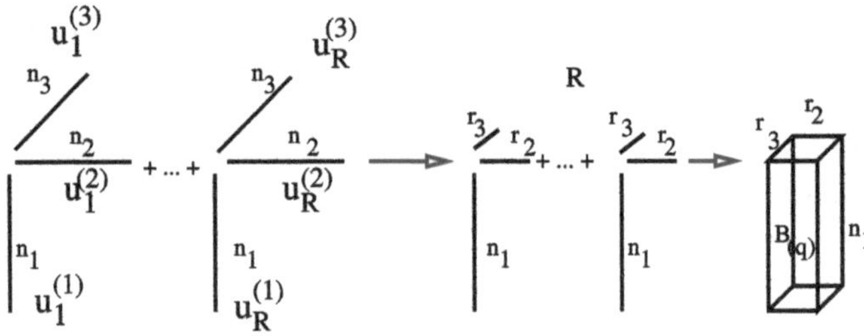

Figure 3.15: Building a single-hole tensor in the C2T algorithm.

- Reshape the tensor $\widetilde{\mathbf{B}}_{(1)} \in \mathbb{R}^{n_1 \times r_2 \times r_3}$ into a matrix $M_{A_1} \in \mathbb{R}^{n_1 \times (r_2 r_3)}$, representing the span of the optimized subset of mode-1 columns of the partially projected tensor $\widetilde{\mathbf{B}}_{(1)}$. Compute the SVD of the matrix M_{A_1}:

$$M_{A_1} = V^{(1)} S^{(1)} W^{(1)},$$

  and truncate the set of singular vectors in $V^{(1)} \mapsto V_{r_1}^{(1)} \in \mathbb{R}^{n_1 \times r_1}$, according to the restriction on the mode-1 Tucker rank, r_1.
- Update the current approximation to the mode-1 dominating subspace, $V_{r_1}^{(1)} \mapsto \widetilde{V}^{(1)}$.
- Implement the single step of the ALS iteration for mode $\ell = 2$ and $\ell = 3$.
- End of the complete ALS iteration sweep.
- Repeat the complete ALS iteration $m_{\max}$ times at most to obtain the optimized Tucker orthogonal side matrices $\widetilde{V}^{(1)}, \widetilde{V}^{(2)}, \widetilde{V}^{(3)}$, and final projected image $\widetilde{\mathbf{B}}_3$.

(IV) Project the final iterated tensor $\widetilde{\mathbf{B}}_3$ in (3.36) using the resultant basis set in $\widetilde{V}^{(3)}$ to obtain the core tensor $\boldsymbol{\beta} \in \mathbb{R}^{r_1 \times r_2 \times r_3}$.

Output data: The Tucker core tensor $\boldsymbol{\beta}$ and the Tucker orthogonal side matrices $\widetilde{V}^{(\ell)}$, $\ell = 1, 2, 3$.

In such a way, it is possible to obtain the Tucker decomposition of a canonical tensor with large mode-size and with rather large ranks, as it may be the case for electrostatic potentials of biomolecules or the electron densities in electronic structure calculations. The *Canonical-to-Tucker* algorithm can be easily modified to use an ε-truncation stopping criterion. Notice that the maximal canonical rank[3] of the core tensor $\boldsymbol{\beta}$ does not exceed $\min_\ell (r_1 r_2 r_3)/r_\ell$; see [161].

Our numerical study indicates that in the case of tensors via grid-based representation of functions describing physical quantities in electronic structure calculations, the ALS step in the C2T transform is usually not required, that is, the RHOSVD approximation is sufficient.

3 Further optimization of the canonical rank in the small-size core tensor $\boldsymbol{\beta}$ can be implemented by applying the ALS iterative scheme in the canonical format, see e. g. [193].

The following remark addresses the complexity issues.

Remark 3.13 ([174]). Algorithm C2T ($\mathcal{C}_{R,\mathbf{n}} \to \mathcal{T}_{\mathcal{C}_R,\mathbf{r}}$) exhibits polynomial cost in R, r, n,

$$O(dRn\min\{n,R\} + dr^{d-1}n\min\{r^{d-1},n\}),$$

with exponential scaling in d. In absence of Step (2) (i. e., if RHOSVD provides a satisfactory approximation), the algorithm does not contain iteration loops, and for any $d \geq 2$, it is a finite SVD-based scheme that is free of the curse of dimensionality.

Numerical tests show that Algorithm C2T($\mathcal{C}_{R,\mathbf{n}} \to \mathcal{T}_{\mathcal{C}_R,\mathbf{r}}$) is efficient for moderate R and n; in particular, it works well in electronic structure calculations on 3D Cartesian grids for moderate grid size $n \lesssim 10^3$ and for $R \leq 10^3$. However, in real life applications the computations may require one-dimension grid sizes in the range $n_\ell \lesssim 3 \cdot 10^4$ ($\ell = 1,2,3$) with canonical ranks $R \leq 10^4$. Therefore, to get rid of a polynomial scaling in R, n, r for 3D applications one can apply the best Tucker approximation methods based on the multigrid acceleration of the nonlinear ALS iteration as described in the following section.

3.3.3 Multigrid canonical-to-Tucker algorithm

The concept of multigrid acceleration (MGA) in tensor calculations can be applied to the multidimensional data obtained as a discretization of some smooth enough functions on a sequence of refined spacial grids [174]. A typical application area is solving integral-differential equations in $\mathbb{R}^d$, approximation of multidimensional operators and functionals, and data-structured representation of physically relevant quantities, such as molecular or electron densities, the Hartree and exchange potentials, and electrostatic potentials of proteins. This concept can be applied to the fully populated and to the canonical rank-R target tensors. In the case of rank-R input data, it can be understood as an adaptive tensor approximation method running over an incomplete set of data in the dual space.

We introduce the equidistant tensor grid $\omega_{\mathbf{d},n} := \omega_1 \times \omega_2 \times \cdots \times \omega_d$, where $\omega_\ell := \{-b + (m-1)h : m = 1,\ldots,n+1\}$ ($\ell = 1,\ldots,d$) with mesh-size $h = 2b/n$. Define a set of collocation points $\{x_\mathbf{m}\}$ in $\Omega \in \mathbb{R}^d$, located at the midpoints of the grid-cells numbered by $\mathbf{m} \in \mathcal{I} := \{1,\ldots,n\}^d$. For fixed n, the target tensor $\mathbf{A}_n = [a_{n,\mathbf{m}}] \in \mathbb{R}^\mathcal{I}$ is defined as the trace of the given continuous multivariate function $f : \Omega \to \mathbb{R}$ on the set of collocation points $\{x_\mathbf{m}\}$ as follows:

$$a_{n,\mathbf{m}} = f(x_\mathbf{m}), \quad \mathbf{m} \in \mathcal{I}.$$

Notice that projected Galerkin discretization method can be applied as well. For further constructions, we also need an "accurate" 1D interpolation operator $\mathcal{I}_{m-1\to m}$ from the coarse to fine grids, acting in each spacial direction. For example, this might be either the interpolation by piecewise linear or cubic splines.

The idea of the *multigrid accelerated* best orthogonal Tucker approximation , see [174], can be described as follows (for $\mathcal{C}_{R,\mathbf{n}}$ initial data):

(1) *General multigrid concept.* Solving a *sequence of nonlinear approximation problems* for $\mathbf{A} = \mathbf{A}_n$ as in (2.18) with $n = n_m := n_0 2^m$, $m = 0, 1, \dots, M$, corresponding to a sequence of (d-adic) refined spacial grids $\omega_{\mathbf{d},n_m}$. The sequence of approximation problems is treated successively in one run from coarse-to-fine grid (reminiscent of the cascadic version of the MG method).

(2) *Coarse initial approximation to the side-matrices $U^{(q)}$.* The initial approximation of $U^{(q)}$ on finer grid $\omega_{\mathbf{d},n_m}$ is obtained by the linear interpolation from coarser grid $\omega_{\mathbf{d},n_{m-1}}$, up to interpolation accuracy $O(n_m^{-\alpha})$, $\alpha > 0$.

(3) *Most important fibers.* We employ the idea of "most important fibers" (MIFs) of the q-mode unfolding matrices $B_{(q)} \in \mathbb{R}^{n \times \overline{r}_q}$, whose positions are extracted from the coarser grids. To identify the location of MIFs, the so-called *maximum energy principle* is applied as follows:

(3a) On the coarse grid, we calculate a projection of the q-mode unfolding matrix $B_{(q)}$ onto the true q-mode orthogonal subspace $\operatorname{Im} U^{(q)} = \operatorname{span}\{\mathbf{u}_1^{(q)}, \dots, \mathbf{u}_{r_q}^{(q)}\}$, which is computed as the matrix product

$$\boldsymbol{\beta}_{(q)} = {U^{(q)}}^T B_{(q)} \in \mathbb{R}^{r_q \times \overline{r}_q}. \tag{3.37}$$

(3b) Now the maximal energy principle specifies the location of MIFs by finding pr columns in $\boldsymbol{\beta}_{(q)}$ with maximal Euclidean norms (supposing that $pr \ll \overline{r}_q$), see Figures 3.16 and 3.17.[4] The positions of MIFs are numbered by the index set $\mathcal{I}_{q,p}$ with $\#\mathcal{I}_{q,p} = pr$, being the subset of the larger index set,

$$\mathcal{I}_{q,p} \subset \mathcal{I}_{\overline{r}_q} := I_{r_1} \times \cdots \times I_{r_{q-1}} \times I_{r_{q+1}} \times \cdots \times I_{r_d}, \quad \#\mathcal{I}_{\overline{r}_q} = \overline{r}_q = O(r^{d-1}).$$

The practical significance of the use the MIFs is justified by the observation that positions of MIFs[5] remain almost independent on the grid parameters $n = n_m$.

(4) *Restricted ALS iteration.* The proposed choice of MIFs allows to accelerate the ALS iteration to solving the problem of best rank-r approximation to the large unfolding matrix $B_{(q)} \in \mathbb{R}^{n \times \overline{r}_q}$ with dominating second dimension $\overline{r}_q = r^{d-1}$ (always the case for large d). This approach allows to reduce the ALS iteration to computation of the r-dimensional dominating subspace of small $n \times pr$ submatrices $B_{(q,p)}$ of $B_{(q)}$ ($q = 1, \dots, d$), where $p = O(1)$ is some fixed small parameter.

4 This strategy allows a "blind search" sampling of a fixed portion of q-mode fibers in the Tucker core that accumulate the maximum part of ℓ^2-energy. The union of selected fibers from every space dimension (specified by the index set $\mathcal{I}_{q,p}$, $q = 1, \dots, d$) accumulates the most important information about the structure of the rank R-tensor in the dual space $\mathbb{R}^{r_1 \times \cdots \times r_d}$. This *knowledge* reduces the amount of computational work on fine grids (SVD with matrix-size $n \times pr$ instead of $n \times \overline{r}_q$).

5 It resembles the multidimensional "adaptive cross approximation" (see, e. g., [231] and [87] related to the 3D case) but now acting on a fixed subset of fibers defined by MIFs.

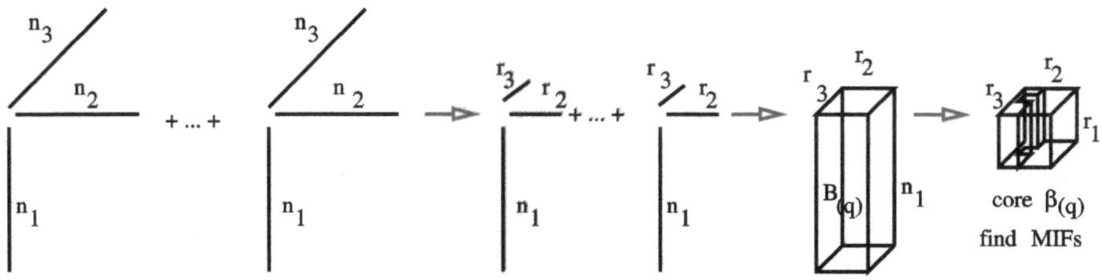

Figure 3.16: Illustration for $d = 3$. Finding MIFs in the "preliminary" core $\boldsymbol{\beta}_{(q)}$ for $q = 1$ for the rank-R initial data on the coarse grid $\mathbf{n} = \mathbf{n}_0 = (n_1, n_2, n_3)$. $B_{(q)}$ is presented in a tensor form for explanatory reasons.

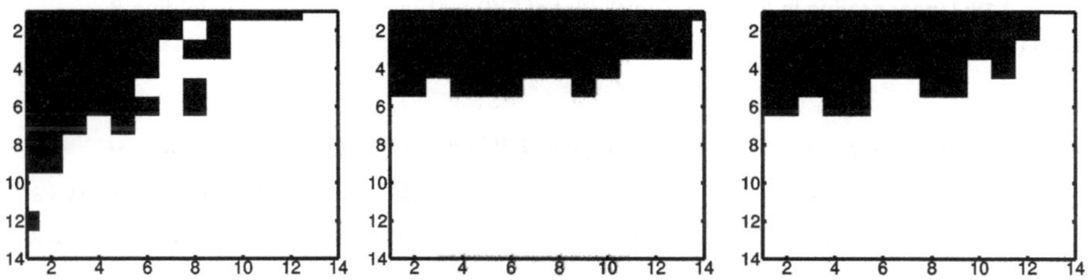

Figure 3.17: MIFs: selected projections of the fibers of the "preliminary" cores for computing $U^{(1)}$ (left), $U^{(2)}$ (middle), and $U^{(3)}$ (right). The example is taken from the multigrid rank compression in the computation of the Hartree potential for the water molecule with the choice $r = 14$, $p = 4$.

The above guidelines lead to considerable complexity reduction of the standard Tucker tensor decomposition algorithms ($\mathbb{V}_\mathbf{n} \to \mathcal{T}_\mathbf{r}$) (discussed in Section 3.2 and for C2T, ($\mathcal{C}_{R,\mathbf{n}} \to \mathcal{T}_{\mathcal{C}_R,\mathbf{r}}$). In the latter case, this approach leads to the efficient tensor approximation method with linear scaling in all governing parameters: d, n, R, and r up to the computational cost on the "very coarse" level.

The algorithm of MG accelerated (MGA) best Tucker approximation for the canonical tensor input $\mathbf{A} \in \mathcal{C}_{R,\mathbf{n}}$ can be outlined as follows [174]:

Algorithm MG-C2T ($\mathcal{C}_{R,\mathbf{n}_M} \to \mathcal{T}_{\mathcal{C}_R,\mathbf{r}}$) (MGA canonical-to-Tucker approximation).

(1) Given $\mathbf{A}_m \in \mathcal{C}_{R,\mathbf{n}_m}$ in the form (2.13), corresponding to a sequence of grid parameters $n_m := n_0 2^m$, $m = 0, 1, \ldots, M$. Fix a reliability threshold parameter $\varepsilon > 0$, a structural constant $p = O(1)$, the critical grid level $m_0 < M$, and the Tucker rank parameter $\mathbf{r}$.

(2) For $m = 0$, solve the approximation problem C2T($\mathcal{C}_{R,\mathbf{n}_M} \to \mathcal{T}_{\mathcal{C}_R,\mathbf{r}}$) and compute the index set $\mathcal{I}_{q,p}(n_0) \subset \mathcal{I}_{\overline{r}_q}$ via identification of MIFs in the matrix unfolding $B_{(q)}$, $q = 1, \ldots, d$, using *the maximum energy principle* applied to the "preliminary core" $\boldsymbol{\beta}_{(q)}$ in (3.37).

(3) For $m = 1, \ldots, m_0$, perform the *cascadic MG* nonlinear ALS iteration:

(3a) Compute initial orthogonal basis by interpolation (say, using cubic splines)

$$\{U^{(1)}, \ldots, U^{(d)}\}_m = \mathcal{I}_{m-1 \to m}(\{U^{(1)}, \ldots, U^{(d)}\}_{m-1}).$$

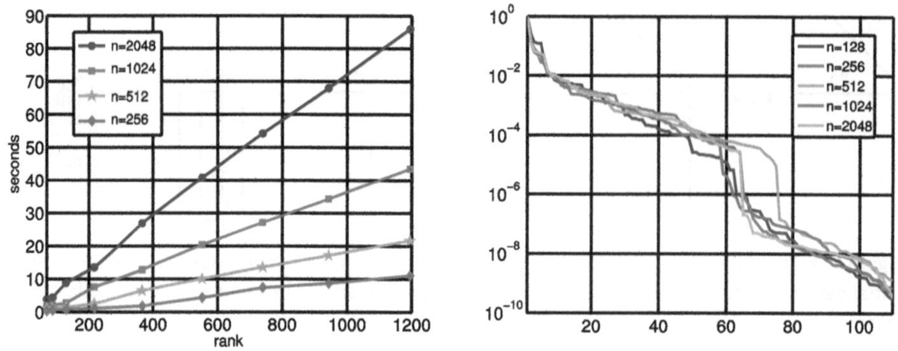

Figure 3.18: Linear scaling in R and in n (left). Plot of SVD for the mode-1 matrix unfolding $B_{(1,p)}$, $p = 4$ (right).

For each $q = 1, \ldots, d$ and with fixed $U^{(\ell)}$ ($\ell = 1, \ldots, d$, $\ell \neq q$), perform:

(3b) Define the index set $\mathcal{I}_{q,p}(n_m) = \mathcal{I}_{q,p}(n_{m-1}) \subset \mathcal{I}_{\bar{r}_q}$ and check the reliability (approximation) criteria by SVD analysis of the small-size matrix (see illustration in Figure 3.18, right)

$$B_{(q,p)} = B_{(q)|\mathcal{I}_{q,p}(n_m)} \in \mathbb{R}^{n_m \times pr},$$

If $\sigma_{\min}(B_{(q,p)}) \leq \varepsilon$, then the index set $\mathcal{I}_{q,p}$ is admissible. If for $m = m_0$ the approximation criteria above is not satisfactory, then choose $p = p + 1$ and repeat steps $m = 0, \ldots, m_0$.

(3c) Determine the orthogonal matrix $U^{(q)} \in \mathbb{R}^{n \times r}$ via computing the r-dimensional dominating subspace for the *"restricted"* matrix unfolding $B_{(q,p)}$.

(4) For levels $m = m_0 + 1, \ldots, M$, perform the MGA Tucker approximation by ALS iteration as in Steps (3a) and (3c), but now with fixed positions of MIFs specified by the index set $\mathcal{I}_{q,p}(n_{m_0})$, i. e., by discarding all fibers in $B_{(q)}$ corresponding to the "less important" index set $\mathcal{I}_{\bar{r}_q} \setminus \mathcal{I}_{q,p}$.

(5) Compute the rank-R core tensor $\boldsymbol{\beta} \in \mathcal{C}_{R,\mathbf{r}}$, as in Step (3) of basic algorithm C2T ($\mathcal{C}_{R,\mathbf{n}} \to \mathcal{T}_{\mathcal{C}_R,\mathbf{r}}$).

Theorem 3.14 ([174]). *Algorithm MG-C2T($\mathcal{C}_{R,\mathbf{n}_M} \to \mathcal{T}_{\mathcal{C}_R,\mathbf{r}}$) amounts to*

$$O(dRrn_M + dp^2r^2n_M)$$

operations per ALS loop, plus extra cost $W_{n_0} = O(dRn_0^2)$ of the coarse mesh solver C2T ($\mathcal{C}_{R,\mathbf{n}_0} \to \mathcal{T}_{\mathcal{C}_R,\mathbf{r}}$). It requires $O(drn_M + drR)$ storage to represent the result.

Proof. Step (3a) requires $O(drn_M)$ operations and memory. Notice that for large M, we have $pr \leq n_M$. Hence, the complexity of Step (3c) is bounded by $O(dRrn_M + prn_M + p^2r^2n_M)$ per iteration loop, and same for Step (3b). Rank-R representation of $\boldsymbol{\beta} \in \mathcal{C}_{R,\mathbf{r}}$

requires $O(drRn_M)$ operations and $O(drR)$-storage. Summing up these costs over levels $m = 0, \dots, M$ proves the result. □

Theorem 3.14 shows that Algorithm MG-C2T realizes the fast rank reduction method that scales linearly in d, n_M, R, and r. Moreover, the complexity and error of the multigrid Tucker approximation can be effectively controlled by the tuning of the governing parameters p, m_0, and n_0.

Figure 3.18 (left) demonstrates linear complexity scaling of the multigrid Tucker approximation in the input rank R, and in the grid size n (electron density for the CH_4 molecule). Figure 3.18 (right) shows the exponentially fast decaying singular values of the mode-1 matrix unfolding $B_{(1,p)}$ with the choice $p = 4$, which demonstrates the reliability of the maximal energy principal in the error control. Similar fast decay of respective singular values is typical in most of our numerical examples in electronic structure calculations considered so far.

3.4 Mixed Tucker-canonical transform

Since the Tucker core still presupposes r^d storage, we consider the approximation methods, using a mixed (two-level) representation [161, 173] that gainfully combines the beneficial features of both the Tucker and canonical models. The main idea of the mixed approximation consists of a rank-structured representation to the Tucker core $\boldsymbol{\beta}$ in certain tensor classes $\mathcal{S} \subset \mathbb{B}_{\mathbf{r}}$. In particular, we consider a class $\mathcal{S} = \mathcal{C}_{R,\mathbf{r}}$ of rank-R canonical tensors, i. e., $\boldsymbol{\beta} \in \mathcal{C}_{R,\mathbf{r}}$.

Definition 3.15 (The mixed two-level Tucker-canonical format). Given the rank parameters $\mathbf{r}, R$, we denote by $\mathcal{T}_{\mathcal{C}_{R,\mathbf{r}}}$ the subclass of tensors in $\mathcal{T}_{\mathbf{r},\mathbf{n}}$ with the core $\boldsymbol{\beta}$ represented in the canonical format, $\boldsymbol{\beta} \in \mathcal{C}_{R,\mathbf{r}} \subset \mathbb{B}_{\mathbf{r}}$. An explicit representation of $\mathbf{A} \in \mathcal{T}_{\mathcal{C}_{R,\mathbf{r}}}$ is given by

$$\mathbf{A} = \left(\sum_{\nu=1}^{R} \xi_\nu \mathbf{u}_\nu^{(1)} \otimes \cdots \otimes \mathbf{u}_\nu^{(d)} \right) \times_1 V^{(1)} \times_2 V^{(2)} \cdots \times_d V^{(d)}, \tag{3.38}$$

with some $\mathbf{u}_\nu^{(\ell)} \in \mathbb{R}^{r_\ell}$. Clearly, we have the embedding $\mathcal{T}_{\mathcal{C}_{R,\mathbf{r}}} \subset \mathcal{C}_{R,\mathbf{n}}$ with the corresponding (non-orthogonal) side-matrices $U^{(\ell)} = [V^{(\ell)}\mathbf{u}_1^{(\ell)} \cdots V^{(\ell)}\mathbf{u}_R^{(\ell)}]$ and scaling coefficients ξ_ν ($\nu = 1, \dots, R$).

A target tensor $\mathbf{A} \in \mathbb{V}_{\mathbf{n}}$ can be approximated by a sum of rank-1 tensors as in (2.15), (2.13), or using the mixed format $\mathcal{T}_{\mathcal{C}_{R,\mathbf{r}}}$ as in (3.38). In what follows, we discuss fast and efficient methods to compute the corresponding rank structured approximations in different problem settings.

In this case, to reduce the ranks of input tensors, we present the two-level canonical-to-Tucker (C2T) approximation with the consequent Tucker-to-canonical (T2C) transform. The corresponding canonical-to-Tucker-to-canonical approximation scheme introduced in [161, 173] can be presented as the following two-level chain:

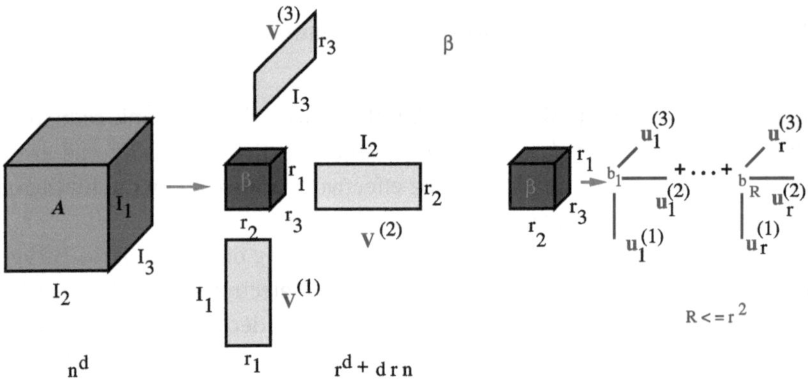

Figure 3.19: Mixed Tucker-canonical decomposition.

$$\mathcal{C}_{R,\mathbf{n}} \xrightarrow{I} \mathcal{T}_{\mathcal{C}_{R},\mathbf{r}} \xrightarrow{II} \mathcal{T}_{\mathcal{C}_{R'},\mathbf{r}} \subset \mathcal{C}_{R',\mathbf{n}}. \tag{3.39}$$

Here, on Level-I, we compute the best orthogonal Tucker approximation applied to the $\mathcal{C}_{R,\mathbf{n}}$-type input, so that the resultant core is represented in the $\mathcal{C}_{R,\mathbf{r}}$ format with the same CP rank R as for the target tensor. On Level-II, the small-size Tucker core in $\mathcal{C}_{R,\mathbf{r}}$ is approximated by a tensor in $\mathcal{C}_{R',\mathbf{r}}$ with $R' < R$. Here we describe the Algorithm on Level-I (which is, in fact, the most laborious part in computational scheme (3.39)), which has a polynomial cost in the size of the input data in $\mathcal{C}_{R,\mathbf{n}}$ (see Remark 3.13).

In the case of full format tensors, the two-level version of Algorithm ALS Tucker ($\mathbb{V}_\mathbf{n} \to \mathcal{T}_{\mathbf{r},\mathbf{n}}$) can be described as the following computation chain:

$$\mathbb{V}_\mathbf{n} \xrightarrow{I} \mathcal{T}_{\mathbf{r},\mathbf{n}} \xrightarrow{II} \mathcal{T}_{\mathcal{C}_{R,\mathbf{r}}} \subset \mathcal{C}_{R,\mathbf{n}},$$

where the Level-I is understood as application of Tucker decomposition Algorithm to full format tensors ($\mathbb{V}_\mathbf{n} \to \mathcal{T}_{\mathbf{r},\mathbf{n}}$) or its multigrid version, and the Level-II includes the rank-R canonical approximation to the small size Tucker core $\boldsymbol{\beta} \in \mathbb{B}_\mathbf{r}$. Figure 3.19 illustrates the computational scheme of the two-level Tucker approximation.

In the case of function-related tensors, it is possible to compute the Level-I approximation with linear cost in the size of the input data (see Section 3.2).

If the input tensor $\mathbf{A}_0$ is already presented in the rank-$\mathbf{r}$ Tucker format, then one can apply the following Lemma 3.16. This lemma presents a simple but useful characterization of the mixed (two-level) Tucker model (cf. [161, 173]) that allows one to approximate the elements in $\mathcal{T}_\mathbf{r}$ via the canonical decomposition applied to the small sized core tensor.

Lemma 3.16 (Mixed Tucker-to-canonical approximation). *Let the target tensor* $\mathbf{A} \in \mathcal{T}_{\mathbf{r},\mathbf{n}}$ *in* (2.18) *have the form* $\mathbf{A} = \boldsymbol{\beta} \times_1 V^{(1)} \times_2 \cdots \times_d V^{(d)}$ *with the orthogonal side-matrices* $V^{(\ell)} = [\mathbf{v}_1^{(\ell)} \cdots \mathbf{v}_{r_\ell}^{(\ell)}] \in \mathbb{R}^{n \times r_\ell}$ *and with* $\boldsymbol{\beta} \in \mathbb{R}^{r_1 \times \cdots \times r_d}$. *Then, for a given* $R \le \min_{1 \le \ell \le d} \overline{r}_\ell$,

$$\min_{Z \in \mathcal{C}_{R,\mathbf{n}}} \|\mathbf{A} - \mathbf{Z}\| = \min_{\boldsymbol{\mu} \in \mathcal{C}_{R,\mathbf{r}}} \|\boldsymbol{\beta} - \boldsymbol{\mu}\|. \tag{3.40}$$

Assume that there exists the best rank-R approximation $\mathbf{A}_{(R)} \in \mathcal{C}_{R,\mathbf{n}}$ *of* $\mathbf{A}$*, then there is the best rank-R approximation* $\boldsymbol{\beta}_{(R)} \in \mathcal{C}_{R,\mathbf{r}}$ *of* $\boldsymbol{\beta}$*, such that*

$$\mathbf{A}_{(R)} = \boldsymbol{\beta}_{(R)} \times_1 V^{(1)} \times_2 \cdots \times_d V^{(d)}. \tag{3.41}$$

Proof. We present the more detailed proof compared with the sketch in Lemma 2.5, [173]. Notice that the canonical vectors $\mathbf{y}_k^{(\ell)}$ of any test element (see (2.13)) in the left-hand side of (3.40),

$$\mathbf{Z} = \sum_{k=1}^{R} \lambda_k \mathbf{y}_k^{(1)} \otimes \cdots \otimes \mathbf{y}_k^{(d)} \in \mathcal{C}_{R,\mathbf{n}}, \tag{3.42}$$

can be chosen in $\operatorname{span}\{\mathbf{v}_1^{(\ell)}, \ldots, \mathbf{v}_{r_\ell}^{(\ell)}\}$, i. e.,

$$\mathbf{y}_k^{(\ell)} = \sum_{m=1}^{r_\ell} \mu_{k,m}^{(\ell)} \mathbf{v}_m^{(\ell)}, \quad k = 1, \ldots, R,\ \ell = 1, \ldots, d. \tag{3.43}$$

Indeed, assuming

$$y_k^{(\ell)} = \sum_{m=1}^{r_\ell} \mu_{k,m}^{(\ell)} \mathbf{v}_m^{(\ell)} + E_k^{(\ell)} \quad \text{with } E_k^{(\ell)} \perp \operatorname{span}\{\mathbf{v}_1^{(\ell)}, \ldots, \mathbf{v}_{r_\ell}^{(\ell)}\},$$

we conclude that $E_k^{(\ell)}$ does not effect the cost function in (3.40) because of the orthogonality of $V^{(\ell)}$. Hence, setting $E_k^{(\ell)} = 0$ and substituting (3.43) into (3.42), we arrive at the desired Tucker decomposition of Z,

$$\mathbf{Z} = \boldsymbol{\beta}_z \times_1 V^{(1)} \times_2 \cdots \times_d V^{(d)}, \quad \boldsymbol{\beta}_z \in \mathcal{C}_{R,\mathbf{r}}.$$

This implies

$$\|\mathbf{A} - \mathbf{Z}\|^2 = \left\|(\boldsymbol{\beta}_z - \boldsymbol{\beta}) \times_1 V^{(1)} \times_2 \cdots \times_d V^{(d)}\right\|^2 = \|\boldsymbol{\beta} - \boldsymbol{\beta}_z\|^2 \geq \min_{\boldsymbol{\mu} \in \mathcal{C}_{R,\mathbf{r}}} \|\boldsymbol{\beta} - \boldsymbol{\mu}\|^2.$$

On the other hand, we have

$$\min_{\mathbf{Z} \in \mathcal{C}_{R,\mathbf{n}}} \|\mathbf{A} - \mathbf{Z}\|^2 \leq \min_{\boldsymbol{\beta}_z \in \mathcal{C}_{R,\mathbf{r}}} \left\|(\boldsymbol{\beta} - \boldsymbol{\beta}_z) \times_1 V^{(1)} \times_2 \cdots \times_d V^{(d)}\right\|^2 = \min_{\boldsymbol{\mu} \in \mathcal{C}_{R,\mathbf{r}}} \|\boldsymbol{\beta} - \boldsymbol{\mu}\|^2.$$

Hence, we arrive at (3.40).

Likewise, for any minimizer $\mathbf{A}_{(R)} \in \mathcal{C}_{R,\mathbf{n}}$ in the right-hand side of (3.40), we obtain

$$\mathbf{A}_{(R)} = \boldsymbol{\beta}_{(R)} \times_1 V^{(1)} \times_2 V^{(2)} \cdots \times_d V^{(d)}$$

with the respective rank-R core tensor

$$\boldsymbol{\beta}_{(R)} = \sum_{k=1}^{R} \lambda_k \mathbf{u}_k^{(1)} \otimes \cdots \otimes \mathbf{u}_k^{(d)} \in \mathcal{C}_{R,\mathbf{r}},$$

where $\mathbf{u}_k^{(\ell)} = \{\mu_{k,m_\ell}^{(\ell)}\}_{m_\ell=1}^{r_\ell} \in \mathbb{R}^{r_\ell}$ are calculated by using representation (3.43). Now changing the order of summation, we have

$$\begin{aligned}\mathbf{A}_{(R)} &= \sum_{k=1}^{R} \lambda_k \mathbf{y}_k^{(1)} \otimes \cdots \otimes \mathbf{y}_k^{(d)} \\ &= \sum_{k=1}^{R} \lambda_k \left(\sum_{m_1=1}^{r_1} \mu_{k,m_1}^{(1)} \mathbf{v}_{m_1}^{(1)} \right) \otimes \cdots \otimes \left(\sum_{m_d=1}^{r_d} \mu_{k,m_d}^{(d)} \mathbf{v}_{m_d}^{(d)} \right) \\ &= \sum_{m_1=1}^{r_1} \cdots \sum_{m_d=1}^{r_d} \left\{ \sum_{k=1}^{R} \lambda_k \prod_{\ell=1}^{d} \mu_{k,m_\ell}^{(\ell)} \right\} \mathbf{v}_{m_1}^{(1)} \otimes \cdots \otimes \mathbf{v}_{m_d}^{(d)}.\end{aligned}$$

The relation (3.41) implies that

$$\|\mathbf{A} - \mathbf{A}_R\| = \|\boldsymbol{\beta} - \boldsymbol{\beta}_{(R)}\|,$$

since the ℓ-mode multiplication with orthogonal side matrices $V^{(\ell)}$ does not change the cost function. Using the already proven relation (3.40), this indicates that $\boldsymbol{\beta}_{(R)}$ is the minimizer in the right-hand side of (3.40). □

Lemma 3.16 means that the corresponding low-rank Tucker-canonical approximation of $\mathbf{A} \in \mathcal{T}_{\mathbf{r},\mathbf{n}}$ can be reduced to the canonical approximation of a small size core tensor.

Lemma 3.16 suggests a two-level dimensionality reduction approach that leads to a sparser data structure compared with the standard Tucker model. Though $\mathbf{A}_{(R)} \in \mathcal{C}_{R,\mathbf{n}}$ can be represented in the mixed Tucker-canonical format, its efficient storage depends on further multilinear operations. In fact, if the resultant tensor is further used in scalar, Hadamard, or convolution products with canonical tensors, it is better to store $\mathbf{A}_{(R)}$ in the canonical format of the complexity rdn.

The numerics for illustrating the performance of the multigrid canonical-to-Tucker algorithm will be presented in Section 8.1 describing calculation of the Hartree potential and Coulomb matrix in the Hartree–Fock equation.

3.5 On Tucker-to-canonical transform

In the rank reduction scheme for the canonical rank-R tensors, we use consequently the canonical-to-Tucker (C2T) transform, and then the Tucker-to-canonical (T2C) tensor approximation. Next, we give two useful remarks which characterize the canonical representation of the full format tensors.

Remark 3.17 applied to the Tucker core tensor of the size $r \times r \times r$ indicates that the ultimate canonical rank of a large-size tensor in $\mathbb{V}_\mathbf{n}$ has the upper bound r^2, as illustrated by Figure 3.20. According to Remark 3.18, its canonical rank can be reduced to a smaller value using the SVD-based truncation procedure up to a fixed tolerance $\varepsilon > 0$.

Denote by $\overline{n}_\ell$ the single-hole product of ℓ-mode dimensions

$$\overline{n}_\ell = n_1 \cdots n_{\ell-1} n_{\ell+1} \cdots n_d. \tag{3.44}$$

Remark 3.17. The canonical rank of a tensor $\mathbf{A} \in \mathbb{V}_\mathbf{n}$ has the upper bound

$$R \leq \min_{1\leq \ell \leq d} \overline{n}_\ell. \tag{3.45}$$

Proof. First, consider the case $d = 3$. Let $n_1 = \max_{1\leq \ell \leq d} n_\ell$ for definiteness. We can represent a tensor $\mathbf{A}$ as

$$\mathbf{A} = \sum_{k=1}^{n_3} B_k \otimes Z_k, \quad B_k \in \mathbb{R}^{n_1 \times n_2},\ Z_k \in \mathbb{R}^{n_3},$$

where $B_k = \mathbf{A}(:, :, k)$ $(k = 1, \ldots, n_3)$ is the $n_1 \times n_2$ matrix slice of A, and $Z_k(i) = 0$, for $i \neq k$, $Z_k(k) = 1$. Let $\operatorname{rank}(B_k) = r_k \leq n_2$, $k = 1, \ldots, n_3$. Then

$$\operatorname{rank}(B_k \otimes Z_k) = \operatorname{rank}(B_k) \leq n_2,$$

and we obtain

$$\operatorname{rank}(A) \leq \sum_{n=1}^{n_3} \operatorname{rank}(B_k) \leq n_2 n_3 = \min_{1\leq \ell \leq 3} \overline{n}_\ell.$$

The general case of $d > 3$ can be proven similarly by induction argument. □

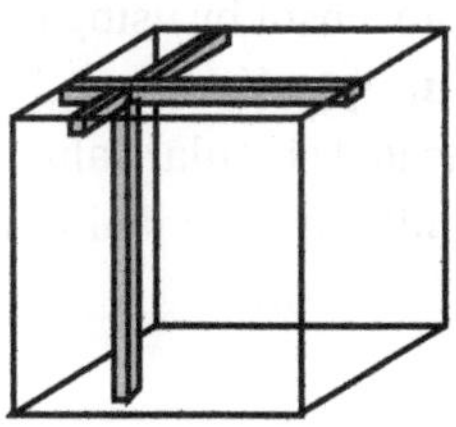

Figure 3.20: Tucker-to-canonical decomposition for a small core tensor.

The next remark shows that the maximal canonical rank of the Tucker core of 3rd-order tensor can be easily reduced to the value $\leq r^2$ by the SVD-based procedure. Though being not practically attractive for arbitrary high-order tensors, the simple algorithm described in Remark 3.18 is proved to be useful for the treatment of small size 3rd-order Tucker core tensors in the rank reduction algorithms described in the previous sections.

Remark 3.18. There is a simple procedure based on SVD to reduce the canonical rank of the core tensor $\boldsymbol{\beta}$ within the accuracy $\varepsilon > 0$. Let $d = 3$ for the sake of clearness.

Denote by $B_m \in \mathbb{R}^{r\times r}$, $m = 1,\dots,r$, the matrix slices of $\boldsymbol{\beta}$ in some fixed mode. Hence, we can represent

$$\boldsymbol{\beta} = \sum_{m=1}^{r} B_m \otimes \mathbf{z}_m, \quad \mathbf{z}_m \in \mathbb{R}^r, \tag{3.46}$$

where $\mathbf{z}_m(m) = 1$, $\mathbf{z}_m(j) = 0$ for $j = 1,\dots,r$, $j \neq m$ (there are exactly d possible decompositions). Let p_m be the minimal integer such that the singular values of B_m satisfy $\sigma_k^{(m)} \leq \frac{\varepsilon}{r^{3/2}}$ for $k = p_m + 1,\dots,r$ (if $\sigma_r^{(m)} > \frac{\varepsilon}{r^{3/2}}$, then set $p_m = r$). Then, denoting by

$$B_{p_m} = \sum_{k_m=1}^{p_m} \sigma_{k_m}^{(m)} \mathbf{u}_{k_m} \otimes \mathbf{v}_{k_m},$$

the corresponding rank-p_m approximation to B_m (by truncation of $\sigma_{p_m+1}^{(m)},\dots,\sigma_r^{(m)}$), we arrive at the rank-R canonical approximation to $\boldsymbol{\beta}$,

$$\boldsymbol{\beta}_{(R)} := \sum_{m=1}^{r} B_{p_m} \otimes \mathbf{z}_m, \quad \mathbf{z}_m \in \mathbb{R}^r, \tag{3.47}$$

providing the error estimate

$$\|\boldsymbol{\beta} - \boldsymbol{\beta}_{(R)}\| \leq \sum_{m=1}^{r} \|B_m - B_{p_m}\| = \sum_{m=1}^{r} \sqrt{\sum_{k_m=p_m+1}^{r} (\sigma_{k_m}^{(m)})^2} \leq \sum_{m=1}^{r} \sqrt{r\frac{\varepsilon^2}{r^3}} = \varepsilon.$$

Representation (3.47) is a sum of rank-p_m terms, so that the total rank is bounded by $R \leq p_1 + \dots + p_r \leq r^2$.

This approach can be easily extended to arbitrary $d \geq 3$ with the bound $R \leq r^{d-1}$.

Figure 3.21 illustrates the canonical decomposition of the core tensor by using the SVD of slices B_m of the core tensor $\boldsymbol{\beta}$, yielding matrices $U_m = \{\mathbf{u}_{k_m}\}_{k=1}^{p_m}$, $V_m = \{\mathbf{v}_{k_m}\}_{k=1}^{p_m}$ and a diagonal matrix of small size $p_m \times p_m$ containing the truncated singular values. It also shows the vector $\mathbf{z}_m = [0,\dots,0,1,0,\dots,0]$, containing all entries equal to 0 except 1 at the mth position.

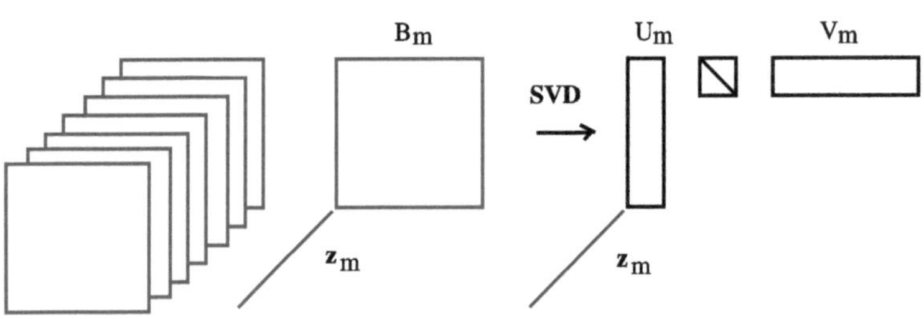

Figure 3.21: Tucker-to-canonical decomposition for a small core tensor, see Remark 3.18.

4 Multiplicative tensor formats in $\mathbb{R}^d$

4.1 Tensor train format: linear scaling in d

The product-type representation of dth-order tensors, which is called the matrix product states (MPS) decomposition in the physical literature, was introduced and successfully applied in DMRG quantum computations [302, 294, 293], and, independently, in quantum molecular dynamics as the multilayer (ML) MCTDH methods [297, 221, 211]. Representations by MPS-type formats in multidimensional problems reduce the complexity of storage to $O(dr^2N)$, where r is the maximal rank parameter.

In recent years, the various versions of the MPS-type tensor format were discussed and further investigated in mathematical literature, including the hierarchical dimension splitting [161], the tensor train (TT) [229, 226], the tensor chain and combined Tucker-TT [167], the QTT-Tucker [66] formats, and the hierarchical Tucker (HT) representation [110], which belongs to the class of ML-MCTDH methods [297], or more generally tensor network states models. The MPS-type tensor approximation was proved by extensive numerics to be efficient in high-dimensional electronic/molecular structure calculations, in molecular dynamics, and in quantum information theory (see survey papers [293, 138, 169, 264]).

Note that although the multiplicative TT and HT parametrizations formally apply to any full format tensor in higher dimensions, they become computationally feasible only when using the RHOSVD-like procedures applied either to the canonical format input or to tensors already given in the TT form. The HOSVD in MPS-type formats was discussed in [294, 100, 226].

The TT format that is the particular case of MPS-type factorization in the case of open boundary conditions, can be defined as follows: for a given rank parameter $\mathbf{r} = (r_0, \ldots, r_d)$ and the respective index sets $J_\ell = \{1, \ldots, r_\ell\}$ ($\ell = 0, 1, \ldots, d$) with the constraint $J_0 = J_d = \{1\}$ (i. e., $r_0 = r_d = 1$), the rank-$\mathbf{r}$ TT format contains all elements $\mathbf{A} = [a(i_1, \ldots, i_d)] \in \mathbb{R}^{n_1 \times \cdots \times n_d}$ that can be represented as the contracted product of 3-tensors over the d-fold product index set $\mathcal{J} := \times_{\ell=1}^{d} J_\ell$ such that

$$\mathbf{A} = \sum_{\alpha \in \mathcal{J}} \mathbf{a}^{(1)}_{\alpha_1} \otimes \mathbf{a}^{(2)}_{\alpha_1, \alpha_2} \otimes \cdots \otimes \mathbf{a}^{(d)}_{\alpha_{d-1}} \equiv \mathbf{A}^{(1)} \bowtie \mathbf{A}^{(2)} \bowtie \cdots \bowtie \mathbf{A}^{(d)},$$

where $\mathbf{a}^{(\ell)}_{\alpha_\ell, \alpha_{\ell+1}} \in \mathbb{R}^{N_\ell}$ ($\ell = 1, \ldots, d$), and $\mathbf{A}^{(\ell)} = [\mathbf{a}^{(\ell)}_{\alpha_\ell, \alpha_{\ell+1}}] \in \mathbb{R}^{N_\ell \times r_\ell \times r_{\ell+1}}$ is the vector-valued $r_\ell \times r_{\ell+1}$ matrix (3-tensor). Here, and in the following (see Definition 4.3), the rank product operation "$\bowtie$" is defined as a regular matrix product of the two core vector-valued matrices, their fibers (blocks) being multiplied by means of tensor product [142]. The particular entry of $\mathbf{A}$ is represented by

$$a(i_1, \ldots, i_d) = \sum_{\alpha_1=1}^{r_1} \cdots \sum_{\alpha_d=1}^{r_d} \mathbf{a}^{(1)}_{\alpha_1}(i_1) \mathbf{a}^{(2)}_{\alpha_1, \alpha_2}(i_2) \cdots \mathbf{a}^{(d)}_{\alpha_{d-1}}(i_d) \equiv \mathbf{A}^{(1)}(i_1) \mathbf{A}^{(2)}(i_2) \cdots \mathbf{A}^{(d)}(i_d),$$

https://doi.org/10.1515/9783110365832-004

so that the latter is written in the *matrix product* form (explaining the notion MPS), where $\mathbf{A}^{(\ell)}(i_\ell)$ is an $r_{\ell-1} \times r_\ell$ matrix.

Example 4.1. Figure 4.1 illustrates the TT representation of a 5th-order tensor; each particular entry $a(i_1, i_2, \dots, i_5)$ is presented as a product of five matrices (and vectors) corresponding to indexes i_ℓ of three-tensors, $i_\ell \in \{1, \dots, n_\ell\}$, $\ell = 1, 2, \dots, 5$.

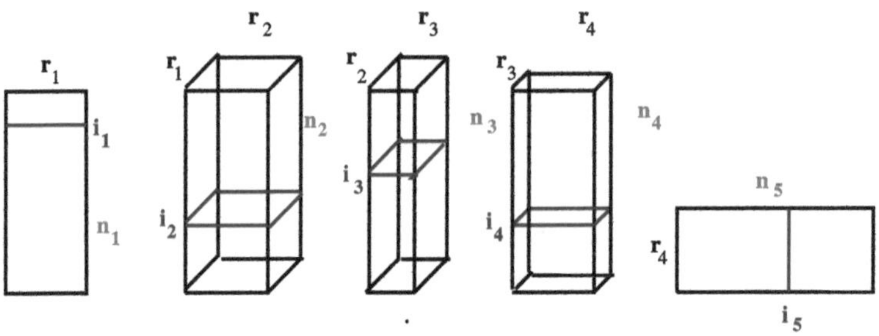

Figure 4.1: Visualizing 5th-order TT tensor.

In case $J_0 = J_d \neq \{1\}$, we arrive at the more general form of MPS, the so-called tensor chain (TC) format [167]. In some cases, TC tensor can be represented as a sum of not more than r_* TT-tensors ($r_* = \min r_\ell$), which can be converted to the TT tensor based on multilinear algebra operations like sum-and-compress. The storage cost for both TC and TT formats is bounded by $O(dr^2N)$, $r = \max r_\ell$.

Clearly, one and the same tensor might have different ranks in different formats (and, hence, different number of representation parameters). The next example considers the Tucker and TT representations of a *function-related* canonical tensor $\mathbf{F} := T(f)$, obtained by sampling of the function $f(x) = x_1 + \cdots + x_d$, $x \in [0,1]^d$, on the Cartesian grid of size $N^{\otimes d}$ and specified by N-vectors $X_\ell = \{ih\}_{i=1}^N$ ($h = 1/N$, $\ell = 1, \dots, d$), and all-ones vector $1 \in \mathbb{R}^N$. The canonical rank of this tensor can be proven to be exactly d [201].

Example 4.2. We have $\mathrm{rank}_{\mathrm{Tuck}}(\mathbf{F}) = 2$, with the explicit representation

$$\mathbf{F} = \sum_{\mathbf{k}=1}^{2} b_{\mathbf{k}} V_{k_1}^{(1)} \otimes \cdots \otimes V_{k_d}^{(d)}, \quad V_1^{(\ell)} = 1,\ V_2^{(\ell)} = X_\ell, \quad [b_{\mathbf{k}}] \in \bigotimes_{\ell=1}^{d} \mathbb{R}^2.$$

Moreover, $\mathrm{rank}_{\mathrm{TT}}(\mathbf{F}) = 2$ in view of the exact decomposition

$$\mathbf{F} = \begin{bmatrix} X_1 & 1 \end{bmatrix} \bowtie \begin{bmatrix} 1 & 0 \\ X_2 & 1 \end{bmatrix} \bowtie \cdots \bowtie \begin{bmatrix} 1 & 0 \\ X_{d-1} & 1 \end{bmatrix} \bowtie \begin{bmatrix} 1 \\ X_d \end{bmatrix}.$$

The rank-structured tensor formats like canonical, Tucker and MPS/TT-type decompositions induce the important concept of *canonical, Tucker or matrix product op-*

erators (CO/TO/MPO) acting between two tensor-product Hilbert spaces, each of dimension d,

$$\mathcal{A} : \mathbb{X} = \bigotimes_{\ell=1}^{d} X^{(\ell)} \to \mathbb{Y} = \bigotimes_{\ell=1}^{d} Y^{(\ell)}.$$

For example, the R-term canonical operator (matrix) takes a form

$$\mathcal{A} = \sum_{\alpha=1}^{R} \bigotimes_{\ell=1}^{d} \mathcal{A}_\alpha^{(\ell)}, \quad \mathcal{A}_\alpha^{(\ell)} : X^{(\ell)} \to Y^{(\ell)}.$$

The action $\mathcal{A}\mathbf{X}$ on rank-R_X canonical tensor $\mathbf{X} \in \mathbb{X}$ is defined as RR_X-term canonical sum in $\mathbb{Y}$,

$$\mathcal{A}\mathbf{X} = \sum_{\alpha=1}^{R} \sum_{\beta=1}^{R_X} \bigotimes_{\ell=1}^{d} \mathcal{A}_\alpha^{(\ell)} \mathbf{x}_\beta^{(\ell)} \in \mathbb{Y}.$$

The rank-$\mathbf{r}$ Tucker matrix can be defined in a similar way.

In the case of rank-$\mathbf{r}$ TT format, the respective matrices are defined as follows.

Definition 4.3. The rank-$\mathbf{r}$ TT-operator (TTO/MPO) decomposition symbolized by a set of factorized operators $\mathcal{A}$ is defined by

$$\mathcal{A} = \sum_{\alpha \in \mathcal{J}} A_{\alpha_1}^{(1)} \otimes A_{\alpha_1\alpha_2}^{(2)} \otimes \cdots \otimes A_{\alpha_{d-1}}^{(d)} \equiv \mathcal{A}^{(1)} \bowtie \mathcal{A}^{(2)} \bowtie \cdots \bowtie \mathcal{A}^{(d)},$$

where $\mathcal{A}^{(\ell)} = [A_{\alpha_\ell\alpha_{\ell+1}}^{(\ell)}]$ denotes the operator-valued $r_\ell \times r_{\ell+1}$ matrix, and where $A_{\alpha_\ell\alpha_{\ell+1}}^{(\ell)}$: $X^{(\ell)} \to Y^{(\ell)}$ ($\ell = 1, \ldots, d$), or, in the index notation,

$$\begin{aligned} \mathcal{A}(i_1, j_1, \ldots, i_d, j_d) = \sum_{\alpha_1=1}^{r_1} \cdots \sum_{\alpha_{d-1}=1}^{r_{d-1}} A_{\alpha_1}^{(1)}(i_1, j_1) A_{\alpha_1\alpha_2}^{(2)}(i_2, j_2) \cdots \\ \cdot A_{\alpha_{d-2}\alpha_{d-1}}^{(d-1)}(i_{d-1}, j_{d-1}) A_{\alpha_{d-1}}^{(d)}(i_d, j_d). \end{aligned} \tag{4.1}$$

Given a rank-$\mathbf{r}_X$ TT-tensor $\mathbf{X} = \mathbf{X}^{(1)} \bowtie \mathbf{X}^{(2)} \bowtie \cdots \bowtie \mathbf{X}^{(d)} \in \mathbb{X}$, the action $\mathcal{A}\mathbf{X} = \mathbf{Y}$ is defined as the TT element $\mathbf{Y} = \mathbf{Y}^{(1)} \bowtie \mathbf{Y}^{(2)} \bowtie \cdots \bowtie \mathbf{Y}^{(d)} \in \mathbb{Y}$,

$$\mathcal{A}\mathbf{X} = \mathbf{Y}^{(1)} \bowtie \mathbf{Y}^{(2)} \bowtie \cdots \bowtie \mathbf{Y}^{(d)} \in \mathbb{Y}, \quad \text{with } \mathbf{Y}^{(\ell)} = [\mathcal{A}_{\alpha_1\alpha_2}^{(\ell)} \mathbf{X}_{\beta_1\beta_2}^{(\ell)}]_{\alpha_1\beta_1, \alpha_2\beta_2},$$

where, in the brackets, we use the standard matrix–vector multiplication. The TT-rank of $\mathbf{Y}$ is bounded by $\mathbf{r}_Y \le \mathbf{r} \odot \mathbf{r}_X$, where $\odot$ means the standard Hadamard (entry-wise) product of two vectors.

To describe the index-free operator representation of the TT matrix–vector product, we introduce the tensor operation denoted by $\bowtie^*$ that can be viewed as dual to $\bowtie$; it is defined as the tensor (Kronecker) product of the two corresponding core matrices, their blocks being multiplied by means of a regular matrix product operation. Now, with the substitution $\mathbf{Y}^{(\ell)} = \mathcal{A}^{(\ell)} \bowtie^* \mathbf{X}^{(\ell)}$, the matrix–vector product in TT format takes the operator form,

$$\mathbf{AX} = (\mathcal{A}^{(1)} \bowtie^* \mathbf{X}^{(1)}) \bowtie \cdots \bowtie (\mathcal{A}^{(d)} \bowtie^* \mathbf{X}^{(d)}).$$

As an example, we consider the finite difference negative d-Laplacian over uniform tensor grid, which is known to have the Kronecker rank-d representation

$$\Delta_d = A \otimes I_N \otimes \cdots \otimes I_N + I_N \otimes A \otimes I_N \otimes \cdots \otimes I_N + \cdots + I_N \otimes I_N \otimes \cdots \otimes A \in \mathbb{R}^{N^{\otimes d} \times N^{\otimes d}}, \quad (4.2)$$

with $A = \Delta_1 = \mathrm{tridiag}\{-1, 2, -1\} \in \mathbb{R}^{N \times N}$ and the $N \times N$ identity matrix I_N.

For the canonical rank we have $\mathrm{rank}_{\mathrm{Can}}(\Delta_d) = d$, whereas the TT-rank of Δ_d is equal to 2 for any dimension due to the explicit representation [142]

$$\Delta_d = \begin{bmatrix} \Delta_1 & I_N \end{bmatrix} \bowtie \begin{bmatrix} I_N & 0 \\ \Delta_1 & I_N \end{bmatrix}^{\otimes(d-2)} \bowtie \begin{bmatrix} I_N \\ \Delta_1 \end{bmatrix},$$

where the rank product operation "$\bowtie$" in the matrix case is defined as above [142]. The similar statement is true concerning the Tucker rank, $\mathrm{rank}_{\mathrm{Tuck}}(\Delta_d) = 2$.

Application of tensor methods for multidimensional PDEs are reported in [65, 67, 68], [212, 214, 213, 21, 257] and in [182, 188]. The basic mathematical models in quantum molecular dynamics have been previously described in [210, 211]. Greedy algorithms for high-dimensional non-symmetric linear problems have been considered in [48].

Basic multilinear algebra operations and solution of linear systems in TT and HT formats have been addressed [10, 228, 11, 21, 195]. The corresponding theoretical analysis can be found in [57, 196, 250, 249, 8] and [136, 137, 250]. Some applications of HT tensor format have been discussed in [261, 262, 242].

Recently the TT and QTT tensor formats were applied in electronic structure calculations for small molecules [240, 239].

4.2 $O(\log n)$-quantics (QTT) tensor approximation

The quantized (or quantics) tensor train (QTT) approximation was introduced and rigorously analyzed by B. Khoromskij in 2009 [165, 167]. It was initiated by the idea to test the TT-ranks of long function-related vectors reshaped to multidimensional hypercubes. For function-generated vectors (tensors), the QTT approximations were proved to provide the logarithmic data compression $O(d \log N)$ on the wide class of functions in $\mathbb{R}^d$, sampled on a tensor grid of size N^d. The basic approximation theory indicates that for a class of function-generated vector of size $N = q^L$, its reshaping into a $q \times \cdots \times q$ hypercube allows a small TT rank decomposition of the resultant L-dimensional tensor, [165, 167]. The storage of vectors of size q^L is reduced to qLr^2, where r is a small QTT rank. Thus, for example, if we have a vector of size $2^L = 2^{20}$, obtained by the grid discretization of an exponential function, then its quantized representation will need only $2 \cdot 20$ numbers, that is, $2 \cdot \log(2^L)$, since the exponential function has a QTT rank equal to 1. Correspondingly, algebraic operations with the QTT images are performed with logarithmic cost.

The QTT- or QCP-type approximation of an N-vector with $N = q^L$, $L \in \mathbb{N}$, is defined as the tensor decomposition (approximation) in the TT or canonical [189] formats applied to a tensor obtained by the q-adic folding (reshaping) of the target vector to an L-dimensional $q \times \cdots \times q$ data array (tensor) that is thought as an element of the L-dimensional quantized tensor space.

In particular, in the vector case, i. e., for $d = 1$, a vector $\mathbf{x} = [\mathbf{x}(i)]_{i \in I} \in \mathbb{V}_{N,1}$, is reshaped to its quantized image in $\mathbb{Q}_{q,L} = \bigotimes_{j=1}^{L} \mathbb{K}^q$, $\mathbb{K} \in \{\mathbb{R}, \mathbb{C}\}$, by q-adic folding,

$$\mathcal{F}_{q,L} : \mathbf{x} \to \mathbf{Y} = [Y(\mathbf{j})] \in \mathbb{Q}_{q,L}, \quad \mathbf{j} = \{j_1, \ldots, j_L\}, \quad \text{with } j_\nu \in \{1, 2, \ldots, q\},\ \nu = 1, \ldots, L,$$

where for fixed i, we have $\mathbf{Y}(\mathbf{j}) := \mathbf{x}(i)$, and $j_\nu = j_\nu(i)$ is defined via q-coding, $j_\nu - 1 = C_{-1+\nu}$, such that the coefficients $C_{-1+\nu}$ are found from the q-adic representation of $i - 1$,

$$i - 1 = C_0 + C_1 q^1 + \cdots + C_{L-1} q^{L-1} \equiv \sum_{\nu=1}^{L} (j_\nu - 1) q^{\nu-1}.$$

For $d > 1$, the construction is similar [167].

Suppose that the quantized image for certain N-d tensor (i. e., an element of D-dimensional quantized tensor space with $D = d \log_q N = dL$) can be effectively represented (approximated) in the low-rank TT (or CP) format living in the higher-dimensional tensor space $\mathbb{Q}_{q,dL}$. In this way, we introduce the QTT approximation of an N-d tensor. For given rank $\{r_k\}$ $(k = 1, \ldots, dL)$, the number of representation parameters for the QTT approximation of an N-d tensor can be estimated by

$$dqr^2 \log_q N \ll N^d, \quad \text{where } r_k \le r,\ k = 1, \ldots, dL,$$

providing log-volume scaling in the size of initial tensor $O(N^d)$. The optimal choice of the base q is shown to be $q = 2$ or $q = 3$ [167]. However, the numerical realizations are usually implemented by using binary coding, i. e., for $q = 2$. Figure 4.2 illustrates the QTT tensor approximation in cases $L = 3$ and $L = 10$.

The *principal question* arises: either there is the rigorous theoretical substantiation of the QTT approximation scheme that establishes it as the *new powerful approximation tool* applicable to the broad class of data, or this is simply the heuristic algebraic procedure that may be efficient in certain numerical examples.

The answer is positive: the power of QTT approximation method is due to the perfect rank-r decomposition discovered in [165, 167] for the wide-ranging class of function-related tensors obtained by sampling a continuous functions over uniform (or properly refined) grid. In particular, we have

- $r = 1$ for complex exponents;

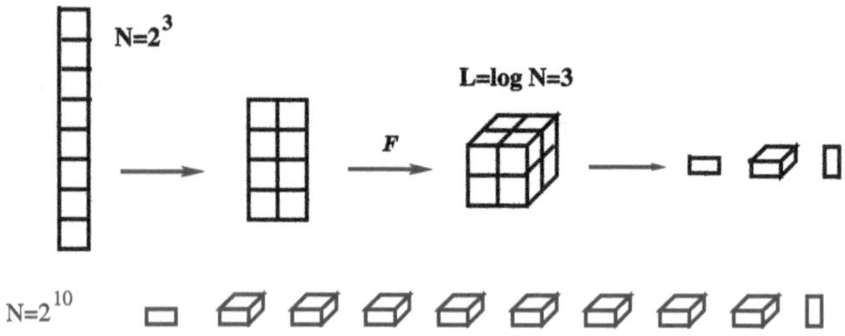

Figure 4.2: Visualizing the QTT tensor approximation in cases $L = 3$ and $L = 10$.

- $r = 2$ for trigonometric functions and for Chebyshev polynomials sampled on Chebyshev–Gauss–Lobatto grid;
- $r \leq m + 1$ for polynomials of degree m;
- r is a small constant for standard wavelet basis functions, etc.

The above rank bounds remain valid independently on the vector size N, and they are applicable to the general case $q = 2, 3, \ldots$.

Approximation of $2^d \times 2^d$ Laplacian-type matrices using TT tensor decomposition was introduced in [225].

Notice that the name quantics (or quantized) tensor approximation (with a shorthand QTT), originally introduced in 2009 [165], is a reminiscent of the entity "quantum of information" that mimics the minimal possible mode size ($q = 2$ or $q = 3$) of the quantized image. Later on, in some publications the QTT approximation method was renamed as "vector tensorization" [101, 110].

4.3 Low-rank representation of functions in quantized tensor spaces

The simple isometric folding of a multi-index data array into the $2 \times 2 \times \cdots \times 2$ format living in the virtual (higher) dimension $D = d \log N$ is the conventional reshaping operation in computer data representation. The most gainful features of numerical computations in the quantized tensor space appear via the remarkable rank-approximation properties figured out for the wide-ranging class of function-related vectors/tensors [167].

The next lemma presents the basic results on the rank-1 (resp. rank-2) q-folding representation of the exponential (resp. trigonometric) vectors.

Lemma 4.4 ([167]). *For given $N = q^L$, with $q = 2, 3, \ldots, L \in \mathbb{N}$, and $z \in \mathbb{C}$, the exponential N-vector $\mathbf{z} := \{x_n = z^{n-1}\}_{n=1}^N$ can be reshaped by the q-folding to the rank-1 $q^{\otimes L}$-tensor,*

$$\mathcal{F}_{q,L} : \mathbf{z} \mapsto \mathbf{Z} = \bigotimes_{p=1}^{L} [1 \quad z^{q^{p-1}} \quad \cdots \quad z^{(q-1)q^{p-1}}]^T \in \mathbb{Q}_{q,L}. \tag{4.3}$$

The number of representation parameters specifying the QTT image is reduced dramatically from N to $qL = q\log N$.

The trigonometric N-vector $\mathbf{t} = \Im m(\mathbf{z}) := \{t_n = \sin(\omega(n-1))\}_{n=1}^{N}$, $\omega \in \mathbb{R}$, can be reshaped by the successive q-adic folding

$$\mathcal{F}_{q,L} : \mathbf{t} \mapsto \mathbf{T} \in \mathbb{Q}_{q,L}$$

to the $q^{\otimes L}$-tensor $\mathbf{T}$ that has both the canonical $\mathbb{C}$-rank and the TT-rank equal exactly to 2. The number of representation parameters does not exceed $4qL$.

Example 4.5. In case $q = 2$, the single sin-vector has the explicit rank-2 QTT-representation in $\{0,1\}^{\otimes L}$ (see [69, 227]) with $k_p = 2^{p-L} i_p - 1$, $i_p \in \{0,1\}$,

$$\mathbf{t} \mapsto \mathbf{T} = \Im m(\mathbf{Z}) = [\sin \omega k_1 \ \cos \omega k_1] \bowtie_{p=2}^{L-1} \begin{bmatrix} \cos \omega k_p & -\sin \omega k_p \\ \sin \omega k_p & \cos \omega k_p \end{bmatrix} \bowtie \begin{bmatrix} \cos \omega k_L \\ \sin \omega k_L \end{bmatrix}.$$

Other results on QTT representation of polynomial, Chebyshev polynomial, Gaussian-type vectors, multivariate polynomials, and their piecewise continuous versions have been derived in [167] and in subsequent papers [177, 227, 68] substantiating the capability of numerical calculus in quantized tensor spaces.

In computational practice the binary coding representation with $q = 2$ is the most convenient choice, though the Euler number $q_* = e \approx 2.7\ldots$ is shown to be the optimal value [167].

The following example demonstrates that the low-rank QTT approximation can be applied for $O(|\log \varepsilon|)$ complexity integration of functions. Given continuous function $f(x)$ and weight function $w(x)$, $x \in [0, A]$, consider the rectangular N-point quadrature I_N, $N = 2^L$, ensuring the error bound $|I - I_N| = O(2^{-\alpha L})$. Assume that the corresponding functional vectors allow low-rank QTT approximation. Then the rectangular quadrature can be implemented as the scalar product on QTT tensors in $O(\log N)$ operations.

$$\int_0^A w(x) f(x) dx \approx I_N(f) := h \sum_{i=1}^{N} w(x_i) f(x_i) = \langle \mathbf{W}, \mathbf{F} \rangle_{\mathrm{QTT}}, \quad \mathbf{W}, \mathbf{F} \in \bigotimes_{\ell=1}^{L} \mathbb{R}^2.$$

Example 4.6 illustrates below the uniform bound on the QTT rank for nontrivial highly oscillating functions. Here and in the following the threshold error like ϵ_{QTT} corresponds to the Euclidean norm.

Example 4.6. Highly oscillating and singular functions on $[0, A]$, $\omega = 100$, $\epsilon_{\mathrm{QTT}} = 10^{-6}$,

$$f_3(x) = \begin{cases} x + a_k \sin(\omega x), & x \in 10(\frac{k-1}{p}; \frac{k-0.5}{p}], \\ 0, & x \in 10(\frac{k-0.5}{p}; \frac{k}{p}], \end{cases}$$

$$f_4(x) = (x+1)\sin(\omega(x+1)^2), \quad x \in [0,1] \quad \text{(Fresnel integral)},$$

where the function $f_3(x)$, $x \in [0, 10]$, $k = 1, \ldots, p$, $p = 16$, $a_k = 0.3 + 0.05(k-1)$, is recognized on three different scales.

Notice that in the following, in all numerical results, we use the average QTT rank $\bar{r}$ defined as

$$\bar{r} := \sqrt{\frac{1}{d-1} \sum_{k=1}^{d-1} r_k r_{k+1}}. \tag{4.4}$$

The average QTT ranks over all directional ranks for the corresponding functional vectors are given in Table 4.1. The maximum rank over all the fibers is nearly the same as the average one.

Table 4.1: Average QTT ranks of N-vectors generated by f_3 and f_4.

$N \setminus \bar{r}$	$r_{QTT}(f_3)$	$r_{QTT}(f_4)$
2^{14}	3.5	6.5
2^{15}	3.6	7.0
2^{16}	3.6	7.5
2^{17}	3.6	7.9

Further examples concerning the low-rank QTT tensor approximation will be presented in sections related to computation of the two-electron integrals and the summation of electrostatic potentials over large lattice structured system of particles.

Notice that 1D and 2D numerical quadratures, based on interpolation by Chebyshev polynomials, have been developed [120]. Taking into account that Chebyshev polynomial, sampled on Chebyshev grid, has the exact rank-2 QTT representation [167] allows us to perform the efficient numerical integration via Chebyshev interpolation by using the QTT approximation.

In application to multidimensional PDEs, the tensor representation of operators in quantized spaces is also important. Several results on the QTT approximation of discretized multidimensional operators (matrices) were presented in [179, 177, 176, 178, 155] and in [142, 66, 67].

Superfast FFT, wavelet and circulant convolution-type data transforms of logarithmic complexity have been introduced [69, 143, 175].

Various applications of the QTT format to the solution of PDEs were reported in [68, 188, 67, 65, 144, 180, 181].

5 Multidimensional tensor-product convolution

The important prerequisites for the grid-based calculation of the convolution integrals in $\mathbb{R}^d$ arising in computational quantum chemistry are the multidimensional tensor-product convolution techniques and the efficient canonical tensor representation of the Green's kernels by using the Laplace transform and sinc-quadrature methods.

The tensor-product approximation of multidimensional convolution transform discretized via collocation-projection scheme on the uniform or composite refined grids was introduced in 2007 (see [173, 166]). In what follows, we present some of the results in [166], where the examples of convolving kernels are given by the classical Newton, Slater (exponential), and Yukawa potentials, $1/\|x\|$, $e^{-\lambda\|x\|}$, and $e^{-\lambda\|x\|}/\|x\|$ with $x \in \mathbb{R}^d$. For piecewise constant elements on the uniform grid of size n^d, the quadratic convergence rate $O(h^2)$ in the mesh parameter $h = 1/n$ is proved in [166], where it was also shown that the Richardson extrapolation method on a sequence of grids improves the order of approximation up to $O(h^3)$. The fast algorithm of complexity $O(dR_1R_2 n \log n)$ is described for tensor-product convolution on the uniform/composite grids of size n^d, where R_1, R_2 are the tensor ranks of convolving functions. We also discuss the tensor-product convolution scheme in the two-level Tucker-canonical format and discuss the consequent rank reduction strategy. The numerical illustrations confirming the approximation theory for convolution schemes of order $O(h^2)$ and $O(h^3)$ can be found in [166]. The linear-logarithmic complexity scaling in n of 1D discrete convolution on large composite grids and for convolution method on $n \times n \times n$ grids in the range $n \leq 16\,384$ was also demonstrated.

5.1 Grid-based discretization of the convolution transform

The multidimensional convolution in $L^2(\mathbb{R}^d)$ is defined by the integral transform

$$w(x) := (f * g)(x) := \int_{\mathbb{R}^d} f(y)g(x-y)dy \quad f,g \in L^2(\mathbb{R}^d), \quad x \in \mathbb{R}^d. \tag{5.1}$$

We are interested in approximate computation of $f * g$ in some fixed box $\Omega = [-A, A]^d$, assuming that the convolving function f has a support in $\Omega' := [-B, B]^d \subset \Omega$ $(B < A)$, i. e., $\operatorname{supp} f \subset \Omega'$. In electronic structure calculations, the convolving function f may represent electron orbitals or electron densities, which normally have an exponential decay, and, hence, they could be truncated beyond some fixed spacial box.

The common example of the convolving kernel g is given by the restriction of the fundamental solution of an elliptic operator in $\mathbb{R}^d$. For example, in the case of the Laplacian in $\mathbb{R}^d$, $d \geq 3$, we have

https://doi.org/10.1515/9783110365832-005

$$g(x) = c(d)/\|x\|^{d-2}, \quad x = (x_1, \dots, x_d) \in \mathbb{R}^d, \quad \|x\| = \sqrt{x_1^2 + \dots + x_d^2},$$

where $c(d) = -2^{\frac{d}{4-d}}/\Gamma(d/2-1)$. This example will be considered in more detail.

There are three commonly used discretization methods for the integral operators the so-called Nyström, collocation and Galerkin-type schemes. Below, we consider the case of uniform grids, referring to [166] for complete theory, including the case of composite grids.

Introduce the equidistant tensor-product lattice $\omega_{\mathbf{d}} := \omega_1 \times \dots \times \omega_d$ of size $h = 2A/n$ by setting $\omega_\ell := \{-A + (k-1)h : k = 1, \dots, n+1\}$, where, for the sake of convenience, $n = 2p$, $p \in \mathbb{N}$, and define the tensor-product index set $\mathcal{I} := \{1, \dots, n\}^d$. Hence $\Omega = \bigcup_{\mathbf{i} \in \mathcal{I}} \Omega_{\mathbf{i}}$ becomes the union of closed boxes $\Omega_{\mathbf{i}} = \bigotimes_{\ell=1}^d \Omega_{i_\ell}$ specified by segments

$$\Omega_{i_\ell} := \{x_\ell : x_\ell \in [-A + (i_\ell - 1)h, -A + i_\ell h]\} \subset \mathbb{R} \quad (\ell = 1, \dots, d). \tag{5.2}$$

The *Nyström-type scheme* leads to simple discretization

$$(f * g)(x_{\mathbf{j}}) \approx h^d \sum_{\mathbf{i} \in \mathcal{I}} f(y_{\mathbf{i}}) g(x_{\mathbf{j}} - y_{\mathbf{i}}), \quad \mathbf{j} \in \mathcal{I},$$

where, for the ease of presentation, the evaluation points $x_{\mathbf{j}}$, and the collocation points $y_{\mathbf{i}}$, $\mathbf{i}, \mathbf{j} \in \mathcal{I}$ are assumed to be located on the same cell-centered tensor-product grid corresponding to $\omega_{\mathbf{d}}$. The Nyström-type scheme applies to the continuous functions f, g, which leads to certain limitations in the case of singular kernels g.

The *collocation-projection discretization* can be applied to a much more general class of integral operators than the Nyström methods, including Green's kernels with the diagonal singularity, say to the Newton potential $g(x) = 1/\|x\|$. We consider the case of tensor-product piecewise constant basis functions $\{\phi_{\mathbf{i}}\}$ associated with $\omega_{\mathbf{d}}$, so that $\phi_{\mathbf{i}} = \chi_{\Omega_{\mathbf{i}}}$ is the characteristic function of $\Omega_{\mathbf{i}}$,

$$\phi_{\mathbf{i}}(x) = \prod_{\ell=1}^d \phi_{i_\ell}(x_\ell), \quad \text{where } \phi_{i_\ell} = \chi_{\Omega_{i_\ell}}. \tag{5.3}$$

Let $x_{\mathbf{m}} \in \omega_{\mathbf{d}}$ be the set of collocation points with $\mathbf{m} \in \mathcal{M}_n := \{1, \dots, n+1\}^d$ (we use the notation $\mathcal{M}_n = \mathcal{M}$ if there is no confusion), and let $f_{\mathbf{i}}$ be the representation coefficients of f in $\{\phi_{\mathbf{i}}\}$,

$$f(y) \approx \tilde{f}(y) := \sum_{\mathbf{i} \in \mathcal{I}} f_{\mathbf{i}} \phi_{\mathbf{i}}(y).$$

In what follows, we specify the coefficients as $f_{\mathbf{i}} = f(y_{\mathbf{i}})$, where $y_{\mathbf{i}}$ is the midpoint of $\Omega_{\mathbf{i}}$, $\mathbf{i} \in \mathcal{I}$. We consider the following *discrete collocation-projection scheme*:

$$f * g \approx \{w_{\mathbf{m}}\}, \quad w_{\mathbf{m}} := \sum_{\mathbf{i} \in \mathcal{I}} f_{\mathbf{i}} \int_{\mathbb{R}^d} \phi_{\mathbf{i}}(y) g(x_{\mathbf{m}} - y) dy, \quad x_{\mathbf{m}} \in \omega_{\mathbf{d}}, \quad \mathbf{m} \in \mathcal{M}. \tag{5.4}$$

The straightforward pointwise evaluation of this scheme requires $O(n^{2d})$ operations. In the case of equidistant grids, the computational complexity can be reduced to $O(n^d \log n)$ by applying the multidimensional FFT. Our goal is to reduce the numerical complexity to the linear scale in the dimension d.

To transform the collocation scheme (5.4) to the discrete convolution, we precompute the collocation coefficients

$$g_{\mathbf{i}} = \int_{\mathbb{R}^d} \phi_{\mathbf{i}}(y) g(-y) dy, \quad \mathbf{i} \in \mathcal{I}, \tag{5.5}$$

define the dth-order tensors $\mathbf{F} = \{f_{\mathbf{i}}\}, \mathbf{G} = \{g_{\mathbf{i}}\} \in \mathbb{R}^{\mathcal{I}}$, and introduce the d-dimensional *discrete convolution*

$$\mathbf{F} * \mathbf{G} := \{z_{\mathbf{j}}\}, \quad z_{\mathbf{j}} := \sum_{\mathbf{i}} f_{\mathbf{i}} g_{\mathbf{j}-\mathbf{i}+\mathbf{1}}, \quad \mathbf{j} \in \mathcal{J} := \{1, \ldots, 2n-1\}^d, \tag{5.6}$$

where the sum is taken over all $\mathbf{i} \in \mathcal{I}$, which leads to legal subscripts for $g_{\mathbf{j}-\mathbf{i}+\mathbf{1}}$, $\mathbf{j} - \mathbf{i} + \mathbf{1} \in \mathcal{I}$. Specifically, for $j_\ell = 1, \ldots, 2n-1$,

$$i_\ell \in [\max(1, j_\ell + 1 - n), \min(j_\ell, n)] \quad \ell = 1, \ldots, d.$$

The discrete convolution can be gainfully applied to fast calculation of $\{w_{\mathbf{m}}\}_{\mathbf{m} \in \mathcal{M}}$ in the collocation scheme (5.4) as shown in the following statement.

Proposition 5.1 ([166]). *The discrete collocation scheme $\{w_{\mathbf{m}}\}$, $\mathbf{m} \in \mathcal{M}$, is obtained by copying the corresponding portion of $\{z_{\mathbf{j}}\}$ from (5.6), centered at $\mathbf{j} = \mathbf{n} = n^{\otimes d}$,*

$$\{w_{\mathbf{m}}\} = \{z_{\mathbf{j}}\}_{|\mathbf{j}=\mathbf{j}_0+\mathbf{m}}, \quad \mathbf{m} \in \mathcal{M}, \ \mathbf{j}_0 = \mathbf{n}/2.$$

Proof. In the 1D case, we have

$$z(1) = f(1) \cdot g(1), \quad z(2) = f(1) \cdot g(2) + f(2) \cdot g(1), \quad \ldots,$$
$$z(n) = f(1) \cdot g(n) + f(2) \cdot g(n-1) + \cdots + f(n) \cdot g(1), \quad \ldots, \quad z(2n-1) = f(n) \cdot g(n).$$

Then we find that elements $\{w_m\}$ coincide with $\{z_j\}_{|j=j_0+m}$, $m \in \mathcal{M}, j_0 = n/2$. The general case $d \geq 1$ can be justified by applying the above argument to each spatial variable. □

Notice that the *Galerkin method* of discretization reads as follows:

$$f * g \approx \sum_{\mathbf{i}, \mathbf{j}-\mathbf{i}+\mathbf{1} \in \mathcal{I}, \mathbf{j} \in \mathbf{j}_0 + \mathcal{M}} f_{\mathbf{i}} g_{\mathbf{j}-\mathbf{i}+\mathbf{1}} \quad \text{with } g_{\mathbf{j}-\mathbf{i}+\mathbf{1}} := \int_{\mathbb{R}^d} \phi_{\mathbf{j}}(x) \phi_{\mathbf{i}}(y) g(x-y) dx dy$$

with the choice $f_{\mathbf{i}} = \langle f, \phi_{\mathbf{i}} \rangle_{L^2}$. The Galerkin scheme is known as the most convenient for theoretical error analysis. However, compared with the collocation method, it has higher implementation cost because of the presence of double integration. Hence, classical discretization methods mentioned above may differ from each other by construction of the tensor-product decompositions. To keep a reasonable compromise between the numerical complexity of the scheme and its generality, in the following we focus on the collocation method by simple low-order finite elements.

Recall that in the case of piecewise constant basis functions the error bound $O(h^2)$ for the collocation scheme is proved in [166], whereas the Richardson extrapolation method on a sequence of grids proved to provide the improved approximation error $O(h^3)$. Such an extrapolation, when available, allows a substantial reduction of the approximation error without extra cost. It is worth noting that the Richardson extrapolation can also be applied to some functionals of the convolution product, say to eigenvalues of the operator that includes the discrete convolution.

5.2 Tensor approximation to discrete convolution on uniform grids

Recall that in the case of uniform grids, the discrete convolution in $\mathbb{R}^d$ can be implemented by d-dimensional FFT with linear cost in the volume size, $O(n^d \log n)$, which preserves the exponential scaling in d. To avoid the curse of dimensionality, we represent the d-dimensional convolution product approximately in the low-rank tensor product formats. This reduces dramatically the computational cost to $O(dn \log n)$. Note that tensor approximation to discrete convolution on non-uniform grids is considered in full detail in [166]; see also [109].

We notice that the multidimensional convolution product appears to be one of the most computationally elaborate multilinear operations. The key idea is to calculate the d-dimensional convolution approximately using rank-structured tensor approximations [166]. Recall that for given d-th order tensors $\mathbf{F}, \mathbf{G} \in \mathcal{T}_{\mathbf{r}}$ in the Tucker format, represented by

$$\mathbf{F} = \boldsymbol{\beta} \times_1 F^{(1)} \times_2 F^{(2)} \cdots \times_d F^{(d)} \quad \text{and} \quad \mathbf{G} = \boldsymbol{\gamma} \times_1 G^{(1)} \times_2 G^{(2)} \cdots \times_d G^{(d)},$$

the convolution product can be represented in the separable form (cf. [173])

$$\mathbf{F} * \mathbf{G} := \sum_{\mathbf{k}=1}^{\mathbf{r}} \sum_{\mathbf{m}=1}^{\mathbf{r}} \beta_{k_1 \ldots k_d} \gamma_{m_1 \ldots m_d} (\mathbf{f}_1^{k_1} * \mathbf{g}_1^{m_1}) \otimes \cdots \otimes (\mathbf{f}_d^{k_d} * \mathbf{g}_d^{m_d}). \tag{5.7}$$

Computing 1D convolution $\mathbf{f}_\ell^{k_\ell} * \mathbf{g}_\ell^{m_\ell} \in \mathbb{R}^{2n-1}$ in $O(n \log n)$ operations leads to the overall linear-logarithmic complexity in n,

$$\mathcal{N}_{T*T} = O(dr^2 n \log n + \#\boldsymbol{\beta} \cdot \#\boldsymbol{\gamma}).$$

In general one might have $\#\boldsymbol{\beta} \cdot \#\boldsymbol{\gamma} = O(r^{2d})$, which may be restrictive even for moderate d.

A significant complexity reduction is observed if at least one of the convolving tensors can be represented in the canonical format. Letting $\mathbf{F} \in \mathcal{T}_{\mathbf{r}}$, $\mathbf{G} \in \mathcal{C}_R$, i. e., $\gamma = \operatorname{diag}\{\gamma_1, \ldots, \gamma_R\}$, we tensorize the convolution product as follows:

$$\mathbf{F} * \mathbf{G} = \sum_{\mathbf{k}=\mathbf{1}}^{\mathbf{r}} \sum_{m=1}^{R} \beta_{k_1 \ldots k_d} \gamma_m (\mathbf{f}_1^{k_1} * \mathbf{g}_1^m) \otimes \cdots \otimes (\mathbf{f}_d^{k_d} * \mathbf{g}_d^m). \tag{5.8}$$

However, the calculation by (5.8) still scales exponentially in d, which leads to certain limitations in the case of higher dimensions.

To get rid of this exponential scaling, it is better to perform the convolution transform using the two-level tensor format, i. e., $\mathbf{F} \in \mathcal{T}_{\mathcal{C}_{R_1,\mathbf{r}}}$ (see Definition 3.15) in such a way that the result $\mathbf{U} = \mathbf{F} * \mathbf{G}$ with $\mathbf{G} \in \mathcal{C}_{R_G}$ is represented in the two-level Tucker format $\mathcal{T}_{\mathcal{C}_{R_1 R_G, \mathbf{r} R_G}}$. Recall that an explicit representation for $\mathbf{F} \in \mathcal{T}_{\mathcal{C}_{R_1,\mathbf{r}}}$ is given by

$$\mathbf{F} = \left(\sum_{\nu=1}^{R_1} \beta_\nu \mathbf{z}_1^\nu \otimes \cdots \otimes \mathbf{z}_d^\nu \right) \times_1 F^{(1)} \times_2 F^{(2)} \cdots \times_d F^{(d)}, \tag{5.9}$$

so that we have the imbedding $\mathcal{T}_{\mathcal{C}_{R_1,\mathbf{r}}} \subset \mathcal{C}_{R_1,\mathbf{n}}$ with the corresponding (non-orthogonal) side-matrices $S^{(\ell)} = [F^{(\ell)} \mathbf{z}_\ell^1 \; \cdots \; F^{(\ell)} \mathbf{z}_\ell^{R_1}] \in \mathbb{R}^{n \times R_1}$ and scaling factors β_ν ($\nu = 1, \ldots, R_1$). Now we represent the tensor-product convolution in the two-level format

$$\mathbf{F} * \mathbf{G} = \sum_{m=1}^{R_G} \gamma_m \left(\sum_{\nu=1}^{R_1} \beta_\nu \mathbf{z}_1^\nu \otimes \cdots \otimes \mathbf{z}_d^\nu \right) \times_1 (F^{(1)} * \mathbf{g}_1^m) \times_2 \cdots \times_d (F^{(d)} * \mathbf{g}_d^m), \tag{5.10}$$

such that the above expansion can be evaluated by the following algorithm.

Algorithm 5.1 (*d*-dimensional tensor convolution of type $\mathcal{T}_{\mathcal{C}_{R_1,\mathbf{r}}} * \mathcal{C}_{R_G,\mathbf{n}} \to \mathcal{T}_{\mathcal{C}_{R_1 R_G, \mathbf{r} R_G}}$).

(1) Given $\mathbf{F} \in \mathcal{T}_{\mathcal{C}_{R_1,\mathbf{r}}}$ with the core $\boldsymbol{\beta} = \sum_{\nu=1}^{R_1} \beta_\nu \mathbf{z}_1^\nu \otimes \cdots \otimes \mathbf{z}_d^\nu \in \mathcal{C}_{R_1,\mathbf{r}}$, and $\mathbf{G} \in \mathcal{C}_{R_G,\mathbf{n}}$.

(2) For $\ell = 1, \ldots, d$, compute the set of 1D convolutions $\mathbf{u}_\ell^{k,m} = \mathbf{f}_\ell^k * \mathbf{g}_\ell^m$ ($k = 1, \ldots, r$, $m = 1, \ldots, R_G$) of size $2n - 1$, restrict the results onto the index set I_ℓ, and form the $n \times rR_G$ side-matrices $U^{(\ell)} = [U_1^{(\ell)} \; \cdots \; U_{R_G}^{(\ell)}]$, composed of the blocks $U_m^{(\ell)}$ with columns $\mathbf{u}_\ell^{k,m}$ as $U_m^{(\ell)} = [\mathbf{f}_\ell^1 * \mathbf{g}_\ell^m \; \cdots \; \mathbf{f}_\ell^r * \mathbf{g}_\ell^m]$, all at the cost $O(drR_G n \log n)$.

(3) Build the core tensor $\boldsymbol{\omega} = \operatorname{blockdiag}\{\gamma_1 \boldsymbol{\beta}, \ldots, \gamma_R \boldsymbol{\beta}\}$ and represent the resultant two-level Tucker tensor in the form (storage demand is $R_G + R_1 + drR_1 + drR_G n$),

$$\mathbf{U} = \boldsymbol{\omega} \times_1 U^{(1)} \times_2 \cdots \times_d U^{(d)} \in \mathcal{T}_{\mathcal{C}_{R_1 R_G, \mathbf{r} R_G}}.$$

In some cases, one may require the consequent rank reduction for the target tensor $\mathbf{U}$ to the two-level format $\mathcal{T}_{\mathcal{C}_{R_0,\mathbf{r}_0}}$ with moderate rank parameters R_0 and $\mathbf{r}_0 = (r_0, \ldots, r_0)$ [166].

If both convolving tensors are given in the canonical format, $\mathbf{F} \in \mathcal{C}_{R_F}$ with coefficients β_k, $k = 1, \ldots, R_F$ and $\mathbf{G} \in \mathcal{C}_{R_G}$, with coefficients γ_m, $m = 1, \ldots, R_G$, then

$$\mathbf{F} * \mathbf{G} = \sum_{k=1}^{R_F} \sum_{m=1}^{R_G} \beta_k \gamma_m (\mathbf{f}_1^k * \mathbf{g}_1^m) \otimes \cdots \otimes (\mathbf{f}_d^k * \mathbf{g}_d^m), \tag{5.11}$$

leading to the reduced cost that scales linearly in dimensionality parameter d and linear-logarithmically in n,

$$\mathcal{W}_{C*C \to C} = O(dR_F R_G n \log n).$$

Algorithm 5.2 (Multidimensional tensor product convolution of type "$C * C \to C$").
(1) Given $\mathbf{F} \in \mathcal{C}_{R_F,\mathbf{n}}$, $\mathbf{G} \in \mathcal{C}_{R_G,\mathbf{n}}$.
(2) For $\ell = 1, \ldots, d$, compute the set of 1D convolutions $\mathbf{f}_\ell^k * \mathbf{g}_\ell^m$ ($k = 1, \ldots, R_F$, $m = 1, \ldots, R_G$) of size $2n - 1$, restrict the results onto the index set I_ℓ, and form the $n \times R_F R_G$ side-matrix $U^{(\ell)}$ (cost $dR_F R_G n \log n$).
(3) Compute the set of scaling factors $\beta_k \gamma_m$.

Complexity bound $O(dR_F R_G n \log n)$ is proven in [166].

The resulting convolution product $\mathbf{F} * \mathbf{G}$ in (5.11) may be approximated in either Tucker or canonical formats, depending on further multi-linear operations applied to this tensor. In the framework of approximate iterations with structured matrices and vectors, we can fix the $\mathcal{C}_{R_0}$-format for the output tensors. Hence, the rank-R_0 canonical approximation (with $R_0 < R_F R_G$) would be the proper choice to represent $\mathbf{F} * \mathbf{G}$. The tensor truncation of the rank-$(R_F R_G)$ auxiliary result to rank-R_0 tensor can be accomplished by fast multigrid C2T plus T2C tensor approximation, and then the result can be stored by $O(dR_0 n)$ reals.

Based on our experience with Algorithms 5.1 and 5.2, applied in electronic structure calculations in 3D, we notice that Algorithm 5.2 is preferable in the case of moderate grid-size (say, $n \le 10^4$), while Algorithm 5.1 is faster for larger grids. For example, both algorithms work perfectly in electronic structure calculations in the framework of the Hartree–Fock model for $d = 3$ [174, 186]. A case in point is that the Hartree potential of moderate size molecules can be calculated on the $n \times n \times n$ 3D Cartesian grids with $n \le 1.6 \cdot 10^4$ in a few minutes providing the relative accuracy about 10^{-7} already with $n = 8192$. Further numerical illustrations will be given in Chapter 8.

5.3 Low-rank approximation of convolving tensors

In applications related to electronic structure calculations, the function-related collocation coefficient tensor $\mathbf{F} = [f_\mathbf{i}]_{\mathbf{i} \in \mathcal{I}}$ can be generated by the electron density $\rho(x)$, by the product of the interaction potential $V(x)$ with the electron orbitals, $V(x)\psi(x)$, or by some related terms. In this way, we make an a priori assumption on the existence of low-rank approximation to the corresponding tensors. In general, this assumption is not easy to justify. However, it works well in practice.

Example 5.1. In the case of Hydrogen atom, we have

$$\rho(x) = e^{-2\|x\|} \quad \text{and} \quad V(x)\psi(x) = \frac{e^{-\|x\|}}{\|x\|} \quad \text{with } V(x) = \frac{1}{\|x\|}, \; x \in \mathbb{R}^3,$$

hence, the existence of corresponding low-rank tensor approximations can be proven along the lines of [161, Lemma 4.3] and [163, Theorem 3].

To construct a low-rank approximation of the convolving tensor $\mathbf{G}$, we consider a class of multivariate spherically symmetric (radial) convolving kernels $g : \mathbb{R}^d \to \mathbb{R}$ parameterized by

$$g = g(\rho(y)) \quad \text{with } \rho \equiv \rho(y) = y_1^2 + \cdots + y_d^2,$$

where the univariate function $g : \mathbb{R}_+ \to \mathbb{R}$ can be represented via the generalized Laplace transform

$$g(\rho) = \int_{\mathbb{R}_+} \widehat{g}(\tau^2) e^{-\rho\tau^2} d\tau. \tag{5.12}$$

Without loss of generality, we introduce one and the same scaling function

$$\phi_i(\cdot) = \phi(\cdot + (i-1)h), \quad i \in I_n,$$

for all spatial dimensions $\ell = 1, \ldots, d$, where $h > 0$ is the mesh parameter, so that the corresponding tensor-product basis function $\phi_{\mathbf{i}}$ is defined by (5.3).

Using *sinc*-quadrature methods, [271], we approximate the collocation coefficient tensor $\mathbf{G} = [g_{\mathbf{i}}]_{\mathbf{i} \in \mathcal{I}}$ in (5.5) via the rank-$(2M+1)$ canonical decomposition

$$g \approx \sum_{k=-M}^{M} w_k \mathbf{E}(\tau_k) \quad \text{with } \mathbf{E}(\tau_k) = [e_{\mathbf{i}}(\tau_k)], \; \mathbf{i} \in \mathcal{I}, \tag{5.13}$$

with suitably chosen coefficients $w_k \in \mathbb{R}$ and quadrature points $\tau_k \in \mathbb{R}_+$, where the rank-1 tensor $\mathbf{E}(\tau_k) \in \mathbb{R}^{\mathcal{I}}$ is given entrywise by

$$e_{\mathbf{i}}(\tau_k) = \widehat{g}(\tau_k^2) \prod_{\ell=1}^{d} \int_{\mathbb{R}} e^{-y_\ell^2 \tau_k^2} \phi_{i_\ell}(y_\ell) dy_\ell. \tag{5.14}$$

For a class of analytic functions the exponentially fast convergence in M of the above quadrature can be proven (see [111, 163]). Notice that the quadrature points τ_k can be chosen symmetrically, i. e., $\tau_k = \tau_{-k}$, hence reducing the number of terms in (5.13) to $r = M + 1$.

In the particular applications in electronic structure calculations, we are interested in fast convolution with the Newton or Yukawa kernels. In the case of the Newton kernel, $g(x) = 1/\|x\|$, the approximation theory can be found in [111]. In the case of the Yukawa potential $e^{-\kappa\|x\|}/\|x\|$ for $\kappa \in [0, \infty)$, we apply the generalized Laplace transform (cf. (5.12))

$$g(\rho) = \frac{e^{-\kappa\sqrt{\rho}}}{\sqrt{\rho}} = \frac{2}{\sqrt{\pi}} \int_{\mathbb{R}_+} \exp(-\rho\tau^2 - \kappa^2/4\tau^2)d\tau, \tag{5.15}$$

corresponding to the choice

$$\hat{g}(\tau^2) = \frac{2}{\sqrt{\pi}} e^{-\kappa^2/4\tau^2}.$$

Approximation theory in the case of Yukawa potential is presented in [163].

In our numerical experiments, the collocation coefficient tensor $\mathbf{G} \in \mathbb{R}^{\mathcal{I}}$ for the Newton kernel is approximated in the rank-R canonical format with $R \in [20, 40]$ providing high accuracies about 10^{-6}–10^{-8} for the grid sizes up to $n^3 = 131\,072^3$.

5.4 Algebraic recompression of the sinc approximation

In the case of large computational grids, the tensor rank of the (problem independent) convolving kernel g can be reduced by an algebraic recompression procedure [166]. For ease of presentation let us consider the case $d = 3$. The idea of our recompression algorithm is based on the observation that a typical feature of the analytic tensor approximation by the sinc quadratures as in (5.13)–(5.14) (for symmetric quadrature points it is agglomerated to the sequence with $k = 0, 1, \ldots, M$) is the presence of many terms all supported only by a few grid points belonging to the small $p \times p \times p$ sub-grid in domain $\Omega^{(p)}$ that is a vicinity of the point-type singularity (say, at $x = 0$). Assume that this group of rank-1 tensors is numbered by $k = 0, \ldots, K < M$. The sum of these tensors, further denoted as $\mathbf{A}_p$, effectively belongs to the low-dimensional space of tri-linear $p \times p \times p$-tensors. Hence, the maximal tensor rank of $\mathbf{A}_p$ does not exceed $r = p^2 \leq K$. Furthermore, we can perform the rank-R_0 canonical approximation of this small tensor with $R_0 < K$ using the ALS or gradient type optimization. The following Algorithm sketches the main steps of the rank recompression scheme described above.

Algorithm 5.3 (Rank recompression for the canonical sinc-based approximation).
(1) Given the canonical tensor $\mathbf{A}$ with rank $R = M + 1$.
(2) Agglomerate all rank-1 terms supported by the only one point, say by $\Omega^{(1)}$, into one rank-1 tensor, further called $\mathbf{A}_1$.

(3) Agglomerate by a summation all terms supported by $\Omega^{(2)} \setminus \Omega^{(1)}$ in one tensor $\mathbf{A}_2$ (with maximal rank 3), approximate with the tensor rank $r_2 \leq 3$, and so on until we end up with tensor $\mathbf{A}_p$ supported by $\Omega^{(p)} \setminus \Omega^{(p-1)} \setminus \cdots \setminus \Omega^{(1)}$.
(4) Approximate the canonical sum $\mathbf{A}_1 + \cdots + \mathbf{A}_p$ by a low-rank tensor.

Notice that in the sinc-quadrature approximations most of these "local" terms are supported by only one point, say by $\Omega^{(1)}$, hence they are all agglomerated in the rank-1 tensor. In approximation of the classical potentials like $1/\|x\|$ or $e^{-\|x\|}/\|x\|$ the usual choice is $p = 1, 2$.

The simple rank recompression procedure described above allows to reduce noticeably the initial rank $R = M + 1$ appearing in the (symmetric) sinc quadratures. Numerical examples on the corresponding rank reduction by Algorithm 5.3 are depicted in [163], Figure 2.

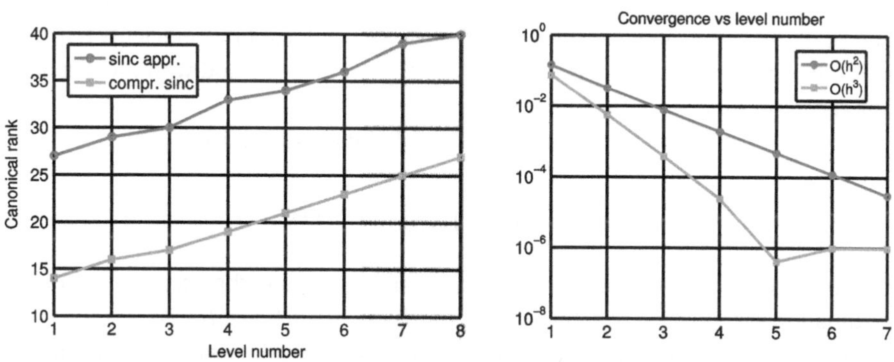

Figure 5.1: Tensor rank of the sinc- and recompressed sinc-approximation for $1/\|x\|$ (left). Convergence history for the $O(h^2)$ and $O(h^3)$ Richardson extrapolated convolution schemes (right).

Figure 5.1 (left) presents the rank parameters obtained from the sinc approximations of $g(x) = 1/\|x\|$ up to threshold $\varepsilon = 0.5 \cdot 10^{-6}$ in max-norm, computed on $n \times n \times n$ grids with $n = 2^{L+3}$ for the level number $L = 1, \ldots, 8$ (upper curve), and the corresponding values obtained by Algorithm 5.3 with $p = 1$ (lower curve). One observes the significant reduction of the tensor rank.

5.5 Numerical verification on quantum chemistry data

We test the approximation error of the tensor-product collocation convolution scheme on practically interesting data arising in electronic structure calculations using the Hartree–Fock equation (see [174] for more detail). We consider the pseudo electron density of the CH_4-molecule represented by the exponential sum

$$f(x) := \sum_{\nu=1}^{M}\left(\sum_{k=1}^{R_0} c_{\nu,k}(x-x_k)^{\beta_k} e^{-\lambda_k(x-x_k)^2}\right)^2, \quad x \in \mathbb{R}^3,\ R_0 = 50,\ M = 4, \tag{5.16}$$

with x_k corresponding to the locations of the C and H atoms. We extract the "principal exponential" approximation of the electron density, f_0, obtained by setting $\beta_k = 0$ $(k = 1, \dots, R_0)$ in (5.16). Using the fast tensor-product convolution method, the Hartree potential of f_0,

$$V_H(x) = \int_{\Omega} \frac{f_0(y)}{\|x-y\|} dy, \quad x \in \Omega = [-A, A]^3,$$

is computed with high accuracy on a sequence of uniform $n \times n \times n$ grids with $n = 2^p + 1$, $p = 5, 6, \dots, 12$, and $A = 9.6$. The initial rank of the input tensor $\mathbf{F} = [f_0(y_{\mathbf{i}})]_{\mathbf{i} \in \mathcal{I}}$, presented in the canonical format, is bounded by $R \le \frac{R_0(R_0+1)}{2}$ (even for simple molecules it normally amounts about several thousands). The collocation coefficients tensor $\mathbf{G}$ in (5.5) for the Newton kernel is approximated by the sinc-method with the algebraic rank-recompression described in Algorithm 5.3.

Note that the Hartree potential has slow polynomial decay, i. e.,

$$V_H(x) = O\left(\frac{1}{\|x\|}\right) \quad \text{as} \quad \|x\| \to \infty.$$

However, the molecular orbitals decay exponentially. Hence, the accurate tensor approximation is computed in some smaller box $\Omega' = [-B, B]^3 \subset \Omega$, $B < A$.

In this numerical example the resultant convolution product with the Newton convolving kernel can be calculated exactly by using the analytic representation for each individual Gaussian,

$$\left(e^{-\alpha\|\cdot\|^2} * \frac{1}{\|\cdot\|}\right)(x) = \left(\frac{\alpha}{\pi}\right)^{-3/2} \frac{1}{\|x\|} \operatorname{erf}(\sqrt{\alpha}\|x\|),$$

where the erf-function is defined by

$$\operatorname{erf}(t) := \frac{2}{\sqrt{\pi}} \int_0^t \exp(-\tau) d\tau, \quad t \ge 0.$$

The Hartree potential $V_H = f_0 * 1/\|\cdot\|$ attains its maximum value at the origin $x = 0$ that is $V_H(0) = 7.19$. Figure 5.1 (right) demonstrates the accuracy $O(h^2)$ of our tensor approximation and $O(h^3)$ of the corresponding improved values, due to the Richardson extrapolation. Here, the grid-size is given by $n = n_\ell = 2^{\ell+4}$ for the level number $\ell = 1, \dots, 7$, with the finest grid-size $n_7 = 2048$. It can be seen that beginning from the level number $\ell = 5$ $(n_5 = 512)$ the extrapolated scheme already achieves the saturation error 10^{-6} of the tensor approximation related to the chosen Tucker rank $r = 22$. This example demonstrates high accuracy of the Richardson extrapolation.

The numerical results on tensor product approximation of the convolution operators in the Hartree–Fock equation compared with the commonly used MOLPRO calculations will be presented in the forthcoming Chapter 11.

6 Tensor decomposition for analytic potentials

Methods of separable approximation of the 3D Newton kernel (electrostatic potential of the Hydrogen atom) using Gaussian sums have been addressed in the chemical and mathematical literature since [38] and [39, 40]. However, these methods were based on non-explicit heuristic approaches, not explaining how to derive such Gaussian sums in an optimal way and with controllable accuracy. A constructive tensor-product approximation to the multivariate Newton kernel was first proposed in [96, 111] based on the sinc approximation [271], and then efficiently implemented and analyzed for a three-dimensional case in [30]. This tensor decomposition has been already successfully applied to assembled tensor-based summation of electrostatic potentials on 3D rectangular lattices invented in [148, 153], and it was one of the basic tools in the construction of the range-separated tensor format introduced in [24].

An alternative method for computation of the convolution transform with the Newton kernel is based on the direct solution of the Poisson equation. The data-sparse elliptic operator inverse based on explicit approximation to the Green function is presented in [159].

6.1 Grid-based canonical/Tucker representation of the Newton kernel

We discuss the grid-based method for the low-rank canonical and Tucker tensor representations of a spherically symmetric kernel function $p(\|x\|)$, $x \in \mathbb{R}^3$ (for example, for the 3D Newton kernel, we have $p(\|x\|) = \frac{1}{\|x\|}$, $x \in \mathbb{R}^3$) by its projection onto the set of piecewise constant basis functions; see [30] for more detail.

In the computational domain $\Omega = [-b/2, b/2]^3$, let us introduce the uniform $n \times n \times n$ rectangular Cartesian grid Ω_n with the mesh size $h = b/n$. Let $\{\psi_{\mathbf{i}}\}$ be a set of tensor-product piecewise constant basis functions $\psi_{\mathbf{i}}(x) = \prod_{\ell=1}^{3} \psi_{i_\ell}^{(\ell)}(x_\ell)$ for the 3-tuple index $\mathbf{i} = (i_1, i_2, i_3)$, $i_\ell \in \{1, \dots, n\}$, $\ell = 1, 2, 3$. The kernel $p(\|x\|)$ can be discretized by its projection onto the basis set $\{\psi_{\mathbf{i}}\}$ in the form of a third-order tensor of size $n \times n \times n$ defined, pointwise, as

$$\mathbf{P} := [p_{\mathbf{i}}] \in \mathbb{R}^{n \times n \times n}, \quad p_{\mathbf{i}} = \int_{\mathbb{R}^3} \psi_{\mathbf{i}}(x) p(\|x\|) dx. \tag{6.1}$$

The low-rank canonical decomposition of the 3rd-order tensor $\mathbf{P}$ can be based on using exponentially convergent sinc-quadratures for approximation of the Laplace–Gauss transform to the analytic function $p(z)$ specified by a certain coefficient $a(t) > 0$,

$$p(z) = \int_{\mathbb{R}_+} a(t) e^{-t^2 z^2} dt \approx \sum_{k=-M}^{M} a_k e^{-t_k^2 z^2} \quad \text{for } |z| > 0, \tag{6.2}$$

https://doi.org/10.1515/9783110365832-006

where the quadrature points and weights are given by

$$t_k = k\mathfrak{h}_M, \quad a_k = a(t_k)\mathfrak{h}_M, \quad \mathfrak{h}_M = C_0 \log(M)/M, \quad C_0 > 0. \tag{6.3}$$

Under the assumption $0 < a \le \|z\| < \infty$, this quadrature can be proven to provide the exponential convergence rate in M for a class of analytic functions $p(z)$; see [271, 111, 163, 166]. For example, in the particular case $p(z) = 1/z$, which can be adapted to the Newton kernel by substitution $z = \sqrt{x_1^2 + x_2^2 + x_3^2}$, we apply the Laplace–Gauss transform

$$\frac{1}{z} = \frac{2}{\sqrt{\pi}} \int_{\mathbb{R}_+} e^{-t^2 z^2} dt.$$

Now for any fixed $x = (x_1, x_2, x_3) \in \mathbb{R}^3$ such that $\|x\| > 0$, we apply the sinc-quadrature approximation to obtain the separable expansion

$$p(\|x\|) = \int_{\mathbb{R}_+} a(t) e^{-t^2\|x\|^2} dt \approx \sum_{k=-M}^{M} a_k e^{-t_k^2\|x\|^2} = \sum_{k=-M}^{M} a_k \prod_{\ell=1}^{3} e^{-t_k^2 x_\ell^2}, \quad a_k = a(t_k). \tag{6.4}$$

Under the assumption $0 < a \le \|x\| \le A < \infty$, this approximation can be proven to provide the exponential convergence rate in M:

$$\left| p(\|x\|) - \sum_{k=-M}^{M} a_k e^{-t_k^2\|x\|^2} \right| \le \frac{C}{a} e^{-\beta\sqrt{M}} \quad \text{with some } C, \beta > 0. \tag{6.5}$$

Combining (6.1) and (6.4) and taking into account the separability of the Gaussian functions, we arrive at the separable approximation for each entry of the tensor $\mathbf{P}$,

$$p_{\mathbf{i}} \approx \sum_{k=-M}^{M} a_k \int_{\mathbb{R}^3} \psi_{\mathbf{i}}(x) e^{-t_k^2\|x\|^2} dx = \sum_{k=-M}^{M} a_k \prod_{\ell=1}^{3} \int_{\mathbb{R}} \psi_{i_\ell}^{(\ell)}(x_\ell) e^{-t_k^2 x_\ell^2} dx_\ell.$$

Define the vector (recall that $a_k > 0$) $\mathbf{p}_k^{(\ell)} = a_k^{1/3} \mathbf{b}^{(\ell)}(t_k) \in \mathbb{R}^{n_\ell}$, where

$$\mathbf{b}^{(\ell)}(t_k) = [b_{i_\ell}^{(\ell)}(t_k)]_{i_\ell=1}^{n_\ell} \in \mathbb{R}^{n_\ell} \quad \text{with } b_{i_\ell}^{(\ell)}(t_k) = \int_{\mathbb{R}} \psi_{i_\ell}^{(\ell)}(x_\ell) e^{-t_k^2 x_\ell^2} dx_\ell.$$

Then the 3rd-order tensor $\mathbf{P}$ can be approximated by the R-term canonical representation

$$\mathbf{P} \approx \mathbf{P}_R = \sum_{k=-M}^{M} a_k \bigotimes_{\ell=1}^{3} \mathbf{b}^{(\ell)}(t_k) = \sum_{q=1}^{R} \mathbf{p}_q^{(1)} \otimes \mathbf{p}_q^{(2)} \otimes \mathbf{p}_q^{(3)} \in \mathbb{R}^{n\times n\times n}, \tag{6.6}$$

where $R = 2M + 1$. For the given threshold $\varepsilon > 0$, M is chosen as the minimal number such that, in the max-norm,

$$\|\mathbf{P} - \mathbf{P}_R\| \le \varepsilon \|\mathbf{P}\|.$$

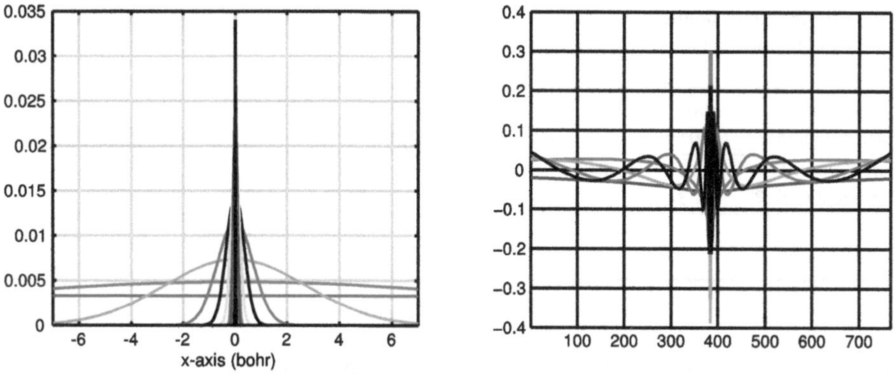

Figure 6.1: Examples of vectors of the canonical $\{\mathbf{p}_q^{(1)}\}_{q=1}^R$ (left) and Tucker $\{\mathbf{t}_k^{(1)}\}_{k=1}^{r_1}$ (right) tensor representations for the single Newton kernel displayed along x-axis.

The canonical skeleton vectors are renumbered by $k \to q = k+M+1$, $\mathbf{p}_q^{(\ell)} \leftarrow \mathbf{p}_k^{(\ell)} \in \mathbb{R}^n$, $\ell = 1,2,3$. The canonical tensor $\mathbf{P}_R$ in (6.6) approximates the discretized 3D symmetric kernel function $p(\|x\|)$ $(x \in \Omega)$ centered at the origin, giving rise to $\mathbf{p}_q^{(1)} = \mathbf{p}_q^{(2)} = \mathbf{p}_q^{(3)}$ $(q = 1,\dots,R)$.

In the following, we also consider the Tucker approximation to the 3rd-order tensor $\mathbf{P}$. Given rank parameters $\mathbf{r} = (r_1, r_2, r_3)$, the rank-$\mathbf{r}$ Tucker tensor approximating $\mathbf{P}$ is defined by the following parameterization: $\mathbf{T_r} = [t_{i_1 i_2 i_3}] \in \mathbb{R}^{n\times n\times n}$ $(i_\ell \in \{1,\dots,n\})$,

$$\mathbf{T_r} := \sum_{\mathbf{k}=1}^{\mathbf{r}} b_{\mathbf{k}} \mathbf{t}_{k_1}^{(1)} \otimes \mathbf{t}_{k_2}^{(2)} \otimes \mathbf{t}_{k_3}^{(3)} \equiv \mathbf{B} \times_1 T^{(1)} \times_2 T^{(2)} \times_3 T^{(3)}, \tag{6.7}$$

where the orthogonal side-matrices $T^{(\ell)} = [\mathbf{t}_1^{(\ell)} \cdots \mathbf{t}_{r_\ell}^{(\ell)}] \in \mathbb{R}^{n\times r_\ell}$, $\ell = 1,2,3$, define the set of Tucker vectors, and $\mathbf{B} \in \mathbb{R}^{r_1\times r_2\times r_3}$ is the Tucker core tensor. Choose the truncation error $\varepsilon > 0$ for the canonical approximation $\mathbf{P}_R$ obtained by the quadrature method, then compute the best orthogonal Tucker approximation of $\mathbf{P}$ with tolerance $O(\varepsilon)$ by applying the canonical-to-Tucker algorithm [174] to the canonical tensor $\mathbf{P}_R \mapsto \mathbf{T_r}$. The latter algorithm is based on the rank optimization via ALS iteration. The rank parameter $\mathbf{r}$ of the resultant Tucker approximand $\mathbf{T_r}$ is minimized subject to the ε-error control,

$$\|\mathbf{P}_R - \mathbf{T_r}\| \le \varepsilon \|\mathbf{P}_R\|.$$

Remark 6.1. Since the maximal Tucker rank does not exceed the canonical one, we apply the approximation results for canonical tensor to derive the exponential convergence in the Tucker rank for a wide class of functions p. This implies the relation $\max\{r_\ell\} = O(|\log \varepsilon|^2)$, which can be observed in all numerical tests implemented so far.

Table 6.1: CPU times (Matlab) to compute with tolerance $\varepsilon = 10^{-6}$ canonical and Tucker vectors of $\mathbf{P}_R$ for the single Newton kernel in a box.

grid size n^3	4608^3	9216^3	$18\,432^3$	$36\,864^3$	$73\,768^3$
mesh size h (Å)	0.0019	0.001	$4.9 \cdot 10^{-4}$	$2.8 \cdot 10^{-4}$	$1.2 \cdot 10^{-4}$
Time (Canon.)	2.	2.7	8.1	38	164
Canonical rank R	34	37	39	41	43
Time (C2T)	17	38	85	200	435
Tucker rank	12	11	10	8	6

Figure 6.1 displays several skeleton vectors of the canonical and Tucker tensor representations for a single Newton kernel along the x-axis from a set $\{\mathbf{p}_q^{(1)}\}_{q=1}^R$. Symmetry of the tensor $\mathbf{P}_R$ implies that the canonical vectors $\mathbf{p}_q^{(2)}$ and $\mathbf{p}_q^{(3)}$ corresponding to y and z-axes, respectively, are of the same shape as $\mathbf{p}_q^{(1)}$. It is clearly seen that there are canonical/Tucker vectors representing the long-, intermediate- and short-range contributions to the total electrostatic potential. This interesting feature will be also recognized for the low-rank lattice sum of potentials (see Section 14.2).

Table 6.1 presents CPU times (sec) for generating a canonical rank-R tensor approximation of the single Newton kernel over $n \times n \times n$ 3D Cartesian grid corresponding to Matlab implementation on a terminal of the 8 AMD Opteron Dual-Core processor. The corresponding mesh sizes are given in Angstroms. We observe the logarithmic scaling of the canonical rank R in the grid size n, whereas the maximal Tucker rank has the tendency to decrease for larger n. The compression rate related to the grid $73\,768^3$, which is the ratio $n^3/(nR)$ for the canonical format and $n^3/(r^3 + rn)$ for the Tucker format, is of orders 10^8 and 10^7, respectively.

Notice that the low-rank canonical/Tucker approximation of the tensor $\mathbf{P}$ is the problem independent task, hence the respective canonical/Tucker vectors can be precomputed at once on large enough 3D $n \times n \times n$ grid, and then stored for the multiple use. The storage size is bounded by Rn or $rn + r^3$ in the case of canonical and Tucker formats, respectively.

6.2 Low-rank representation for the general class of kernels

Along with Coulombic systems corresponding to $p(\|x\|) = \frac{1}{\|x\|}$, the tensor approximation described above can be also applied to a wide class of commonly used long-range potentials $p(\|x\|)$ in $\mathbb{R}^3$, for example, to the Slater, Yukawa, Lennard-Jones or Van der Waals, and dipole–dipole interactions potentials defined as follows:

$$\text{Slater function:} \quad p(\|x\|) = \exp(-\lambda\|x\|), \quad \lambda > 0;$$

$$\text{Yukawa kernel:} \quad p(\|x\|) = \frac{\exp(-\lambda\|x\|)}{\|x\|}, \quad \lambda > 0;$$

$$\text{Lennard-Jones potential:} \quad p(\|x\|) = 4\epsilon\left[\left(\frac{\sigma}{\|x\|}\right)^{12} - \left(\frac{\sigma}{\|x\|}\right)^{6}\right].$$

The simplified version of the Lennard-Jones potential is the so-called Buckingham function:

$$\text{Buckingham potential:} \quad p(\|x\|) = 4\epsilon\left[e^{\|x\|/r_0} - \left(\frac{\sigma}{\|x\|}\right)^{6}\right].$$

The electrostatic potential energy for the dipole–dipole interaction due to Van der Waals forces is defined by

$$\text{Dipole–dipole interaction energy:} \quad p(\|x\|) = \frac{C_0}{\|x\|^3}.$$

The existence of quasi-optimal low-rank decompositions based on the sinc-quadrature approximation to the Laplace transform of the above-mentioned functions can be rigorously proven for a wide class of generating kernels. In particular, the following Laplace (or Laplace–Gauss) integral transforms [309] with parameter $\rho > 0$ can be combined with the sinc-quadrature approximation to obtain the low-rank representation to the corresponding function generated tensor:

$$e^{-2\sqrt{\kappa\rho}} = \frac{\sqrt{\kappa}}{\sqrt{\pi}} \int_{\mathbb{R}_+} t^{-3/2} e^{-\kappa/t} e^{-\rho t}\, dt, \tag{6.8}$$

$$\frac{e^{-\kappa\sqrt{\rho}}}{\sqrt{\rho}} = \frac{2}{\sqrt{\pi}} \int_{\mathbb{R}_+} e^{-\kappa^2/t^2} e^{-\rho t^2}\, dt, \tag{6.9}$$

$$\frac{1}{\sqrt{\rho}} = \frac{2}{\sqrt{\pi}} \int_{\mathbb{R}_+} e^{-\rho t^2}\, dt, \tag{6.10}$$

$$\frac{1}{\rho^n} = \frac{1}{(n-1)!} \int_{\mathbb{R}_+} t^{n-1} e^{-\rho t}\, dt, \quad n = 1, 2, \ldots. \tag{6.11}$$

This approach is combined with the subsequent substitution of a parameter ρ by the appropriate function $\rho(x) = \rho(x_1, x_2, x_3)$, usually by using an additive representation $\rho(x) = c_1 x_1^2 + c_2 x_2^2 + c_3 x_3^2$. In cases (6.11) ($n = 1$) and (6.10), the convergence rate for the sinc-quadrature approximations of type (6.3) has been estimated in [39, 40] and later analyzed in more detail in [95, 111]. The case of the Yukawa and Slater kernel has been investigated in [161, 163]. The exponentially fast error decay for the general transform (6.11) can be derived by minor modification of the above-mentioned results.

Remark 6.2. The idea behind the low-rank tensor representation for a sum of spherically symmetric potentials on a 3D lattice can be already recognized on the continuous level by introducing the Laplace transform of the generating kernel. For example, in representation (6.9) with the particular choice $\kappa = 0$, given by (6.10), we can set up

$\rho = x_1^2 + x_2^2 + x_3^2$, i. e., $p(\|x\|) = 1/\|x\|$ $(1 \leq x_\ell < \infty)$, and apply the sinc-quadrature approximation as in (6.2)–(6.3),

$$p(z) = \frac{2}{\sqrt{\pi}} \int_{\mathbb{R}_+} e^{-t^2 z^2} dt \approx \sum_{k=-M}^{M} a_k e^{-t_k^2 z^2} \quad \text{for } |z| > 0. \tag{6.12}$$

Now the simple sum

$$\Sigma_L(x) = \sum_{i_1,i_2,i_3=1}^{L} \frac{1}{\sqrt{(x_1 + i_1 b)^2 + (x_2 + i_2 b)^2 + (x_3 + i_3 b)^2}}$$

on a rectangular $L \times L \times L$ lattice of width $b > 0$ can be represented by the agglomerated integral transform

$$\begin{aligned} \Sigma_L(x) &= \frac{2}{\sqrt{\pi}} \int_{\mathbb{R}_+} \left[\sum_{i_1,i_2,i_3=1}^{L} e^{-[(x_1+i_1 b)^2 + (x_2+i_2 b)^2 + (x_3+i_3 b)^2] t^2} \right] dt \\ &= \frac{2}{\sqrt{\pi}} \int_{\mathbb{R}_+} \sum_{i_1=1}^{L} e^{-(x_1+i_1 b)^2 t} \sum_{i_2=1}^{L} e^{-(x_2+i_2 b)^2 t} \sum_{i_3=1}^{L} e^{-(x_3+i_3 b)^2 t} dt, \end{aligned} \tag{6.13}$$

where the integrand is separable. Representation (6.13) indicates that applying the same quadrature approximation to the lattice sum integral (6.13) as that for the single kernel (6.12) leads to the decomposition of the total sum of potentials with the same canonical rank as for the single one.

In the following, we construct the low-rank canonical and Tucker decompositions to the lattice sum of long range interaction potentials discretized on the fine 3D-grid and applied to the general class of kernel functions and to more general configuration of a lattice, including the case of lattices with vacancies.

7 The Hartree–Fock equation

The Hartree–Fock (HF) equation governed by the 3D integral-differential operator is the basic model in ab initio calculations of the ground state energy and electronic structure of molecular systems [123, 277, 128]. It is a strongly nonlinear eigenvalue problem for which one should find the solution when the part of the governing operator depends on the eigenfunctions. This dependence is expressed by the convolution of the electron density, which is a function of the solution (molecular orbitals) with the Newton kernel in $\mathbb{R}^3$. Multiple strong singularities, due to nuclear cusps in the electron density of a molecule, impose strong requirements on the accuracy of Hartree–Fock calculations. Finally, the eigenvalues and the ground state energy should be computed with high accuracy to be suitable for more precise post-Hartree–Fock computations.

7.1 Electronic Schrödinger equation

The Hartree–Fock equation provides the model reduction to the electronic Schrödinger equation

$$\mathcal{H}_e \Psi = E\Psi, \tag{7.1}$$

with the Hamiltonian

$$\mathcal{H}_e = -\frac{1}{2}\sum_{i=1}^{N} \Delta_i + \sum_{i=1}^{N}\sum_{A=1}^{M} \frac{Z_A}{x_i - a_A} + \sum_{\substack{i,j=1 \\ i \neq j}}^{N} \frac{1}{|x_i - x_j|}, \quad a_A, x_i, x_j \in \mathbb{R}^3, \tag{7.2}$$

which describes the energy of an N-electron molecular system in the framework of the so-called Born–Oppenheimer approximation, implying a system with clapped nuclei. Here, M is the number of nuclei, Z_A are nuclei charges located at the distinct points a_A, $A = 1, \ldots, M$. Since the nuclei are much heavier than electrons, and their motion is much slower, the nuclei and electronic parts of the energy can be considered separately. Thus, the electronic Schrödinger equation specifies the energy of a molecular system at a fixed nuclei geometry. The Hamiltonian (7.2) includes the kinetic energy of electrons, the potential energy of the interaction between nuclei and electrons, and the electron correlation energy. The electronic Schrödinger equation is a multidimensional problem in $\mathbb{R}^{3N}$, and it is computationally unfeasible except for the simple Hydrogen or Hydrogen-like atoms.

The Hartree–Fock equation is a 3D eigenvalue problem in space variables obtained as a result of the minimization of the energy functional for the electronic Schrödinger equation [277, 128]. The underlying condition for the wavefunction is that it should be a single Slater determinant containing the products of molecular

https://doi.org/10.1515/9783110365832-007

orbitals. For fermions the wavefunction Ψ should be antisymmetric, therefore, it is parameterized using a Slater determinant representation,

$$\Psi(x_1,\dots,x_N) = \frac{1}{N!}\begin{vmatrix} \varphi_1(x_1) & \varphi_2(x_1) & \dots & \varphi_N(x_1) \\ \varphi_1(x_2) & \varphi_2(x_2) & \dots & \varphi_N(x_2) \\ \dots & \dots & \dots & \dots \\ \varphi_1(x_N) & \varphi_2(x_N) & \dots & \varphi_N(x_N) \end{vmatrix},$$

where $\varphi_i(x_j)$ are the one-electron wavefunctions, $i,j = 1,\dots N$. We refer to the literature on electronic structure calculations for the derivation of the Hartree–Fock equation [277, 128].

The Hartree–Fock equations are orbital equations obtained within a mean-field approximation to the many-electron problem [128]. They are derived from application of the variational principle to the expectation value of the many-electron Hamiltonian over a configuration state function (CSF) characterizing the desired state of the many-electron system under study. In simple cases, like the ground state of a closed-shell system (N even) to which we restrict ourselves here, this CSF reduces to a single Slater determinant built up from the orbitals.

7.2 The Hartree–Fock eigenvalue problem

Here, we consider the Hartree–Fock problem for the closed shell systems, where the number of molecular orbitals equals the number of electron pairs, $N_{\text{orb}} = N/2$. The Hartree–Fock equation is a nonlinear eigenvalue problem,

$$\mathcal{F}\varphi_i(x) = \lambda_i\varphi_i(x), \quad x \in \mathbb{R}^3, \tag{7.3}$$

with respect to the (orthogonal) molecular orbitals $\varphi_i(x)$,

$$\int_{\mathbb{R}^3} \varphi_i\varphi_j = \delta_{ij}, \quad i = 1,\dots,N_{\text{orb}},\ x \in \mathbb{R}^3,$$

and the Fock operator is given by

$$\mathcal{F} = H_c + V_H - \mathcal{K}. \tag{7.4}$$

The core Hamiltonian part H_c of the Fock operator consists of the kinetic energy of electrons specified by the Laplace operator and the nuclear potential energy of interaction of electrons and nuclei,

$$H_c(x) = -\frac{1}{2}\Delta - \sum_{A=1}^{M}\frac{Z_A}{\|x - a_A\|}, \quad Z_A > 0,\ x, a_A \in \mathbb{R}^3, \tag{7.5}$$

where M is the number of nuclei in a molecule, and Z_A and a_A are their charges and positions, respectively. Here,

$$V_c(x) = -\sum_{A=1}^{M} \frac{Z_A}{\|x - a_A\|}$$

is the nuclear potential operator. The electron correlation parts of the Fock operator are described by the Hartree potential

$$V_H(x) := \int_{\mathbb{R}^3} \frac{\rho(y)}{\|x - y\|} dy \tag{7.6}$$

with the electron density

$$\rho(y) = 2 \sum_{i=1}^{N_{\mathrm{orb}}} (\varphi_i(y))^2, \quad x,\ y \in \mathbb{R}^3, \tag{7.7}$$

and the exchange operator

$$(\mathcal{K}\varphi)(x) := \int_{\mathbb{R}^3} \frac{\tau(x,y)}{\|x - y\|} \varphi(y) dy, \quad \tau(x,y) = \sum_{i=1}^{N_{\mathrm{orb}}} \varphi_i(x)\varphi_i(y), \quad x \in \mathbb{R}^3, \tag{7.8}$$

where $\tau(x,y)$ is the density matrix. Since both operators V_H and $\mathcal{K}$ depend on the solution of the eigenvalue problem (7.3), the nonlinear Hartree–Fock equation is solved iteratively by using self-consistent field (SCF) iteration [238, 44].

The Hartree–Fock model is often called a mean-field approximation, since the energy of electrons in a molecule is computed with respect to the mean field created by all electrons in a molecular system, including the target electrons.

7.3 The standard Galerkin scheme for the Hartree–Fock equation

The standard Galerkin approach to the numerical solution of the Hartree–Fock problem [277, 128] is based on the expansion of the molecular orbitals in a separable Gaussian-type basis $\{g_\mu\}_{1 \le \mu \le N_b}$,

$$\varphi_i(x) = \sum_{\mu=1}^{N_b} c_{i\mu} g_\mu(x), \quad i = 1, \dots, N_{\mathrm{orb}},\ x \in \mathbb{R}^3, \tag{7.9}$$

which yields the system of nonlinear equations for the coefficients matrix $C = \{c_{i\mu}\} \in \mathbb{R}^{N_{\mathrm{orb}} \times N_b}$ (and the density matrix $D = 2CC^* \in \mathbb{R}^{N_b \times N_b}$),

$$F(C)C = SC\Lambda, \quad \Lambda = \mathrm{diag}(\lambda_1, \dots, \lambda_{N_b}), \quad C^T SC = I_{N_b}, \tag{7.10}$$

where $S = \{s_{\mu\nu}\}$ is the overlap matrix for the chosen Galerkin basis, where $s_{\mu\nu} = \int_{\mathbb{R}^3} g_\mu g_\nu dx$. The Galerkin counterpart of the Fock operator

$$F(C) = H + J(C) + K(C) \tag{7.11}$$

includes the core Hamiltonian H discretizing the Laplacian and the nuclear potential operators (7.5), and the matrices $J(C)$ and $K(C)$ corresponding to the Galerkin projections of the operators V_H and $\mathcal{K}$, respectively.

In this way, one can precompute the one-electron integrals in the core Hamiltonian $H = \{h_{\mu\nu}\}_{\mu,\nu=1}^{N_b}$,

$$h_{\mu\nu} = \frac{1}{2} \int_{\mathbb{R}^3} \nabla g_\mu \cdot \nabla g_\nu dx + \int_{\mathbb{R}^3} V_c(x) g_\mu g_\nu dx \quad 1 \le \mu, \nu \le N_b, \tag{7.12}$$

and the so-called two-electron integrals (TEI) tensor, also known as electron repulsion integrals,

$$b_{\mu\nu\kappa\lambda} = \int_{\mathbb{R}^3} \int_{\mathbb{R}^3} \frac{g_\mu(x) g_\nu(x) g_\kappa(y) g_\lambda(y)}{\|x - y\|} dxdy, \quad 1 \le \mu, \nu \le N_b,\ x, y \in \mathbb{R}^3, \tag{7.13}$$

since they depend only on the choice of the basis functions in (7.9).

Then, the solution is sought by the self-consistent fields (SCF) iteration using the core Hamiltonian H as the initial guess, and by updating the Coulomb

$$J(C)_{\mu\nu} = \sum_{\kappa,\lambda=1}^{N_b} b_{\mu\nu,\kappa\lambda} D_{\kappa\lambda}, \tag{7.14}$$

and the exchange Galerkin matrices

$$K(C)_{\mu\nu} = -\frac{1}{2} \sum_{\kappa,\lambda=1}^{N_b} b_{\mu\lambda,\nu\kappa} D_{\kappa\lambda}, \tag{7.15}$$

at every iteration step. The direct inversion of iterative subspaces (DIIS) method, introduced in 1982 by Pulay [238], provides stable convergence of iteration. The DIIS method is based on defining the weights of the previous solutions to be used as the initial guess for the current step of iteration.

Finally, the Hartree–Fock energy (or electronic energy, [277]) is computed as

$$E_{\mathrm{HF}} = 2 \sum_{i=1}^{N_{\mathrm{orb}}} \lambda_i - \sum_{i=1}^{N_{\mathrm{orb}}} (\widetilde{J}_i - \widetilde{K}_i),$$

where

$$\widetilde{J}_i = (\varphi_i, V_H \varphi_i)_{L^2} = \langle C_i, J C_i \rangle$$

and

$$\widetilde{K}_i = (\varphi_i, K\varphi_i)_{L^2} = \langle C_i, KC_i \rangle, \quad i = 1, \ldots, N_{\text{orb}},$$

are the Coulomb and exchange integrals in the basis of Hartree–Fock orbitals φ_i.

Given the geometry of nuclei, the resulting ground state energy E_0 of the molecule is defined by

$$E_0 = E_{\text{HF}} + E_{\text{nuc}}, \tag{7.16}$$

where the so-called nuclear shift

$$E_{\text{nuc}} = \sum_{k=1}^{M} \sum_{m<k}^{M} \frac{Z_k Z_m}{\|a_k - a_m\|} \tag{7.17}$$

describes the repulsion energy of nuclei in a molecule.

Commonly used numerical methods for solving the Hartree–Fock equation are based on the analytical computation of the two-electron integrals (7.13) in the problem adapted naturally separable Gaussian-type bases [3] by using *erf*-function expansions. This rigorous approach is a well-established standard for the ab initio Hartree–Fock calculations, and there is a number of efficient programm packages which required years of development by large scientific groups.

The success of the analytical integration methods stems from the big amount of precomputed information based on the physical insight, including the construction of problem adapted atomic orbitals basis sets and elaborate nonlinear optimization for calculation of density fitting basis. The known limitations of this approach appear due to a strong dependence of the numerical efficiency on the size and quality of the chosen Gaussian basis sets that might be crucial for larger molecular clusters and heavier atoms.

7.4 Rank-structured grid-based approximation of the Hartree–Fock problem

The tensor-structured numerical methods, both the name and the concept, first appeared as efficient grid-based rank-structured algorithms for calculation of the multidimensional convolution operators in the Hartree–Fock equation [174, 145], supplemented by the approximation theory on low-rank approximation of the multivariate functions and operators [166]. These numerical studies were particularly convenient for testing novel tensor approaches since the results of calculations could be easily checked by comparison with the standard software packages in computational chemistry like MOLPRO [299], which present the accurate results from analytical evaluation of the same multidimensional operators.

In the next chapters, we describe the two ab initio Hartree–Fock solvers using the tensor-structured grid-based calculation of all quantities involved, including the rank-structured calculation of the core Hamiltonian introduced in [156]. We briefly summarize the basic approaches as follows:

- In Section 8, we describe the multilevel Hartree–Fock solver using the nontraditional concept for the numerical solution of the eigenvalue problem, which avoids the computation of TEI. Instead, it employes the grid-based rank-structured computation of the Coulomb and exchange matrices on the fly in the course of SCF iterations. This solver was introduced in [174, 145, 146, 187]. Though this approach eliminates the requirement for the challenging computation of the two-electron integrals, it exhibits time limitations in the loops for computation of the exchange operator $\mathcal{K}$. Therefore, its MATLAB implementation is non-competitive with the standard packages based on the analytical calculations. However, it may be good for parallel implementations.
- In Section 11, we present the fast TESC[1] Hartree–Fock solver[2] introduced in [157, 147], which is comparable in time and accuracy (in MATLAB implementation) with the benchmark packages. It is based on efficient rank-structured calculation of the TEI tensor in a factorized form by using an algebraic "1D density fitting" scheme and the truncated Cholesky decomposition algorithm. Due to the rank-structured representation of the two-electron integrals, this tensor-based solver proved to be attractive as a starting point for computation of the excitation energies of molecules.

Note that the grid-based approaches are not restricted to Gaussian-type basis functions and may be applied for the construction of new well-separable grid-based basis functions, for example, combination of Gaussians with plane waves or/and the Slater-type orbitals.

1 TESC is the abbreviation for the tensor-based electronic structure calculations.

2 It was first called the "black-box" Hartree–Fock solver.

8 Multilevel grid-based tensor-structured HF solver

In this chapter, we describe the nontraditional concept for the numerical solution of the Hartree–Fock (HF) equation, which avoids computation of the two-electron integrals. Alternatively, it employs the grid-based rank-structured computation of the Coulomb and exchange matrices on the fly in the framework of SCF iterations [174, 145]. This solver was introduced in [146, 187] and called the "multilevel HF solver"since both the 3D-dimensional integrals in the Hartree potential and the nonlocal six-dimensional integral exchange potential are calculated on a sequence of dyadically refined $n \times n \times n$ Cartesian grids in $\mathbb{R}^3$.

The important ingredients of grid-based approach to the 3D Hartree–Fock equation include many classical numerical techniques, in particular, the discretization of differential and integral operators, error analysis, fast data transforms, algebraic solvers for linear systems, and eigenvalue problems, which are widely presented in the literature. We refer to some of many existing monographs addressing these issues, [273, 246, 104, 29, 282, 98, 281, 275, 53, 184, 160].

8.1 Calculation of the Hartree and exchange operators

The rank-structured grid-based representation of the multivariate functions and operators was first applied in evaluation of the convolution operators in the Hartree–Fock equation, where both accuracy and timing were essential and calculations had to be performed on large 3D grids to be competitive with analytically based methods [174, 145]. The efficient low-rank canonical tensor representation to the Newton kernel was also an essential ingredient [161, 30]. To reduce the initial rank of the electron density that is quadratically proportional to the number of Gaussian-type basis functions, the canonical-to-Tucker tensor transform was applied [174, 173], which made computations tractable for large tensor grids and for extended molecules even when using MATLAB. In particular, it was shown that the rank-structured tensor approach allows reducing the calculation of the three-dimensional (3D) convolution integrals to sequences of one-dimensional (1D) convolution transforms via 1D-FFT, 1D Hadamard, and scalar products.

The idea to replace or assist the analytical computations for the Hartree–Fock problem by a data-sparse grid-based approach is not new. In particular, the wavelet multiresolution schemes [122, 305, 83, 80, 267, 35], as well as the sparse grids approach in [43, 308, 106] have been proposed, though the entirely wavelet-based method is successful only for small atomic systems with one or two electrons [105, 7]. The multiresolution approach in density functional computations of the electronic structure is considered in [97, 263]. The grid-based numerical method in Hartree–Fock calculations for diatomic molecules proposed in [276, 200] is hardly extendable to compact (3D)

https://doi.org/10.1515/9783110365832-008

molecules. The domain decomposition approach to problems with "linear geometry" of molecules is discussed in [12]. Recently it was shown that the QTT tensor approximation can be applied for Hartree–Fock calculations for small molecules [239, 240]. Rank decomposition of the Schrödinger operator and the corresponding wavefunctions have been first discussed in [31, 32, 33]. The results on the regularity properties of the wave functions in the Schrödinger equation, which motivate the data-sparse approaches in electronic structure calculations, are presented in [307, 308, 82].

8.1.1 Agglomerated representation of the Galerkin matrices

For the multilevel Hartree–Fock solver [146, 187], fast and accurate evaluation of the Galerkin matrices $J(D)$ and $K(D)$ is based on a certain reorganization of the standard iterative computational scheme for the eigenvalue problem (7.10) given in Section 7.3. Specifically, instead of precomputing the full set of two-electron integrals $b_{\mu\nu,\kappa\lambda}$ and computation of matrices in (7.14) and (7.15) employing the updated elements of the density matrix D, we use the explicit integral representations for $J(D)$ and $K(D)$. In particular, the Galerkin representation of the Hartree operator (the Coulomb matrix) is now calculated by the grid-based quadrature integration of

$$J(D)_{\mu\nu} = \int_{\mathbb{R}^3} g_\mu(x) V_H(x) g_\nu(x)dx, \quad 1 \leq \mu, \nu \leq N_b, \tag{8.1}$$

including a single convolution transform in $\mathbb{R}^3$ to compute the Hartree potential in (8.1),

$$V_H = \rho * \frac{1}{\|\cdot\|},$$

where the electron density is given by

$$\rho(y) = 2\sum_{a=1}^{N_{\mathrm{orb}}} \left(\sum_{\kappa,\lambda=1}^{N_b} C_{\kappa a} C_{\lambda a} g_\kappa(y) g_\lambda(y) \right). \tag{8.2}$$

In turn, as proposed in [145], we represent the matrix entries of the exchange operator $K(D)$ by the following loops. For $a = 1, \ldots, N_{\mathrm{orb}}$, compute the convolution integrals

$$W_{a\nu}(x) = \int_{\mathbb{R}^3} \frac{g_\nu(y) \sum_{\kappa=1}^{N_b} C_{\kappa a} g_\kappa(y)}{\|x-y\|} dy, \quad \nu = 1, \ldots, N_b, \tag{8.3}$$

and then the scalar products

$$K_{\mu\nu,a} = \int_{\mathbb{R}^3} \left[\sum_{\kappa=1}^{N_b} C_{\kappa a} g_\kappa(x) \right] g_\mu(x) W_{a\nu}(x) dx, \quad \mu, \nu = 1, \ldots, N_b. \tag{8.4}$$

Finally, the entries of the exchange matrix are given by sums over all orbitals,

$$K(C)_{\mu\nu} = \sum_{a=1}^{N_{orb}} K_{\mu\nu,a}, \quad \mu,\nu = 1,\dots,N_b. \tag{8.5}$$

The advantage of above representations is in the minimization of the number of convolution products that have to be computed by numerical quadratures. What is even more important is that we have the possibility of efficient low-rank separable approximation of the discretized density $\rho(x)$ and of the auxiliary potentials $W_{a\nu}(x)$ at step (8.3).

Effective realization of this concept is based on certain separability assumptions on the Galerkin basis functions g_μ. First, we suppose that the initial problem is posed in a finite volume box $\Omega = [-b,b]^3 \in \mathbb{R}^3$ subject to the homogeneous Dirichlet boundary conditions on $\partial\Omega$ (due to the exponential decay of the orbitals $\psi_i(x)$ as $\|x\| \to \infty$). For a given discretization parameter $n \in \mathbb{N}$, introduce the equidistant tensor grid

$$\omega_{3,n} := \omega_1 \times \omega_2 \times \omega_3, \quad \omega_\ell := \{-b + (m-1)h : m = 1,\dots,n+1\}, \quad \ell = 1,\dots,3, \tag{8.6}$$

with the mesh-size $h = 2b/n$. Define the set of piecewise constant basis functions $\{\phi_\mathbf{i}\}$, $\mathbf{i} \in \mathcal{I} := \{1,\dots,n\}^3$, associated with the respective grid-cells in $\omega_{3,n}$ (indicator functions), and the corresponding set $\{\chi_\mathbf{j}\}$, $\mathbf{j} \in \mathcal{J} := \{1,\dots,n-1\}^3$, of tensor-product continuous piecewise linear polynomials in each spacial variable. We denote the corresponding finite element spaces as

$$\mathcal{V}_n = \operatorname{span}\{\phi_\mathbf{i}\} \quad \text{and} \quad \mathcal{W}_n = \operatorname{span}\{\chi_\mathbf{j}\} \in H_0^1(\Omega). \tag{8.7}$$

Now the basis set $\{g_\mu\}$ is supposed to satisfy the following properties:

- The Galerkin approximation error over the reduced basis set $\{g_\mu\}$ is physically admissible, presupposing the sufficient approximation quality.
- Each basis function $g_\mu(x) \in H_0^1(\Omega)$ can be represented (approximated) by the R_G-term separable expansion in $x = (x_1,x_2,x_3)$ with moderate number of terms R_G,

$$g_\mu(x) = \sum_{k=1}^{R_G} g_{\mu,k}^{(1)}(x_1) g_{\mu,k}^{(2)}(x_2) g_{\mu,k}^{(3)}(x_3), \quad \mu = 1,\dots,N_b. \tag{8.8}$$

- The property of discrete separability. The functions $g_\mu(x)$ (and $g_{\mu,k}^{(\ell)}$) allow the approximate representation in either basis sets $\{\phi_\mathbf{i}\}$ and $\{\chi_\mathbf{j}\}$ by the rank-R_G coefficients tensors $\mathbf{G}_\mu = [G_{\mu,\mathbf{i}}] \in \mathbb{R}^\mathcal{I}$ and $\mathbf{X}_\mu = [X_{\mu,\mathbf{j}}] \in \mathbb{R}^\mathcal{J}$, respectively.
- We suppose that the Galerkin integrals for $J(D)$ and $K(D)$, given by (8.1)–(8.5), can be accurately represented by the well-separable numerical quadratures in the discretized basis sets $\{\mathbf{G}_\mu\}$ and $\{\mathbf{X}_\mu\}$, providing asymptotic convergence as $h \to 0$. The representation with respect to both $g_\mu(x)$ and its piecewise linear version provides satisfactory accuracy, which may require large grid parameter n.

8.1.2 On the choice of the Galerkin basis functions

The examples of *problem-independent* grid-oriented basis sets are given by plane waves, wavelets, and by the piecewise polynomial *finite element* (FE) basis functions. The practically tractable grids of size $n \times n \times n$ for representing problem-independent basis sets are presently limited by the value of about $n \approx 500$.

Several efficient "meshless" basis sets $\{g_\mu\}$ are known in the literature on computational quantum chemistry. In particular, we mention the *linear combination of atomic orbitals* (LCAO) and their successors, *Slater-type orbitals* (STOs). The most popular basis sets for representation of molecular orbitals are the so-called *Gaussian-type orbitals* (GTOs),

$$\varphi(x) = \sum_{k=1}^{N_b} c_k g_k(x), \quad x \in \mathbb{R}^3, \tag{8.9}$$

defined by Gaussians scaled by polynomials,

$$g_k(x) = \prod_{\ell=1}^{3} (x^\ell - A_k^\ell)^{\beta_k^\ell} e^{-\alpha_k \|x^\ell - A_k\|^2}, \quad x = \{x^{(\ell)}\}_{\ell=1}^3, \tag{8.10}$$

where $A_k = (A_k^1, A_k^2, A_k^3) \in \mathbb{R}^3$ $(k = 1, \ldots, N_b)$ correspond to the locations of atoms in a molecule, $\beta_k^\ell \in \mathbb{N}_0$, and N_b is the number of GTO basis functions. In the case of larger molecules, the so-called *contracted Gaussian functions* are widely used, which probably constitutes the best compromise between STOs and GTOs (cf. [203] for detailed discussion). The construction of such *problem-dependent* basis sets is distinctively based on the precomputed electronic orbitals for single atoms.

An alternative to the analytically given GTO-type basis functions are the so-called *fully numerical atomic orbitals* [203], which are solely specified by their numerical values on a grid. Such a choice of basis functions fits well the nature of our tensor-structured numerical method. This allows utilizing the already existing problem adapted basis sets taking advantage of the important physical information that is well known for the individual atoms. The (low-rank) separable representation of functions and operators reduces the 3D calculations to fast numerical operations implemented only on the univariate grids (1D calculations) [173, 145]. In this way, the computation of the volume integrals, convolution transforms, scalar products, and function-function multiplications can be simplified dramatically.

The particular requirements on the approximating basis set to be fulfilled in the framework of our tensor-structured numerical scheme were formulated at the end of the previous subsection 8.1.1. The systematic construction of the high-quality low tensor rank approximating basis can be established based on the following techniques:

- Algebraic optimization of the conventional "meshless" GTO-type basis sets ($R_G = 1$);
- Rank reduction of the Slater-type orbitals ($R_G = O(\log \varepsilon^{-1})$ up to the tolerance $\varepsilon > 0$), cf. [161];

– Using the orthogonal Tucker vectors computed for simplified problems, whose Tucker rank is supposed to be weakly dependent on the particular molecule and the grid parameters [173, 174, 186].

All these concepts still require further theoretical and numerical analysis.

The main advantage of the low tensor rank approximating basis sets is the linear scaling of the resultant algorithms in the univariate grid size n, which already allows employing huge $n \times n \times n$-grids in $\mathbb{R}^3$ (specifically, $n \leq 2 \cdot 10^4$ for the current computations in the framework of the multilevel Hartree–Fock solver). This could be beneficial in the FEM-DFT computations applied to large molecular clusters.

8.1.3 Tensor computation of the Galerkin integrals in matrices $J(D)$ and $K(D)$

The beneficial feature of our method is that functions and operators involved in the computational scheme for the Coulomb and exchange matrices (8.1)–(8.5) are efficiently evaluated using (approximate) low-rank tensor-product representations in the discretized basis sets $\{\mathbf{G}_\mu\}$ and $\{\mathbf{X}_\mu\}$ at the expense that scales linear-logarithmic in n, $O(n \log n)$.

To that end, we introduce some interpolation/prolongation operators interconnecting the continuous functions on Ω and their discrete representation on the grid via the coefficient tensors in $\mathbb{R}^{\mathcal{I}}$ (or in $\mathbb{R}^{\mathcal{J}}$). Note that the coefficient space of tri-tensors in

$$\mathbb{V}_n = \mathbb{R}^{\mathcal{I}} := V_1 \otimes V_2 \otimes V_3$$

is the tensor-product space with $V_\ell = \mathbb{R}^n$ ($\ell = 1,2,3$). Conventionally, we use the canonical isomorphism between $\mathcal{V}_n$ and $\mathbb{V}_n$,

$$\mathcal{V}_n \ni f(x) = \sum_{\mathbf{i}} f_{\mathbf{i}} \phi_{\mathbf{i}}(x) \quad \Longleftrightarrow \quad \mathbf{F} := [f_{\mathbf{i}}]_{\mathbf{i} \in \mathcal{I}} \in \mathbb{V}_n.$$

We make use of similar entities for the pair $\mathcal{W}_n$ and $\mathbb{W}_n = \mathbb{R}^{\mathcal{J}} := W_1 \otimes W_2 \otimes W_3$ with $W_\ell = \mathbb{R}^{n-1}$ ($\ell = 1,2,3$).

Now we define the collocation and L^2-projection mappings onto $\mathbb{V}_n$. For the continuous function f, we introduce the collocation "projection" operator by

$$\mathcal{P}_C : f \mapsto \sum_{\mathbf{i}} f(y_{\mathbf{i}}) \phi_{\mathbf{i}}(x) \quad \Longleftrightarrow \quad \mathbf{F} := [f(y_{\mathbf{i}})]_{\mathbf{i} \in \mathcal{I}} \in \mathbb{V}_n,$$

where $\{y_{\mathbf{i}}\}$ is the set of cell-centered points with respect to the grid $\omega_{3,n}$. Furthermore, for functions $f \in L^2(\Omega)$, we define the L^2-projection by

$$\mathcal{P}_0 : f \mapsto \sum_{\mathbf{i}} \langle f, \phi_{\mathbf{i}} \rangle \phi_{\mathbf{i}}(x) \quad \Longleftrightarrow \quad \mathbf{F} := [\langle f, \phi_{\mathbf{i}} \rangle]_{\mathbf{i} \in \mathcal{I}} \in \mathbb{V}_n.$$

Likewise, we denote by $\mathcal{Q}_0$ the L^2-projection onto $\mathbb{W}_n$.

Using the discrete representations as above, we are able to rewrite all functional and integral transforms in (8.1)–(8.5) in terms of tensor operations in $\mathbb{V}_n$. In particular, for the continuous targets, the function-times-function and the L^2-scalar product can be discretized by tensor operations as

$$f \cdot g \mapsto \mathbf{F} \odot \mathbf{G} \in \mathbb{V}_n \quad \text{and} \quad \langle f, g \rangle \mapsto h^3 \langle \mathbf{F}, \mathbf{G} \rangle$$

with

$$\mathbf{F} = \mathcal{P}_C(f), \quad \mathbf{G} = \mathcal{P}_C(g),$$

and $\odot$ means the Hadamard (entrywise) product of tensors.

The convolution product is represented by

$$f * g \mapsto \mathbf{F} *_T \mathbf{G} \in \mathbb{V}_n, \quad \text{with } \mathbf{F} = \mathcal{P}_C(f) \in \mathbb{V}_n, \; \mathbf{G} = \mathcal{P}_0(g) \in \mathbb{V}_n,$$

where the tensor operation $*_T$ stands for the tensor-structured convolution transform in $\mathbb{V}_n$ described in [166] (see also [186, 174] for application of fast $*_T$ transform in electronic structure calculations). We notice that under certain assumptions on the regularity of the input functions (see Section 5) the tensor product convolution $*_T$ can be proven to provide an approximation error of order $O(h^2)$, whereas the two-grid version via the Richardson extrapolation leads to the improved error bound $O(h^3)$ (cf. [166]). Tensor-structured calculation of the multidimensional convolution integral operators with the Newton kernel have been introduced and implemented in [174, 187, 145], see also [108].

Representations (8.1)–(8.2) for the Coulomb operator can be now rewritten (approximately) in terms of the discretized basis functions by using tensor operations:

$$\rho \approx \Theta := \sum_{a=1}^{N_{\mathrm{orb}}} \left(\sum_{\kappa,\lambda=1}^{N_b} C_{\kappa a} C_{\lambda a} \mathbf{G}_\kappa \odot \mathbf{G}_\lambda \right), \quad \text{where } \mathbf{G}_\kappa = \mathcal{P}_C(g_\kappa),$$

implying

$$V_H = \rho * g \approx \Theta *_T \mathbf{P}_N, \quad \text{where } \mathbf{P}_N = \mathcal{P}_0(g), \; g = \frac{1}{\|\cdot\|}, \tag{8.11}$$

with $\mathbf{P}_N \in \mathbb{V}_n$ being the collocation tensor for the Coulomb potential. This implies the tensor representation of the Coulomb matrix,

$$J(D)_{\mu\nu} \approx \langle \mathbf{G}_\mu \odot \mathbf{G}_\nu, \Theta *_T \mathbf{P}_N \rangle, \quad 1 \le \mu, \nu \le N_b. \tag{8.12}$$

The separability property of basis functions ensures that $\mathrm{rank}(\mathbf{G}_\mu) \le R_G$, whereas tensors Θ and $\mathbf{P}_N$ are to be approximated by low-rank tensors. Hence, in our method, the corresponding tensor operations are implemented using fast multilinear algebra equipped with the corresponding rank optimization (tensor truncation) [173, 174, 186].

The numerical examples of other rank decompositions to electron density (not including the calculation of the three-dimensional convolution operator) have been presented in [52, 81]. The tensor product convolution was introduced in [173, 174] and also discussed in [108, 109, 166].

Likewise, tensor representations (8.3)–(8.5) for the exchange operator realized in [145] now look as follows:

$$W_{a\nu} \approx \Upsilon_{a\nu} := \left[\mathbf{G}_\nu \odot \sum_{\kappa=1}^{N_b} C_{\kappa a} \odot \mathbf{G}_\kappa \right] *_T \mathbf{P}_N, \quad \nu = 1, \ldots, N_b, \tag{8.13}$$

with the tensor $\mathbf{P}_N \in \mathbb{V}_n$ defined by (8.11). Now we proceed with

$$K_{\mu\nu,a} \approx \chi_{\mu\nu,a} := \left\langle \left[\sum_{\kappa=1}^{N_b} C_{\kappa a} \mathbf{G}_\kappa \right] \odot \mathbf{G}_\mu, \Upsilon_{a\nu} \right\rangle, \quad \mu, \nu = 1, \ldots, N_b, \tag{8.14}$$

finally providing the entries of the exchange matrix by summation over all orbitals

$$K(D)_{\mu\nu} = \sum_{a=1}^{N_{\text{orb}}} \chi_{\mu\nu,a}, \quad \mu, \nu = 1, \ldots, N_b. \tag{8.15}$$

Again, the auxiliary tensors and respective algebraic operations have to be implemented with the truncation to low-rank tensor formats.

8.2 Numerics on three-dimensional convolution operators

Here we discuss the algorithms for grid-based calculation of the Coulomb and exchange operators by using the tensor-structured numerical method introduced in [174, 145], where it was demonstrated that calculation of the three- and six-dimensional convolution integrals with the Newton kernel can be reduced to a combination of one-dimensional Hadamard and scalar products and one-dimensional convolutions.

In the following, for numerical illustrations, we use the Gaussian basis sets, which are convenient for verification of the computational results (the corresponding Galerkin Fock matrix) with the standard MOLPRO output [299]. The univariate Gaussians $g_k^{(\ell)}(x_\ell) = g_{k,1}^{(\ell)}(x_\ell)$, $\ell = 1, 2, 3$, are the functions with infinite support given by

$$g_k^{(\ell)}(x_\ell) = (x_\ell - A_{\ell,k})^{p_{\ell,k}} \exp(-\alpha_k (x_\ell - A_{\ell,k})^2), \quad x_\ell \in \mathbb{R}, \; \alpha_k > 0,$$

where $p_{\ell,k} = 0, 1, \ldots$ is the polynomial degree, and the points $(A_{1,k}, A_{2,k}, A_{3,k}) \in \mathbb{R}^3$ specify the positions of nuclei in a molecule.

The molecule is embedded in a certain fixed computational box $\Omega = [-b, b]^3 \in \mathbb{R}^3$, as in Figure 11.1.[1] For a given discretization parameter $n \in \mathbb{N}$, we use the equidistant $n \times n \times n$ tensor grid $\omega_{3,n} = \{x_\mathbf{i}\}$, $\mathbf{i} \in \mathcal{I} := \{1, \ldots, n\}^3$, with the mesh-size $h = 2b/(n+1)$.

1 In the case of small to moderate size molecules, usually, we use the computational box of size 40^3 bohr.

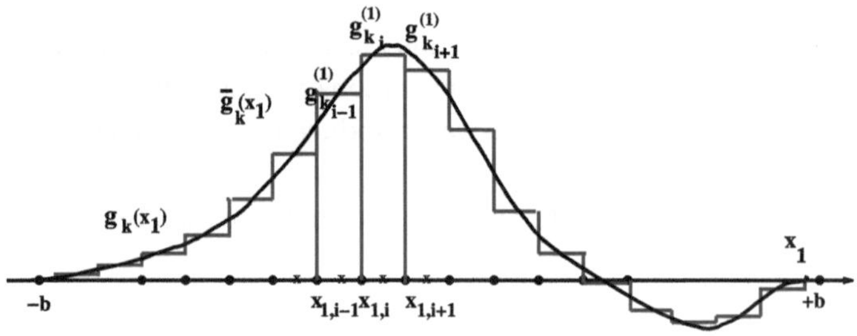

Figure 8.1: Approximation of the Gaussian-type basis function by a piecewise constant function.

The Gaussian-type basis functions are used for the representation of orbitals (8.9). In calculations of integral terms, the separable type basis functions $g_k(x)$, $x \in \mathbb{R}^3$ are approximated by sampling their values at the centers of discretization intervals, as in Figure 8.1, using the product of univariate piecewise constant basis functions $g_k(x) \approx \overline{g}_k(x) = \prod_{\ell=1}^{3} \overline{g}_k^{(\ell)}(x^{(\ell)})$, $\ell = 1, 2, 3$, yielding their rank-1 tensor representation,

$$g_k \mapsto \mathbf{G}_k = \mathbf{g}_k^{(1)} \otimes \mathbf{g}_k^{(2)} \otimes \mathbf{g}_k^{(3)} \in \mathbb{R}^{n \times n \times n}, \quad k = 1, \dots, N_b. \tag{8.16}$$

For the tensor-based calculation of the Hartree potential

$$V_H(x) := \int_{\mathbb{R}^3} \frac{\rho(y)}{\|x - y\|} dy$$

and of the corresponding Coulomb matrix

$$J_{km} := \int_{\mathbb{R}^3} g_k(x) g_m(x) V_H(x) dx, \quad k, m = 1, \dots, N_b, \; x \in \mathbb{R}^3,$$

we use the discrete tensor representation of basis functions (8.16). Then the electron density is approximated by using 1D Hadamard products of skeleton vectors in rank-1 tensors (instead of product of Gaussians)

$$\rho \approx \Theta = 2 \sum_{a=1}^{N_{\text{orb}}} \sum_{k=1}^{N_b} \sum_{m=1}^{N_b} c_{a,m} c_{a,k} (\mathbf{g}_k^{(1)} \odot \mathbf{g}_m^{(1)}) \otimes (\mathbf{g}_k^{(2)} \odot \mathbf{g}_m^{(2)}) \otimes (\mathbf{g}_k^{(3)} \odot \mathbf{g}_m^{(3)}) \in \mathbb{R}^{n \times n \times n}.$$

Further, the representation of the Newton convolving kernel $\frac{1}{\|x-y\|}$ by a canonical rank-R_N tensor [30] is used (see Section 6.1 for details):

$$\mathbf{P}_N \mapsto \mathbf{P}_R = \sum_{q=1}^{R_N} \mathbf{p}_q^{(1)} \otimes \mathbf{p}_q^{(2)} \otimes \mathbf{p}_q^{(3)} \in \mathbb{R}^{n \times n \times n}. \tag{8.17}$$

Since large ranks make tensor operations inefficient, the multigrid canonical-to-Tucker and Tucker-to-canonical algorithms (see Sections 3.3.3 and 3.5) should be applied to reduce the initial rank of $\Theta \mapsto \Theta'$ by several orders of magnitude, from $N_b^2/2$ to essentially smaller value $R_\rho \ll N_b^2/2$. For sufficient accuracy, the ε-threshold is chosen of the order of 10^{-7}.

Tensor approximation to the Hartree potential is calculated by using the 3D tensor product convolution, which is a sum of tensor products of 1D convolutions:

$$V_H \approx \mathbf{V}_H = \Theta' * \mathbf{P}_R = \sum_{j=1}^{R_\rho} \sum_{q=1}^{R_N} c_j (\mathbf{u}_j^{(1)} * \mathbf{p}_q^{(1)}) \otimes (\mathbf{u}_j^{(2)} * \mathbf{p}_q^{(2)}) \otimes (\mathbf{u}_j^{(3)} * \mathbf{p}_q^{(3)}).$$

Finally, the entries of the Coulomb matrix J_{km} are computed by 1D scalar products of the canonical vectors of $\mathbf{V}_H$ with the Hadamard products of the rank-1 tensors representing the Galerkin basis:

$$J_{km} \approx \langle \mathbf{G}_k \odot \mathbf{G}_m, \mathbf{V}_H \rangle, \quad k, m = 1, \ldots N_b.$$

The cost of 3D tensor product convolution is $O(n \log n)$ instead of $O(n^3 \log n)$ for the standard benchmark 3D convolution using the 3D FFT. Table 8.1 shows CPU times (sec) for the Matlab computation of V_H for H_2O molecule [174] on a SUN station using a cluster with 4 Intel Xeon E7-8837/32 cores/2.67 GHz and 1024 GB storage (times for 3D FFT for $n \geq 4096$ are obtained by extrapolation). It is easy to notice cubic scaling of the 3D FFT time in dyadic increasing of the grid size n and approximately linear-logarithmic scaling for 3D convolution on the same grids (see $C * C$ row). C2T shows the time for the canonical-to-Tucker rank reduction.

Following [166], we apply the Richardson extrapolation technique (see [218]) to obtain higher accuracy approximations of order $O(h^3)$ without extra computational cost. The numerical gain of using an extrapolated solution is achieved due to the fact that the approximation error $O(h^3)$ on the single grid would require the univariate grid size $n_1 = n^{3/2} \gg n$. The corresponding Richardson extrapolant $V_{H,\text{Rich}}^{(n)}$ approximating $V_H(x)$ over a pair of nested grids $\omega_{3,n}$ and $\omega_{3,2n}$, and defined on the "coarse" $n^{\otimes 3}$-grid, is given by

$$V_{H,\text{Rich}}^{(n)} = (4 \cdot V_H^{(2n)} - V_H^{(n)})/3 \quad \text{in the grid-points on } \omega_{3,n}.$$

The next numerical results show the accuracy of the tensor-based calculations using $n \times n \times n$ 3D Cartesian grids with respect to the corresponding output from the MOLPRO package [299].

Table 8.1: Times (sec) for the 3D tensor product convolution vs. convolution by 3D FFT in computation of V_H for H_2O molecule.

n^3	1024^3	2048^3	4096^3	8192^3	$16\,384^3$
FFT_3	10	81	640	5120	~11 hours
$C * C$	8.8	20.0	61.0	157.5	299.2
C2T	6.9	10.9	20.0	37.9	86.0

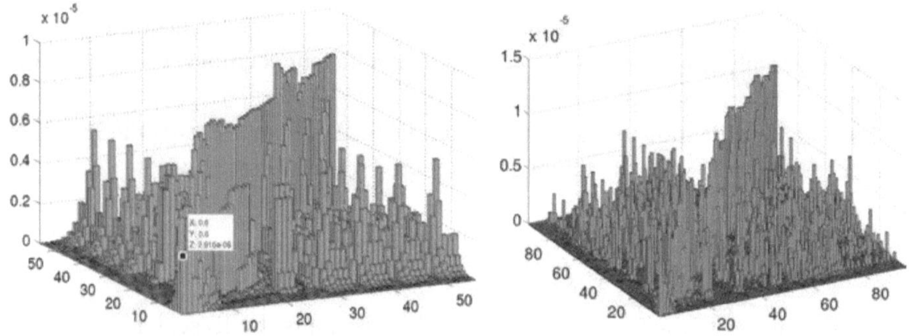

Figure 8.2: Left: Absolute error in tensor computation of the Coulomb matrix for CH_4 and C_2H_6 molecules.

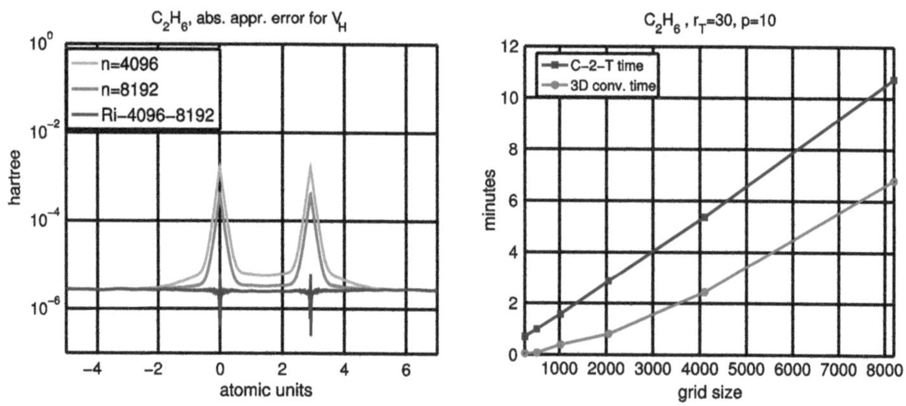

Figure 8.3: Left: Absolute approximation error (blue line: $\approx 10^{-6}$ au) in the tensor-product computation of the Hartree potential of C_2H_6, measured in the grid line $\Omega = [-5, 7] \times \{0\} \times \{0\}$. Right: Times versus n in MATLAB for computation of V_H for C_2H_6 molecule.

Figure 8.2 demonstrates the accuracy ($\sim 10^{-5}$) of the calculation of the Coulomb matrix for CH_4 and C_2H_6 molecules using the Richardson extrapolation on a sequence of grids $\omega_{3,n}$ with $n = 4096$ and $n = 8192$.

Figure 8.3 (left) shows the accuracy in calculation of the Hartree potential (in comparison with the benchmark calculations from MOLPRO) for the C_2H_6 molecule computed on $n \times n \times n$ grids of size $n = 4096$ and $n = 8192$ (dashed lines). The solid line in Figure 8.3 shows the accuracy of the Richardson extrapolation of the results from two grids of size $n = 4096$ and $n = 8192$. One can observe essential improvement of accuracy for the Richardson extrapolation. Figure 8.3 (right) shows the CPU times versus n in MATLAB indicating the linear complexity scaling in the univariate grid size n. See also Figure 8.4 illustrating accuracy for the exchange matrix $K = K_{ex}$.

In a similar way, the algorithm for 3D grid-based tensor-structured calculation of 6D integrals in the exchange potential operator was introduced in [145], $K_{km} =$

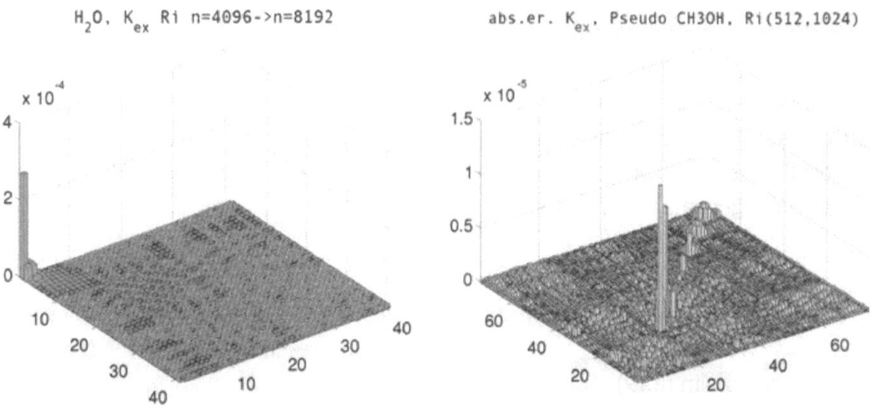

Figure 8.4: L^∞-error in $K_{ex} = K$ for the density of H_2O and pseudodensity of CH_3OH.

$\sum_{a=1}^{N_{orb}} K_{km,a}$ with

$$K_{km,a} := \int_{\mathbb{R}^3}\int_{\mathbb{R}^3} g_k(x)\frac{\varphi_a(x)\varphi_a(y)}{|x-y|}g_m(y)dxdy, \quad k,m = 1,\dots N_b.$$

The contribution from the ath orbital are approximated by tensor anzats,

$$K_{km,a} \approx \left\langle \mathbf{G}_k \odot \left[\sum_{\mu=1}^{N_b} c_{\mu a}\mathbf{G}_\mu\right], \left[\mathbf{G}_m \odot \sum_{\nu=1}^{N_b} c_{\nu a}\mathbf{G}_\nu\right] * \mathbf{P}_R\right\rangle.$$

Here, the tensor product convolution is first calculated for each ath orbital, and then scalar products in canonical format yield the contributions to entries of the exchange Galerkin matrix from the a-th orbital. The algorithm for tensor calculation of the exchange matrix is described in detail in [145].

These algorithms were introduced in the first tensor-structured Hartree–Fock solver using 3D grid-based evaluation of the Coulomb and exchange matrices in 1D complexity at every step of self-consistent field (SCF) iteration [146, 187].

8.3 Multilevel rank-truncated self-consistent field iteration

In the following sections we discuss the first grid-based Hartree–Fock solver, which was developed in 2009 and published in [146] and [187].

The standard self-consistent field iteration (SCF) algorithm can be formulated as the following "fixed-point" iteration [203, 44]: Starting from initial guess C_0, perform iterations of the form

$$\begin{aligned} \widetilde{F}_k C_{k+1} &= S C_{k+1}\Lambda_{k+1}, \quad \Lambda_{k+1} = \operatorname{diag}(\lambda_1^{k+1},\dots,\lambda_{N_{orb}}^{k+1}), \\ C_{k+1}^T S C_{k+1} &= I_{N_{orb}}, \end{aligned} \tag{8.18}$$

where the current Fock matrix $\widetilde{F}_k = \Phi(C_k, C_{k-1}, \ldots, C_0)$, $k = 0, 1, \ldots$, is specified by the particular relaxation scheme. For example, for the simplest approach, called the *Roothaan algorithm*, one has $\widetilde{F}_k = F(C_k)$. In practically interesting situations this algorithm usually leads to "flip-flop" stagnation [203].

Recall that $\lambda_1^{k+1} \leq \lambda_2^{k+1} \leq \cdots \leq \lambda_{N_{\mathrm{orb}}}^{k+1}$ are N_{orb} negative eigenvalues of the linear generalized eigenvalue problem

$$\widetilde{F}_k U = \lambda S U, \tag{8.19}$$

and the $N_b \times N_{\mathrm{orb}}$ matrices C_{k+1} contain the respective N_{orb} orthonormal eigenvectors $\mathbf{u}_1, \ldots, \mathbf{u}_{N_{\mathrm{orb}}}$. We denote by $\widetilde{C}_{k+1} \in \mathbb{R}^{N_b \times N_b}$ the matrix representing the full set of orthogonal eigenvectors in (8.19).

We use the particular choice of $\widetilde{F}_k$, $k = 0, 1, \ldots$, via the DIIS-algorithm (cf. [238]), with the starting value $\widetilde{F}_0 = F(C_0) = H$, where the matrix H corresponds to the core Hamiltonian.

In [146, 187] a modification to the standard DIIS iteration was proposed by carrying out the iteration on a sequence of successively refined grids with the grid-dependent stopping criteria. The multilevel implementation provides robust convergence from the zero initial guess for the Hartree and exchange operators. The coarse-to-fine grids iteration, in turn, accelerates the solution process dramatically due to low cost of the coarse grid calculations.

The principal feature of the tensor-truncated iteration is revealed on the fast update of the Fock matrix $F(C)$ by using tensor-product multilinear algebra of 3-tensors accomplished with the rank truncation. Moreover, the multilevel implementation provides a simple scheme for constructing good initial guess on the fine grid-levels.

8.3.1 SCF iteration by using modified DIIS scheme

For each fixed discretization, we use the original version of DIIS scheme (cf. [128]), defined by the following choice of the *residual error vectors* (matrices):

$$E_i := [\widetilde{C}_{i+1}^T F(C_i) \widetilde{C}_{i+1}]_{|\{1 \leq \mu \leq N_{\mathrm{orb}}; N_{\mathrm{orb}}+1 \leq \nu \leq N_b\}} \in \mathbb{R}^{N_{\mathrm{orb}} \times (N_b - N_{\mathrm{orb}})} \tag{8.20}$$

for iteration number $i = 0, 1, \ldots, k$, which should vanish on the exact solutions of the Hartree–Fock Galerkin equation due to the orthogonality property. Hence, some stopping criterion applies to residual error vector E_i for $i = 0, 1, 2, \ldots$. Here the subindexes μ and ν specify the relevant range of entries in the coefficients for molecular orbitals $\widetilde{C}_{i+1}$.

The minimizing coefficient vector $\widetilde{\mathbf{c}} := (c_0, \ldots, c_k)^T \in \mathbb{R}^{k+1}$ is computed by solving the constrained quadratic minimization problem for the respective cost functional (the averaged residual error vector over previous iterands):

$$f(\widetilde{\mathbf{c}}) := \frac{1}{2} \left\| \sum_{i=0}^{k} c_i E_i \right\|_F^2 \equiv \frac{1}{2} \langle B\widetilde{\mathbf{c}}, \widetilde{\mathbf{c}} \rangle \to \min, \quad \text{provided that } \sum_{i=0}^{k} c_i = 1,$$

where

$$B = \{B_{ij}\}_{i,j=0}^{k} \quad \text{with } B_{ij} = \langle E_i, E_j \rangle,$$

with E_i defined by (8.20). Introducing the Lagrange multiplier $\xi \in \mathbb{R}$, the problem is reduced to minimization of the Lagrangian functional

$$L(\tilde{\mathbf{c}}, \xi) = f(\tilde{\mathbf{c}}) - \xi(\langle \mathbf{1}, \tilde{\mathbf{c}} \rangle - 1),$$

where $\mathbf{1} = (1, \ldots, 1)^T \in \mathbb{R}^{k+1}$, which leads to the linear augmented system of equations

$$B\tilde{\mathbf{c}} - \xi\mathbf{1} = 0, \qquad (8.21)$$
$$\langle \mathbf{1}, \tilde{\mathbf{c}} \rangle = 1.$$

Finally, the updated Fock operator $\tilde{F}_k$ is built up by

$$\tilde{F}_k = \sum_{i=0}^{k-1} c_i^{\text{opt}} \tilde{F}_i + c_k^{\text{opt}} F(C_k), \quad k = 0, 1, 2, \ldots, \qquad (8.22)$$

where the minimizing coefficients $c_i^{\text{opt}} = \tilde{c}_i$ $(i = 0, 1, \ldots, k)$ solve the linear system (8.21). For $k = 0$, the first sum in (8.22) is assumed to be zero, hence providing $c_0^{\text{opt}} = 1$ and $\tilde{F}_0 = F(C_0)$.

Recall that if the stopping criterion on C_k, $k = 1, \ldots$, is not satisfied, then one updates $\tilde{F}_k$ by (8.22) and solves the eigenvalue problem (8.18) for C_{k+1}.

Note that in practice one can use the averaged residual vector only on a reduced subsequence of iterands, $E_k, E_{k-1}, \ldots, E_{k-k_0}$, $k - k_0 > 0$. In our numerical examples below, we usually set $k_0 = 4$.

8.3.2 Unigrid and multilevel tensor-truncated DIIS iteration

In this section, we describe the resultant numerical algorithm. Recall that the discrete nonlinear Fock operator is specified by a matrix

$$F(C) = H + J(C) + K(C), \qquad (8.23)$$

where H corresponds to the core Hamiltonian (fixed in our scheme), and the discrete Hartree and exchange operators are given by tensor representations (8.12) and (8.4), respectively.

First, we describe the unigrid tensor-truncated DIIS scheme [146, 187].

I. Algorithm U_DIIS (unigrid tensor-truncated DIIS iteration).

(1) Given the core Hamiltonian matrix H, the grid parameter n, and the termination parameter $\varepsilon > 0$.

(2) Set $C_0 = 0$ (i. e., $J(C_0) = 0$, $K(C_0) = 0$) and $\widetilde{F}_0 = H$.
(3) For $k = 0, 1, \ldots$, perform
   (a) Solve the full linear eigenvalue problem of size $N_b \times N_b$, given by (8.19), and define C_{k+1} as the matrix containing the N_{orb} eigenvectors corresponding to N_{orb} minimal eigenvalues.
   (b) Terminate the iteration by checking the stopping criterion

$$\|C_{k+1} - C_k\|_F \le \varepsilon.$$

   (c) If $\|C_{k+1} - C_k\|_F > \varepsilon$, then compute the Fock matrix

$$F(C_{k+1}) = H + J(C_{k+1}) + K(C_{k+1})$$

   by the tensor-structured calculations of $J(C_{k+1})$ and $K(C_{k+1})$ using grid-based basis functions with expansion coefficients C_{k+1} (see Section 8.1). Update the Fock matrix $\widetilde{F}_{k+1}$ by (8.22) and switch to Step (a).
(4) Returns: Eigenvalues $\lambda_1, \ldots, \lambda_{N_{\mathrm{orb}}}$ and eigenvectors $C \in \mathbb{R}^{N_b \times N_{\mathrm{orb}}}$.

Numerical illustration on the convergence of Algorithm U_DIIS for solving the Hartree–Fock equation in the pseudopotential case of CH_4 have been presented in [187]. It demonstrates that the convergence history is almost independent of the grid size on the examples with $n = 64$ and $n = 256$.

To enhance the unigrid DIIS iteration, we apply the multilevel version of Algorithm U_DIIS defined on a sequence of discrete Hartree–Fock equations specified by a sequence of grid parameters $n_p = n_0, 2n_0, \ldots, 2^M n_0$, with $p = 0, \ldots, M$, corresponding to the succession of dyadically refined spacial grids. To that end, for ease of exposition, we also introduce the incomplete version of Algorithm U_DIIS, further called Algorithm U_DIIS($\widetilde{k}$), where the DIIS correction starts only after the iteration number $k = \widetilde{k} \ge 1$. The input data for Algorithm U_DIIS($\widetilde{k}$) include the current approximation $C_{\widetilde{k}}$ and a sequence of all already precomputed Fock matrices, $\widetilde{F}_0, \widetilde{F}_1, \ldots, \widetilde{F}_{\widetilde{k}-1}$.

We sketch this algorithm as follows:

II. Algorithm U_DIIS($\widetilde{k}$) (incomplete unigrid tensor-truncated DIIS iteration).
(1) Given the core Hamiltonian matrix H, the grid parameter n, the termination parameter $\varepsilon > 0$, $C_{\widetilde{k}}$, and a sequence of Fock matrices $\widetilde{F}_0, \widetilde{F}_1, \ldots, \widetilde{F}_{\widetilde{k}-1}$.
(2) Compute $J(C_{\widetilde{k}})$, $K(C_{\widetilde{k}})$, $F(C_{\widetilde{k}}) = H + J(C_{\widetilde{k}}) + K(C_{\widetilde{k}})$, and $\widetilde{F}_{\widetilde{k}}$ by (8.22).
(3) For $k = \widetilde{k}, \widetilde{k} + 1, \ldots$, perform steps (a)–(c) in Algorithm M_DIIS.

Next, we consider the multilevel tensor-truncated DIIS scheme [146, 187].

III. Algorithm M_DIIS (multilevel tensor-truncated DIIS scheme).
(1) Given the core Hamiltonian matrix H, the coarsest grid parameter n_0, the termination parameter $\varepsilon_0 > 0$, and the number of grid refinements M.

(2) For $p = 0$, apply the unigrid Algorithm U_DIIS with $n = n_0$, $\varepsilon_p = \varepsilon_0$, and return the number of iterations k_0, matrix C_{k_0+1}, and a sequence of Fock matrices $\widetilde{F}_0, \widetilde{F}_1, \ldots, \widetilde{F}_{k_0}$.

(3) For $p = 1, \ldots, M$, apply successively Algorithm U_DIIS($k_{p-1} + 1$), with the input parameters $n_p := 2^p n_0$, $\varepsilon_p := \varepsilon_0 2^{-2p}$, $C_{k_{p-1}+1}$. Keep continuous numbering of the DIIS iterations through all levels such that the maximal iteration number at level p is given by

$$k_p = \sum_{p=0}^{p} m_p$$

with m_p being the number of iterative steps at level p.

(4) Returns: k_M, C_{k_M+1}, and a sequence of Fock matrices $\widetilde{F}_0, \widetilde{F}_1, \ldots, \widetilde{F}_{k_M}$.

In numerical practice, usually, we start calculations on a small $n_0 \times n_0 \times n_0$ 3D Cartesian grid with $n_0 = 64$ and end up with maximum $n_M = 8192$ for all electron case computations, or $n_M = 1024$ for the pseudopotential case. Further, in Section 11.6, we show by numerical examples that in large-scale computations the multilevel Algorithm M_DIIS allows us to perform most of the iterative steps on coarse grids, thus reducing dramatically the computational cost and, at the same time providing a good initial guess for the DIIS iteration on nonlinearity at each consequent approximation level.

The rest of this section addresses the complexity estimate of the multilevel tensor-truncated iteration in terms of R_N, N_b, n, and other governing parameters of the algorithm. For the ease of discussion we suppose that $\operatorname{rank}(G_\mu) = 1$, $\mu = 1, \ldots, N_b$ (see [187] concerning the more detailed discussion on the general case of $\operatorname{rank}(G_\mu) \geq 1$).

Lemma 8.1 ([187]). *Let* $\operatorname{rank}(G_\mu) = 1$, $\mu = 1, \ldots N_b$, *and* $\operatorname{rank}(\mathbf{P}_N) = R_N \leq C N_{\text{orb}}$. *Suppose that the rank reduction procedure applied to the convolution products* Υ_{av} *in (8.3) provides the rank estimate* $\operatorname{rank}(\Upsilon_{av}) \leq r_0$. *Then the numerical cost of one iterative step in Algorithm* M_DIIS *at level p can be bounded by*

$$W_p = O(N_b R_N n_p \log n_p + N_b^3 r_0 N_{\text{orb}} n_p).$$

Assume that the number of multigrid DIIS iterations at each level is bounded by the constant I_0. *Then the total cost of Algorithm* M_DIIS *does not exceed the double cost at the finest level* $n = n_M$, $2W_M = O(I_0 N_b^3 r_0 N_{\text{orb}} n)$.

Proof. The rank bound $\operatorname{rank}(G_k) = 1$ implies $\operatorname{rank}(\sum_{m=1}^{N_b} c_{ma} G_m) \leq N_b$. Hence, the numerical cost to compute the tensor-product convolution Υ_{av} in (8.3) amounts to

$$W(\Upsilon_{av}) = O(N_b R_N n_p \log n_p).$$

Since the initial canonical rank of Υ_{av} is estimated by $\operatorname{rank}(\Upsilon_{av}) \leq N_b R_N$, the multigrid rank reduction algorithm, having linear scaling in $\operatorname{rank}(\Upsilon_{av})$, see Section 3, provides the complexity bound $O(r_0 N_b R_N n_p)$. Hence the total cost to compute scalar products in $\chi_{\mu\nu,a}$ (see (8.4)) can be estimated by

$$W(\chi_{\mu\nu,a}) = O(N_b^3 r_0 N_{\text{orb}} n_p),$$

which completes the first part of our proof. The second assertion follows due to linear scaling in n_p of the unigrid algorithm, which implies the following bound:

$$n_0 + 2n_0 + \cdots + 2^p n_0 \leq 2^{p+1} n_0 = 2n_M,$$

hence completing the proof. □

Remark 8.2. In the case of large molecules and $R_G = \operatorname{rank}(G_\mu) \geq 1$, further optimization of the algorithm up to $O(R_N N_b^2 n_p)$-complexity may be possible on the base of rank reduction applied to the rank-$R_G N_b$ orbitals and by using an iterative eigenvalue solver instead of currently employed direct solver via matrix diagonalization, or by using direct minimization schemes [263].

Our algorithm for *ab initio* solution of the Hartree–Fock equation in *tensor-structured format* was examined numerically on some moderate size molecules [146, 187]. In particular, we consider the all-electron case of H_2O and the case of pseudopotential of CH_4 and CH_3OH molecules. In the presented numerical examples, we use the discretized GTO basis functions for reasons of convenient comparison of the results with the output from the standard MOLPRO package based on the analytical evaluation of the integral operators in the GTO basis.

The size of the computational box $[-b, b]^3$ introduced in Section 8.1.1 varies from $2b = 11.2$ Å for H_2O up to $2b = 16$ Å for small organic molecules. The smallest step-size of the grid $h = 0.0013$ Å is reached in the SCF iterations for the H_2O molecule, using the finest level grid with $n = 8192$, whereas the average step size for the computations using the pseudopotentials for small organic molecules is about $h = 0.015$ Å, corresponding to the grid size $n = 1024$.

We solve numerically the ab initio Hartree–Fock equation by using Algorithms U_DIIS and M_DIIS presented in Section 8.3.2. Starting with the zero initial guess for matrices $J(C) = 0$ and $K(C) = 0$ in the Galerkin Fock matrix, the eigenvalue problem at the first iterative step ($p = 0$) is solved by using only the H part of the Fock matrix in (8.23), which does not depend on the solution and hence can be precomputed beforehand.

Thus, the SCF iteration starts with the expansion coefficients $c_{\mu i}$ for orbitals in the GTO basis, computed using only the core Hamiltonian H. At every iteration step, the Hartree and exchange potentials and the corresponding Galerkin matrices are computed using the updated coefficients $c_{\mu i}$. The renewed Coulomb and exchange matrices generate the updated Fock matrix to be used for the solution of the eigenvalue

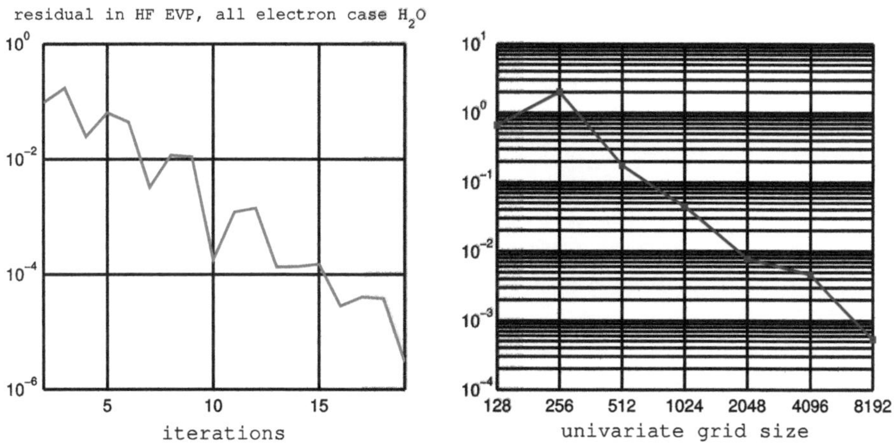

Figure 8.5: Multilevel convergence of the DIIS iteration applied to the all electron case of H_2O (left), and convergence in the energy in n (right).

problem. The minimization of the Frobenius norm of the virtual block of the Fock operator evaluated on eigenvectors of the consequent iterations, $\widetilde{C}_k, \widetilde{C}_{k-1}, \ldots$, is utilized for the DIIS scheme.

The multilevel solution of the nonlinear eigenvalue problem (8.18) is realized via the SCF iteration on a sequence of uniformly refined grids, beginning from the initial coarse grid, say, with $n_0 = 64$, and proceeding on the dyadically refined grids $n_p = n_0 2^p$, $p = 1, \ldots, M$. We use the grid-dependent termination criterion $\varepsilon_{n_p} := \varepsilon_0 2^{-2p}$, keeping a continuous numbering of the iterations.

Figure 8.5 (left) shows the convergence of the iterative scheme in the case of H_2O molecule. Figure 8.5 (right) illustrates the convergence in the total Hartree–Fock energy reaching the absolute error about 10^{-4}, which implies the relative error $9 \cdot 10^{-6}$ in the case of grid size $n = 1024$. The total energy is calculated by

$$E_{\mathrm{HF}} = 2 \sum_{a=1}^{N_{\mathrm{orb}}} \lambda_a - \sum_{a=1}^{N_{\mathrm{orb}}} (\widetilde{J}_a - \widetilde{K}_a)$$

with $\widetilde{J}_a = \langle \psi_a, V_H \psi_a \rangle_{L^2}$, and $\widetilde{K}_a = \langle \psi_a, \mathcal{V}_{\mathrm{ex}} \psi_a \rangle_{L^2}$, being the so-called Coulomb and exchange integrals, respectively, computed in the molecular orbital basis ψ_a $(a = 1, \ldots, N_{\mathrm{orb}})$.

The detailed discussion of the multilevel DIIS iteration, including various numerical tests, can be found in [187, 146].

9 Grid-based core Hamiltonian

In this section, following [156], we discuss the grid-based method for calculating the core Hamiltonian part in the Fock operator (7.4)

$$\mathcal{H} = -\frac{1}{2}\Delta + V_c$$

with respect to the Galerkin basis $\{g_m(x)\}_{1\le m\le N_b}$, $x \in \mathbb{R}^3$, where $V_c(x)$ is given by (7.4), and Δ represents the 3D Laplacian subject to Dirichlet boundary conditions.

9.1 Tensor approach for multivariate Laplace operator

The initial eigenvalue problem is posed in the finite volume box $\Omega = [-b, b]^3 \in \mathbb{R}^3$ subject to the homogeneous Dirichlet boundary conditions on $\partial\Omega$. For given discretization parameter $N \in \mathbb{N}$, we use the equidistant $N \times N \times N$ tensor grid $\omega_{3,N} = \{x_{\mathbf{i}}\}$, $\mathbf{i} \in \mathcal{I} := \{1, \ldots, N\}^3$, with the mesh-size $h = 2b/(N+1)$, which may be different from the grid $\omega_{3,n}$ introduced in Section 8.1.1 (usually, $n \le N$).

Now, similar to Section 8.1.1, define a set of piecewise linear basis functions $\overline{g}_k := \mathbf{I}_1 g_k$, $k = 1, \ldots, N_b$, by linear tensor-product interpolation via the set of product hat functions $\{\xi_{\mathbf{i}}\} = \xi_{i_1}(x_1)\xi_{i_2}(x_2)\xi_{i_3}(x_3)$, $\mathbf{i} \in \mathcal{I}$, associated with the respective grid-cells in $\omega_{3,N}$. Here, the linear interpolant $\mathbf{I}_1 = I_1 \times I_1 \times I_1$ is a product of 1D interpolation operators $\overline{g}_k^{(\ell)} = I_1 g_k^{(\ell)}$, $\ell = 1, \ldots, 3$, where $I_1 : C^0([-b, b]) \to W_h := \operatorname{span}\{\xi_i\}_{i=1}^N$ is defined over the set of piecewise linear basis functions by

$$(I_1\, w)(x_\ell) := \sum_{i_\ell=1}^{N} w(x_{i_\ell})\xi_{i_\ell}(x_\ell), \quad x_{\mathbf{i}} \in \omega_{3,N},\ \ell = 1, 2, 3.$$

This leads to the separable grid-based approximation of the initial Gaussian-type basis functions $g_k(x)$,

$$g_k(x) \approx \overline{g}_k(x) = \prod_{\ell=1}^{3} \overline{g}_k^{(\ell)}(x_\ell) = \prod_{\ell=1}^{3} \sum_{i=1}^{N} g_k^{(\ell)}(x_{i_\ell})\xi_i(x_\ell), \tag{9.1}$$

where the rank-1 coefficients tensor $\mathbf{G}_k$ is given by $\mathbf{G}_k = \mathbf{g}_k^{(1)} \otimes \mathbf{g}_k^{(2)} \otimes \mathbf{g}_k^{(3)}$ with the canonical vectors $\mathbf{g}_k^{(\ell)} = \{g_k^{(\ell)}(x_{i_\ell})\}$ (see Figure 9.1 illustrating the construction of $\overline{g}_k(x_1)$).

We approximate the exact Galerkin matrix $A_g \in \mathbb{R}^{N_b \times N_b}$,

$$A_g = \{a_{km}\} := \{\langle -\Delta g_k, g_m\rangle\} \equiv \{\langle \nabla g_k, \nabla g_m\rangle\}, \quad k, m = 1, \ldots N_b,$$

by using the piecewise linear representation of the basis functions $\overline{g}_k(x) \in \mathbb{R}^3$ (see (9.1)) constructed on $N \times N \times N$ Cartesian grid (see [41] for general theory of finite element

https://doi.org/10.1515/9783110365832-009

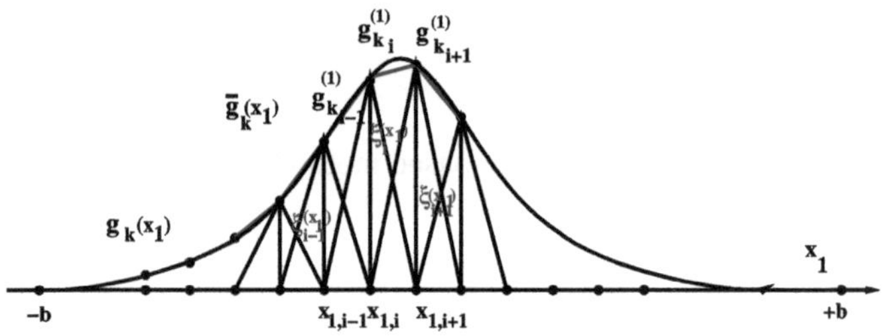

Figure 9.1: Using hat functions $\xi_i(x_1)$ for a single-mode basis function $g_k(x_1)$ yielding the piecewise linear representation $\overline{g}_k(x_1)$ of a continuous function $g_k(x_1)$.

approximation). Here, ∇ denotes the 3D gradient operator. The approximating matrix A_G is now defined by

$$A_g \approx A_G = \{\overline{a}_{km}\} := \{\langle -\Delta \overline{g}_k, \overline{g}_m\rangle\} \equiv \{\langle \nabla \overline{g}_k, \nabla \overline{g}_m\rangle\}, \quad A_G \in \mathbb{R}^{N_b \times N_b}. \tag{9.2}$$

The accuracy of this approximation is of order $\|a_{km} - \overline{a}_{km}\| = O(h^2)$, where h is the mesh size (see [156], Theorem A.4, and numerics in Section 9.3).

Recall that the Laplace operator applies to a separable function $\eta(x)$, $x = (x_1, x_2, x_3) \in \mathbb{R}^3$, having a representation $\eta(x) = \eta_1(x_1)\eta_2(x_2)\eta_3(x_3)$ as follows:

$$\Delta \eta(x) = \frac{d^2\eta_1(x_1)}{dx_1^2}\eta_2(x_2)\eta_3(x_3) + \frac{d^2\eta_2(x_2)}{dx_2^2}\eta_1(x_1)\eta_3(x_3) + \frac{d^2\eta_3(x_3)}{dx_3^2}\eta_1(x_1)\eta_2(x_2), \tag{9.3}$$

which ensures the standard Kronecker rank-3 tensor representation of the respective Galerkin FEM stiffness matrix A_Δ in the tensor basis $\{\xi_i(x_1)\xi_j(x_2)\xi_k(x_3)\}$, $i, j, k = 1, \dots N$,

$$A_\Delta := A^{(1)} \otimes S^{(2)} \otimes S^{(3)} + S^{(1)} \otimes A^{(2)} \otimes S^{(3)} + S^{(1)} \otimes S^{(2)} \otimes A^{(3)} \in \mathbb{R}^{N^{\otimes 3} \times N^{\otimes 3}}.$$

Here, the 1D stiffness and mass matrices $A^{(\ell)}, S^{(\ell)} \in \mathbb{R}^{N \times N}$, $\ell = 1, 2, 3$, are given by

$$A^{(\ell)} := \{\langle \nabla_{(\ell)}\xi_i(x_\ell), \nabla_{(\ell)}\xi_j(x_\ell)\rangle\}_{i,j=1}^N = \frac{1}{h}\,\mathrm{tridiag}\{-1, 2, -1\},$$

$$S^{(\ell)} = \{\langle \xi_i, \xi_j\rangle\}_{i,j=1}^N = \frac{h}{6}\,\mathrm{tridiag}\{1, 4, 1\},$$

respectively, and $\nabla_{(\ell)} = \frac{d}{dx_\ell}$. Since $\{\xi_i\}_{i=1}^N$ are the same for all modes $\ell = 1, 2, 3$, (for simplicity of notation) we further denote $A^{(\ell)} = A_1$ and $S^{(\ell)} = S_1$.

Lemma 9.1 (Galerkin matrix A_G, [156]). *Assume that the basis functions $\{\overline{g}_k(x)\}$, $x \in \mathbb{R}^3$, $k = 1, \dots N_b$, are rank-1 separable, i. e., $\overline{g}_k(x) = \overline{g}_k^{(1)}(x_1)\overline{g}_k^{(2)}(x_2)\overline{g}_k^{(3)}(x_3)$. Then matrix entries of the Laplace operator A_G can be represented by*

$$\begin{aligned}\overline{a}_{km} &= \langle A_1 \mathbf{g}_k^{(1)}, \mathbf{g}_m^{(1)} \rangle \langle S_1 \mathbf{g}_k^{(2)}, \mathbf{g}_m^{(2)} \rangle \langle S_1 \mathbf{g}_k^{(3)}, \mathbf{g}_m^{(3)} \rangle \\ &\quad + \langle S_1 \mathbf{g}_k^{(1)}, \mathbf{g}_m^{(1)} \rangle \langle A_1 \mathbf{g}_k^{(2)}, \mathbf{g}_m^{(2)} \rangle \langle S_1 \mathbf{g}_k^{(3)}, \mathbf{g}_m^{(3)} \rangle \\ &\quad + \langle S_1 \mathbf{g}_k^{(1)}, \mathbf{g}_m^{(1)} \rangle \langle S_1 \mathbf{g}_k^{(2)}, \mathbf{g}_m^{(2)} \rangle \langle A_1 \mathbf{g}_k^{(3)}, \mathbf{g}_m^{(3)} \rangle \\ &= \langle A_\Delta \mathbf{G}_k, \mathbf{G}_m \rangle, \end{aligned} \tag{9.4}$$

where $\mathbf{g}_k^{(\ell)}, \mathbf{g}_m^{(\ell)} \in \mathbb{R}^N (k, m = 1, \ldots, N_b)$ *are the vectors of collocation coefficients of* $\{\overline{g}_k^{(\ell)}(x_\ell)\}$, $\ell = 1, 2, 3$, *and* $\mathbf{G}_k$ *are the corresponding 3-tensors* $\mathbf{G}_k = \mathbf{g}_k^{(1)} \otimes \mathbf{g}_k^{(2)} \otimes \mathbf{g}_k^{(3)}$ *of rank*-1.

Proof. By definition, we have

$$\overline{a}_{km} = \langle \nabla \overline{g}_k, \nabla \overline{g}_m \rangle = \langle \nabla (\overline{g}_k^{(1)} \overline{g}_k^{(2)} \overline{g}_k^{(3)}), \nabla \overline{g}_m^{(1)} \overline{g}_m^{(2)} \overline{g}_m^{(3)} \rangle.$$

Taking into account the representation (9.3), this implies

$$\begin{aligned}\overline{a}_{km} &= \langle \nabla_{(1)} \overline{g}_k^{(1)}, \nabla_{(1)} \overline{g}_m^{(1)} \rangle \langle \overline{g}_k^{(2)}, \overline{g}_m^{(2)} \rangle \langle \overline{g}_k^{(3)}, \overline{g}_m^{(3)} \rangle \\ &\quad + \langle \overline{g}_k^{(1)}, \overline{g}_m^{(1)} \rangle \langle \nabla_{(1)} \overline{g}_k^{(2)}, \nabla_{(1)} \overline{g}_m^{(2)} \rangle \langle \overline{g}_k^{(3)}, \overline{g}_m^{(3)} \rangle \\ &\quad + \langle \overline{g}_k^{(1)}, \overline{g}_m^{(1)} \rangle \langle \overline{g}_k^{(2)}, \overline{g}_m^{(2)} \rangle \langle \nabla_{(1)} \overline{g}_k^{(3)}, \nabla_{(1)} \overline{g}_m^{(3)} \rangle. \end{aligned} \tag{9.5}$$

Simple calculations show that for $\ell = 1$,

$$\begin{aligned}\langle -\Delta_{(1)} \overline{g}_k^{(1)}, \overline{g}_m^{(1)} \rangle &= \left\langle \nabla_{(1)} \sum_{i=1}^N g_{ki} \xi_i(x_1), \nabla_{(1)} \sum_{j=1}^N g_{kj} \xi_j(x_1) \right\rangle \\ &= \left\langle \sum_{i=1}^N g_{ki} \nabla_{(1)} \xi_i(x_1), \sum_{j=1}^N g_{kj} \nabla_{(1)} \xi_j(x_1) \right\rangle \\ &= \sum_{i=1}^N g_{ki} \sum_{j=1}^N g_{kj} \langle \nabla_{(1)} \xi_i(x_1), \nabla_{(1)} \xi_j(x_1) \rangle = \langle A_1 \mathbf{g}_k^{(1)}, \mathbf{g}_m^{(1)} \rangle, \end{aligned}$$

and

$$\langle \overline{g}_k^{(1)}, \overline{g}_m^{(1)} \rangle = \langle S_1 \mathbf{g}_k^{(1)}, \mathbf{g}_m^{(1)} \rangle,$$

and similarly for the remaining modes $\ell = 2, 3$. These representations imply

$$\overline{a}_{km} = \langle A_\Delta \mathbf{G}_k, \mathbf{G}_m \rangle,$$

which completes the proof. □

Remark 9.2. Agglomerating rank-1 vectors $\mathbf{G}_k \in \mathbb{R}^{N^{\otimes 3}}$ $(k = 1, \ldots, N_b)$ into a matrix $G \in \mathbb{R}^{N^{\otimes 3} \times N_b}$, the entrywise representation (9.4) can be written in a matrix form

$$A_G = G^T A_\Delta G \in \mathbb{R}^{N_b \times N_b},$$

corresponding to the standard matrix–matrix transform under the change of basis.

Lemma 9.1 implies that in case of basis functions having ranks larger than one

$$g_m(x) = \sum_{p=1}^{R_m} \eta_p(x), \quad R_m \geq 1, \tag{9.6}$$

where $\eta_p(x)$ is the rank-1 separable function, representation (9.4) takes the following form:

$$\begin{aligned}\overline{a}_{km} = \sum_{p=1}^{R_k}\sum_{q=1}^{R_m} [&\langle A_1 \mathbf{g}_{k,p}^{(1)}, \mathbf{g}_{m,q}^{(1)}\rangle \langle S_1 \mathbf{g}_{k,p}^{(2)}, \mathbf{g}_{m,q}^{(2)}\rangle \langle S_1 \mathbf{g}_{k,p}^{(3)}, \mathbf{g}_{m,q}^{(3)}\rangle \\ &+ \langle S_1 \mathbf{g}_{k,p}^{(1)}, \mathbf{g}_{m,q}^{(1)}\rangle \langle A_1 \mathbf{g}_{k,p}^{(2)}, \mathbf{g}_{m,q}^{(2)}\rangle \langle S_1 \mathbf{g}_{k,p}^{(3)}, \mathbf{g}_{m,q}^{(3)}\rangle \\ &+ \langle S_1 \mathbf{g}_{k,p}^{(1)}, \mathbf{g}_{m,q}^{(1)}\rangle \langle S_1 \mathbf{g}_{k,p}^{(2)}, \mathbf{g}_{m,q}^{(2)}\rangle \langle A_1 \mathbf{g}_{k,p}^{(3)}, \mathbf{g}_{m,q}^{(3)}\rangle],\end{aligned} \tag{9.7}$$

where R_m, $m = 1, \dots, N_b$, denote the rank parameters of the Galerkin basis functions $\overline{g}_m$.

Representation (9.4) can be simplified by the standard lumping procedure preserving the same approximation error $O(h^2)$:

$$\begin{aligned}a_{km} \mapsto \overline{a}_{km} &= \langle A_1 \mathbf{g}_k^{(1)}, \mathbf{g}_m^{(1)}\rangle \langle \mathbf{g}_k^{(2)}, \mathbf{g}_m^{(2)}\rangle \langle \mathbf{g}_k^{(3)}, \mathbf{g}_m^{(3)}\rangle \\ &\quad + \langle \mathbf{g}_k^{(1)}, \mathbf{g}_m^{(1)}\rangle \langle A_1 \mathbf{g}_k^{(2)}, \mathbf{g}_m^{(2)}\rangle \langle \mathbf{g}_k^{(3)}, \mathbf{g}_m^{(3)}\rangle \\ &\quad + \langle \mathbf{g}_k^{(1)}, \mathbf{g}_m^{(1)}\rangle \langle \mathbf{g}_k^{(2)}, \mathbf{g}_m^{(2)}\rangle \langle A_1 \mathbf{g}_k^{(3)}, \mathbf{g}_m^{(3)}\rangle \\ &= \langle A_{\Delta,\mathrm{FD}} \mathbf{G}_k, \mathbf{G}_m\rangle,\end{aligned}$$

where $A_{\Delta,\mathrm{FD}}$ denotes the finite difference (FD) discrete Laplacian

$$A_{\Delta,\mathrm{FD}} := \frac{1}{h}[A^{(1)} \otimes I^{(2)} \otimes I^{(3)} + I^{(1)} \otimes A^{(2)} \otimes I^{(3)} + I^{(1)} \otimes I^{(2)} \otimes A^{(3)}],$$

with $I^{(\ell)}$ being the $N \times N$ identity matrix.

It is worth noting that the extension of Lemma 9.1 to the case of d-dimensional Laplacian

$$\overline{a}_{km} = \langle A_{\Delta,d} \mathbf{G}_k, \mathbf{G}_m\rangle$$

leads to a similar d-term sum representation.

9.2 Nuclear potential operator by direct tensor summation

The method of direct tensor summation of long-range electrostatic potentials [156, 147] described below is based on the use of low-rank canonical representation to the single

Newton kernel $\mathbf{P}_R$ in the bounding box, translated and restricted according to coordinates of the nuclei in a box. The approach is applicable, for example, in tensor-based calculation of the nuclear potential operator describing the Coulombic interaction of electrons with the nuclei in a molecular system in a box or in a (cubic) unit cell. It is defined by the function $V_c(x)$ in the scaled unit cell $\Omega = [-b/2, b/2]^3$,

$$V_c(x) = \sum_{\nu=1}^{M_0} \frac{Z_\nu}{\|x - a_\nu\|}, \quad Z_\nu > 0,\ x, a_\nu \in \Omega \subset \mathbb{R}^3, \tag{9.8}$$

where M_0 is the number of nuclei in Ω, and a_ν and Z_ν represent their coordinates and charges, respectively.

We start with approximating the non-shifted 3D Newton kernel $\frac{1}{\|x\|}$ on the auxiliary extended box $\widetilde{\Omega} = [-b, b]^3$ by its projection onto the basis set $\{\psi_{\mathbf{i}}\}$ of piecewise constant functions defined on the uniform $2n \times 2n \times 2n$ tensor grid Ω_{2n} with the mesh size h described in Section 6.1. This defines the "reference" rank-R canonical tensor as above:

$$\widetilde{\mathbf{P}}_R = \sum_{q=1}^{R} \mathbf{p}_q^{(1)} \otimes \mathbf{p}_q^{(2)} \otimes \mathbf{p}_q^{(3)} \in \mathbb{R}^{2n\times 2n\times 2n}. \tag{9.9}$$

Here, we recall the grid-based approximate summation of nuclear potentials in (9.8) by using the shifted reference canonical tensor in (9.9) once precomputed on fine 3D Cartesian grid. For ease of exposition, we make the technical assumption that each nuclei coordinate a_ν is located exactly at a grid-point $a_\nu = (i_\nu h - b/2, j_\nu h - b/2, k_\nu h - b/2)$ with some $1 \le i_\nu, j_\nu, k_\nu \le n$. Our approximate numerical scheme is designed for nuclei positioned arbitrarily in the computational box, where approximation error of order $O(h)$ is controlled by choosing large enough grid size n. Indeed, 1D computational cost $O(n)$ enables usage of fine grids of size $n^3 \approx 10^{15}$, yielding mesh size $h \approx 10^{-4}$–10^{-5} Å in our MATLAB calculations (h is of the order of the atomic radii). This grid-based tensor calculation scheme for the nuclear potential operator was tested numerically in Hartree–Fock calculations [156], where it was compared with the analytical evaluation of the same operator by benchmark packages.

Let us introduce the rank-1 windowing operator

$$\mathcal{W}_\nu = \mathcal{W}_\nu^{(1)} \otimes \mathcal{W}_\nu^{(2)} \otimes \mathcal{W}_\nu^{(3)}$$

for $\nu = 1, \ldots, M_0$ by

$$\mathcal{W}_\nu \widetilde{\mathbf{P}}_R := \widetilde{\mathbf{P}}_R(i_\nu + n/2 : i_\nu + 3/2n; j_\nu + n/2 : j_\nu + 3/2n; k_\nu + n/2 : k_\nu + 3/2n) \in \mathbb{R}^{n\times n\times n}. \tag{9.10}$$

With this notation, the total electrostatic potential $V_c(x)$ in the computational box Ω is approximately represented by a direct canonical tensor sum

$$\mathbf{P}_c = \sum_{\nu=1}^{M_0} Z_\nu \mathcal{W}_\nu \widetilde{\mathbf{P}}_R$$
$$= \sum_{\nu=1}^{M_0} Z_\nu \sum_{q=1}^{R} \mathcal{W}_\nu^{(1)} \mathbf{p}_q^{(1)} \otimes \mathcal{W}_\nu^{(2)} \mathbf{p}_q^{(2)} \otimes \mathcal{W}_\nu^{(3)} \mathbf{p}_q^{(3)} \in \mathbb{R}^{n\times n\times n} \tag{9.11}$$

with the canonical rank bound

$$\operatorname{rank}(\mathbf{P}_c) \le M_0 R, \tag{9.12}$$

where every rank-R canonical tensor $\mathcal{W}_\nu \widetilde{\mathbf{P}}_R \in \mathbb{R}^{n\times n\times n}$ is thought of as a sub-tensor of the reference tensor $\widetilde{\mathbf{P}}_R \in \mathbb{R}^{2n\times 2n\times 2n}$ obtained by its shifting and restriction (windowing) onto the $n\times n\times n$ grid in the box $\Omega_n \subset \Omega_{2n}$. Here, a shift from the origin is specified according to the coordinates of the corresponding nuclei a_ν counted in the h-units.

For example, the electrostatic potential centered at the origin, i. e., with $a_\nu = 0$, corresponds to the restriction of $\widetilde{\mathbf{P}}_R$ onto the initial computational box Ω_n, i. e., restricted to the index set (assume that n is even)

$$\{[n/2 + i] \times [n/2 + j] \times [n/2 + k]\}, \quad i, j, k \in \{1, \ldots, n\}.$$

Remark 9.3. The rank estimate (9.12) for the sum of arbitrarily positioned electrostatic potentials in a box (unit cell) $R_c = \operatorname{rank}(\mathbf{P}_c) \le M_0 R$ is usually too pessimistic. Our numerical tests for moderate size molecules indicate that the rank of the $(M_0 R)$-term canonical sum in (9.11) can be reduced considerably. This rank optimization can be implemented by the multigrid version of the canonical rank-reduction algorithm, canonical-Tucker-canonical [174] (see also Section 3.3). The resultant canonical tensor will be denoted by $\widehat{\mathbf{P}}_c$.

The described grid-based representation of the exact sum of electrostatic potentials $v_c(x)$ in a form of a tensor in a canonical format enables its easy projection to some separable basis set, like GTO-type atomic orbital basis often used in quantum chemical computations. The following example illustrates calculation of the nuclear potential operator matrix in tensor format for molecules [156, 147]. We show that the projection of a sum of electrostatic potentials of atoms onto a given set of basis functions is reduced to a combination of 1D Hadamard and scalar products.

Let us consider tensor-structured calculation of the nuclear potential operator (9.8) in a molecule [156, 147]. Given the set of continuous basis functions

$$\{g_\mu(x)\}, \quad \mu = 1, \ldots, N_b, \tag{9.13}$$

each of them can be discretized by a third-order tensor

$$\mathbf{G}_\mu = [g_\mu(x_1(i), x_2(j), x_3(k))]_{i,j,k=1}^{n} \in \mathbb{R}^{n\times n\times n}$$

obtained by sampling of $g_\mu(x)$ at the midpoints $(x_1(i), x_2(j), x_3(k))$ of the grid-cells indexed by (i, j, k). Suppose, for simplicity, that it is a rank-1 canonical tensor $\operatorname{rank}(\mathbf{G}_\mu) = 1$, i. e.

$$\mathbf{G}_\mu = \mathbf{g}_\mu^{(1)} \otimes \mathbf{g}_\mu^{(2)} \otimes \mathbf{g}_\mu^{(3)} \in \mathbb{R}^{n\times n\times n}$$

with the canonical vectors $\mathbf{g}_\mu^{(\ell)} \in \mathbb{R}^n$ associated with modes $\ell = 1, 2, 3$.

The sum of potentials in a box $V_c(x)$ (9.8) is represented in the given basis set (9.13) by a matrix $V_g = [v_{km}] \in \mathbb{R}^{N_b\times N_b}$. The entries of the nuclear potential operator matrix are calculated (approximated) by the simple tensor operation (see [156, 147])

$$v_{km} = \int_{\mathbb{R}^3} V_c(x) g_k(x) g_m(x) dx \approx \overline{v}_{km} := \langle \mathbf{G}_k \odot \mathbf{G}_m, \mathbf{P}_c \rangle, \quad 1 \le k, m \le N_b. \tag{9.14}$$

We further denote $V_G = [\overline{v}_{km}]$. Here $\mathbf{P}_c$ is the sum of shifted/windowed canonical tensors (9.11) representing the total electrostatic potential of atoms in a molecule. Recall that

$$\mathbf{G}_k \odot \mathbf{G}_m := (\mathbf{g}_k^{(1)} \odot \mathbf{g}_m^{(1)}) \otimes (\mathbf{g}_k^{(2)} \odot \mathbf{g}_m^{(2)}) \otimes (\mathbf{g}_k^{(3)} \odot \mathbf{g}_m^{(3)})$$

denotes the Hadamard (entrywise) product of tensors representing the basis functions (9.13), which is reduced to 1D products. The scalar product $\langle \cdot, \cdot \rangle$ in (9.14) is also reduced to 1D scalar products due to separation of variables.

We notice that the approximation error $\varepsilon > 0$ caused by a separable representation of the nuclear potential is controlled by the rank parameter $R_c = \operatorname{rank}(\mathbf{P}_c) \approx CR$, where C weakly depends on the number of nuclei M_0. Now letting $\operatorname{rank}(\mathbf{G}_m) = 1$ implies that each matrix element is to be computed with linear complexity in n, $O(Rn)$. The exponential convergence of the canonical approximation in the rank parameter R allows us the optimal choice $R = O(|\log \varepsilon|)$ adjusting the overall complexity bound $O(|\log \varepsilon| n)$ almost independent on M_0.

Remark 9.4. It should be noted that since we remain in the concept of global basis functions for the Galerkin approximation to the HF eigenvalue problem, the sizes of the grids used in discretized representation of these basis functions can be different in the calculation of the kinetic and potential parts in the Fock operator. The corresponding choice is only controlled by the respective approximation error and by the numerical efficiency.

Finally, we note that the Galerkin tensor representation of the identity operator leads to the following mass matrix: $S = \{\overline{s}_{km}\}$, where

$$\overline{s}_{km} = \int_{\mathbb{R}^3} \overline{g}_k(x) \overline{g}_m(x) dx \approx \langle \mathbf{G}_k, \mathbf{G}_m \rangle, \quad 1 \le k, m \le N_b.$$

To conclude this section, we note that the error bound $\|V_g - V_G\| \le Ch^2$ can be proven along the line of the discussion in [166].

9.3 Numerical verification for the core Hamiltonian

First, following [156] we consider the evaluation of a Galerkin matrix entry for the identity and Laplace operators, that is,

$$\langle g, g\rangle = \int_{\mathbb{R}^3} g(x)^2 dx \quad \text{and} \quad \langle -\Delta g, g\rangle = \int_{\mathbb{R}^3} \nabla g(x)\cdot \nabla g(x) dx, \quad g(x) = e^{-\alpha\|x\|^2}, \quad x \in \mathbb{R}^3,$$

for a single Gaussian with sufficiently large $\alpha > 0$ and using large $N \times N \times N$ Cartesian grids. Functions are discretized with respect to the basis set (9.1) in the computational box $[-b, b]^3$ with $b = 14.6$ au ≈ 8 Å.

For a single Gaussian, we compare $\mathcal{J}_h$ computed as in Lemma 9.1 with the exact expression

$$\mathcal{J} = \int_{\mathbb{R}^3} \nabla g(x)\cdot \nabla g(x) dx = 3J_1 J_{01}^2,$$

where

$$J_1 = 4\alpha^2 \int_{-\infty}^{\infty} x^2 e^{-2\alpha x^2} dx = \sqrt{\frac{\pi}{2}}\sqrt{\alpha}, \quad J_{01} = \int_{-\infty}^{\infty} e^{-\alpha x^2} dx = \frac{\sqrt{\pi}}{\sqrt{\alpha}}.$$

Table 9.1 shows the approximation error $|\mathcal{J} - \mathcal{J}_h|$ versus the grid size, where $\mathcal{J}_h$ corresponds to the grid-based evaluation of the matrix element on the corresponding grid for $\alpha = 2500$, $4\cdot 10^4$, and $1.2\cdot 10^5$, which exceed the largest exponents α in the conventional Gaussian sets for hydrogen ($\alpha = 1777$), carbon ($\alpha = 6665$), oxygen ($\alpha = 11\,720$), and mercury ($\alpha = 10^5$) atoms.

Computations confirm the results of Theorem A4 in [156] on the error bound $O(h^2)$. It can be seen that the errors reduce by a distinct factor of 4 for the diadically refined

Table 9.1: Approximation error $|\mathcal{J} - \mathcal{J}_h|$ for the grid-based evaluation of the Laplacian Galerkin matrix entry for a Gaussian $g(x) = e^{-\alpha\|x\|^2}$, $x \in \mathbb{R}^3$, $N = 2^p - 1$.

p	N^3	$\alpha = 2.5\cdot 10^3$		$\alpha = 4\cdot 10^4$		$\alpha = 1.2\cdot 10^5$	
		$\mathcal{J} - \mathcal{J}_h$	RE	$\mathcal{J} - \mathcal{J}_h$	RE	$\mathcal{J} - \mathcal{J}_h$	RE
12	4095^3	0.0037	–	0.0058	–	0.025	–
13	8191^3	$9.3\cdot 10^{-4}$	$1.0\cdot 10^{-5}$	0.0034	0.0026	$2.4\cdot 10^{-5}$	–
14	$16\,383^3$	$2.3\cdot 10^{-4}$	$1.2\cdot 10^{-6}$	$9.1\cdot 10^{-4}$	$9.1\cdot 10^{-5}$	0.0015	–
15	$32\,767^3$	$5.8\cdot 10^{-5}$	$7.6\cdot 10^{-8}$	$2.3\cdot 10^{-4}$	$4.8\cdot 10^{-6}$	$4.03\cdot 10^{-4}$	$3.8\cdot 10^{-5}$
16	$65\,535^3$	$1.4\cdot 10^{-5}$	$4.7\cdot 10^{-9}$	$5.8\cdot 10^{-5}$	$3.0\cdot 10^{-7}$	$1.0\cdot 10^{-4}$	$1.6\cdot 10^{-6}$
17	$131\,071^3$	$3.6\cdot 10^{-5}$	$2.4\cdot 10^{-10}$	$1.5\cdot 10^{-5}$	$1.9\cdot 10^{-8}$	$5.5\cdot 10^{-5}$	$1.0\cdot 10^{-7}$
18	$262\,143^3$	$9.1\cdot 10^{-7}$	$3.1\cdot 10^{-11}$	$3.6\cdot 10^{-6}$	$1.2\cdot 10^{-9}$	$6.4\cdot 10^{-6}$	$6.5\cdot 10^{-9}$
19	$524\,287^3$	$2.2\cdot 10^{-7}$	$5.4\cdot 10^{-13}$	$9.1\cdot 10^{-7}$	$7.3\cdot 10^{-11}$	$1.6\cdot 10^{-6}$	$4.0\cdot 10^{-10}$

grids. Therefore, in spite of sharp "needles" of Gaussians due to large α, the Richardson extrapolation [218] (RE column) on a sequence of large grids provides a higher accuracy of order $O(h^3)$–$O(h^4)$.

In Table 9.1, the largest grid size $N = 2^{19} - 1$ corresponds to the computational box $\Omega \in \mathbb{R}^3$ with the huge number of entries of order $2^{57} \approx 10^{17}$. The corresponding mesh size is of order $h \sim 10^{-5}$ Å. Computing times in MATLAB range from several milliseconds up to 1.2 sec for the largest grid.

Notice that the integral $\langle g, g\rangle = \int_{\mathbb{R}^3} e^{-2\alpha\|x\|^2} dx = J_{01}^3(\alpha)$ involved in the calculation of the mass-matrix S_g is approximated with the same accuracy.

In the following, we consider an example on the grid-based approximation to the Schrödinger equation for the hydrogen atom (see [156]), that is, we verify the proposed algorithms for the Hartree–Fock equation in the simplest case of the hydrogen atom

$$\mathcal{H}\psi = \lambda\psi, \quad \mathcal{H} = -\frac{1}{2}\Delta + \frac{1}{\|x\|}, \qquad x \in \mathbb{R}^3, \tag{9.15}$$

which has the exact solution $\psi = e^{-\|x\|}/\sqrt{\pi}$, $\lambda = -1/2$.

Example 9.1. Consider the traditional expansion of the solution using the ten s-type primitive Gaussian functions from the cc-pV6Z basis set [234, 265]

$$\psi(x) \approx \sum_{k=1}^{N_b} c_k \varphi_k(x), \quad N_b = 10, \; x \in \mathbb{R}^3,$$

which leads to the Galerkin equation corresponding to (9.15) with

$$F = \langle \mathcal{H}\overline{g}_k, \overline{g}_m\rangle := -\frac{1}{2}\langle \Delta\overline{g}_k, \overline{g}_m\rangle + \left\langle \frac{1}{\|x\|}\overline{g}_k, \overline{g}_m \right\rangle, \qquad k, m = 1, \ldots N_b,$$

with respect to the Galerkin basis $\{\overline{g}_k\}$. We choose the appropriate size of the computational box as $b \approx 8$ Å and discretize $\{\overline{g}_k\}$ using $N \times N \times N$ Cartesian grid, obtaining the canonical rank-1 tensor representation $\mathbf{G}_k$ of the basis functions. Then, the kinetic energy and the nuclear potential parts of the Fock operator are computed by (9.4) and (9.14).

Table 9.2, line (1), presents numerical errors in energy $|\lambda - \lambda_h|$ of the grid-based calculations using the cc-pV6Z basis set of $N_b = 10$ Gaussians generated by MOLPRO [299], providing an accuracy of order $\sim 10^{-6}$. Notice that this accuracy is achieved already at the grid-size $N = 8192$, hence, further grid refinement does not improve the results.

Example 9.2. Here, we study the effect of basis optimization by adding an auxiliary basis function to the Gaussian basis set from the previous example, thus increasing the number of basis functions to $N_b = 11$. The second line (2) in Table 9.2 shows improvement of accuracy for the basis augmented by a rank-1 approximation to the

Table 9.2: Examples 9.1–9.3 for hydrogen atom: $|\lambda - \lambda_h|$ vs. grid size N^3 for (1) the discretized basis of $N_b = 10$ Gaussians, (2) 11 basis functions consisting of Gaussians augmented by a rank-1 function φ_0, (3) discretized single rank-R_b Slater function.

N^3	1024^3	2048^3	4096^3	8192^3	$16\,384^3$	$32\,768^3$
(1) $\lvert\lambda - \lambda_h\rvert$	$4.1 \cdot 10^{-4}$	$1.0 \cdot 10^{-4}$	$2.7 \cdot 10^{-5}$	$7.5 \cdot 10^{-6}$	$2.4 \cdot 10^{-6}$	$1.0 \cdot 10^{-6}$
(2) $\lvert\lambda - \lambda_h\rvert$	$1.5 \cdot 10^{-5}$	$7.2 \cdot 10^{-6}$	$2.7 \cdot 10^{-6}$	$1.1 \cdot 10^{-6}$	$8.0 \cdot 10^{-7}$	$7.8 \cdot 10^{-7}$
(3) $\lvert\lambda - \lambda_h\rvert$	$1.0 \cdot 10^{-4}$	$2.7 \cdot 10^{-5}$	$6.8 \cdot 10^{-6}$	$1.7 \cdot 10^{-6}$	$4.3 \cdot 10^{-7}$	–

Slater function given by the grid representation of $\varphi_0 = e^{-(|x_1|+|x_2|+|x_3|)}$. Augmenting by a piecewise linear hat function of the type ξ_i centered at the origin gives similar results as for φ_0.

Example 9.3. In this example, we present computations with the controlled accuracy using a single rank-R_b basis function generated by the sinc-approximation to the Slater function. Using the Laplace transform

$$G(\rho) = e^{-2\sqrt{\alpha\rho}} = \frac{\sqrt{\alpha}}{\sqrt{\pi}} \int_0^\infty \tau^{-3/2} \exp(-\alpha/\tau - \rho\tau) d\tau,$$

the Slater function can be represented as a rank-R canonical tensor by computing the sinc-quadrature decomposition [161, 163] and setting $\rho = x_1^2 + x_2^2 + x_3^2$:

$$G(\rho) \approx \frac{\sqrt{\alpha}}{\sqrt{\pi}} \sum_{k=-L}^{L} w_k \tau_k^{-3/2} \exp(-\alpha/\tau_k) \prod_{\ell=1}^{3} \exp(-\tau_k x_\ell^2),$$

where $\tau_k = e^{kh_L}$, $w_k = h_L \tau_k$, $h_L = C_0 \log L/L$. The accuracy of approximation is controlled by choosing the number of quadrature points L. In this example, we have only one basis function in a set, an approximate Slater function, but represented by the canonical tensor of rank $R_b \leq 2L + 1$. Thus, each of the matrices A_G computed by (9.7) and V_G is of size 1×1. Table 9.2 (3) shows accuracy of the solution to the Hartree–Fock equation for the hydrogen atom using one approximate Slater basis function.

Table 6.3 in [156] presents the Richardson extrapolation for Examples 9.1 and 9.3. Due to noticeable convergence rate of order $O(h^2)$, the Richardson extrapolation (RE) gives further improvement of the accuracy up to $O(h^3)$. It can be seen in Table 6.3, [156], that the Richardson extrapolation for the results of Example 9.3 gives accuracy of order 10^{-7}, beginning from the grid size 4096. Note that with the choice $L = 60$, the accuracy is improved on one order of magnitude compared to those obtained for the standard Gaussian basis set in Example 9.1.

Table 9.3 presents numerical examples of the grid-based approximation to the Galerkin matrices for the Laplace operator A_G and nuclear potential V_G using (9.4) and (9.14) for C_2H_5OH molecule. The mesh size of the $N \times N \times N$ Cartesian grid ranges from $h = 0.0036$ au (atomic units) corresponding to $N = 8192$ up to $h = 2.2 \cdot 10^{-4}$ au for $N = 131\,072$.

Throughout the tables we show the relative Frobenius norms of the differences $\mathrm{Er}(A_G)$ and $\mathrm{Er}(V_G)$ in the corresponding Galerkin matrix elements for the Laplace and nuclear potential operators, respectively, where

$$\mathrm{Er}(A_G) = \frac{\|A_g - A_G\|}{\|A_g\|}, \quad \mathrm{Er}(V_G) = \frac{\|V_g - V_G\|}{\|V_g\|}.$$

The quadratic convergence of both quantities along the line of dyadic grid refinement is in good agreement with the theoretical error estimates $O(h^2)$. Therefore, the employment of the Richardson approximation providing the error

$$E_{\mathrm{Ri},2h,h} = \mathrm{Er}\left(\frac{4 \cdot V_{G,h} - V_{G,2h}}{3}\right)$$

suggests further improvement of the accuracy up to order $O(h^4)$ for the Laplace operator. The "RE" lines in Table 9.3 demonstrate the results of the Richardson extrapolation applied to corresponding quantities at the adjacent grids.

Note that for the grid-based representation of the collective nuclear potential $\mathbf{P}_c$, the univariate grid size n can be noticeably smaller than the size of the grid used for the piecewise linear discretization for the Laplace operator.

Table 9.3: Ethanol (C_2H_5OH): accuracy $\mathrm{Er}(A_G)$ and $\mathrm{Er}(V_G)$ of the Galerkin matrices A_G and V_G corresponding to the Laplace and the nuclear potential operators, respectively, using the discretized basis of 123 primitive Gaussians (from the cc-pVDZ set [75, 265]).

p	13	14	15	16	17
$N^3 = 2^{3p}$	8192^3	$16\,384^3$	$32\,768^3$	$65\,536^3$	$131\,072^3$
$\mathrm{Er}(A_G)$	0.032	0.0083	0.0021	$5.2 \cdot 10^{-4}$	$1.3 \cdot 10^{-4}$
RE	–	$4.0 \cdot 10^{-4}$	$3.3 \cdot 10^{-5}$	$6.0 \cdot 10^{-6}$	$5.0 \cdot 10^{-8}$
$\mathrm{Er}(V_G)$	0.024	0.0083	0.0011	$3.1 \cdot 10^{-4}$	
RE	–	0.0031	0.0013	$5.9 \cdot 10^{-5}$	

Figure 9.2 displays the nuclear potential for the molecule C_2H_5OH (ethanol) computed in a box $[-b, b]^3$ with $b = 16$ au. We show two cross-sections of the 3D function at the level $x = 0.0625$ au and of the permuted function at the level $y = -0.3125$ au. It can be seen from the left figure that three non-hydrogen atoms with the largest charges (two Carbon atoms with $Z = 6$ and one Oxygen atom with $Z = 8$) are placed on the plane $x = 0$. The right figure shows the location close to one of Hydrogen atoms.

The error $\varepsilon > 0$ arising due to the separable approximation of the nuclear potential is controlled by the rank parameter of the nuclear potential $R_P = \mathrm{rank}(\mathbf{P}_c)$. Now letting $\mathrm{rank}(\mathbf{G}_m) = R_m$ implies that each matrix element is to be computed with

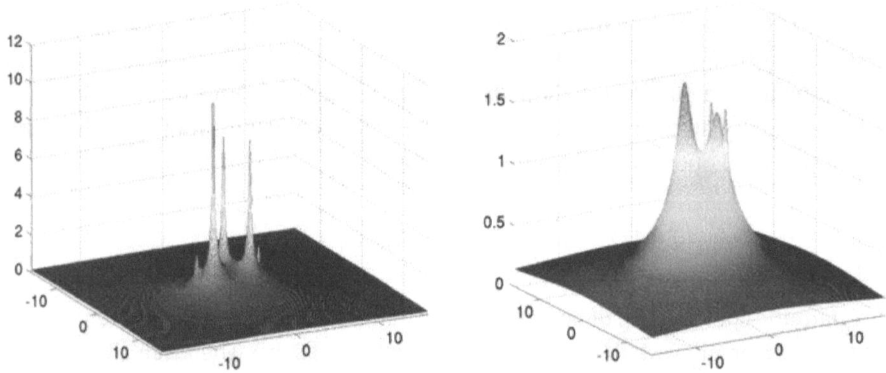

Figure 9.2: Nuclear potential $\mathbf{P}_c$ for the C_2H_5OH molecule, shown for the cross sections along x-axis at the level $x = 0.0625$ au and along y-axis at level $y = 1.6$ au.

linear complexity in n, $O(R_k R_m R_P n)$. The almost exponential convergence of the rank approximation in R_P allows us the choice $R_P = O(|\log \varepsilon|)$.

The maximum computational time for A_G with $N^3 = 131\,072^3$ is of the order of hundred seconds in MATLAB. For the coarser grid with $N^3 = 8192^3$, CPU times are in the range of several seconds for both A_G and V_G.

Comprehensive error estimates for the grid-based calculations of the core Hamiltonian are formulated in [156], where a number of numerical experiments for various molecules is presented as well.

10 Tensor factorization of grid-based two-electron integrals

10.1 General introduction

The efficient tensor-structured method for the grid-based calculation of the *two-electron integrals* (TEI) tensor was introduced by V. Khoromskaia, B. Khoromskij, and R. Schneider in 2012 (see [157]). In this chapter, following [157, 150], we describe the fast algorithm for the grid-based computation of the fourth-order TEI tensor in a form of the Cholesky factorization by using the grid-based algebraic 1D "density fitting" scheme, which applies to the products of basis functions. It is worth to note, that the described approach does not require calculation of the full TEI matrix, but only relies on computation of its few selected columns evaluated by using 1D density fitting factorizations (see Remark 10.3).

Imposing the low-rank tensor representation of the product basis functions and the Newton convolving kernel, all discretized on large $n \times n \times n$ Cartesian grid, the 3D integral transforms are calculated in $O(n \log n)$ complexity. This scheme provides the storage for TEI of the order of $O(N_b^3)$ in the number of basis functions N_b.

The TEI tensor, also known as the Fock integrals or electron repulsion integrals, is the principal ingredient in electronic and molecular structure calculations. In particular, the corresponding coefficient tensor arises in ab initio Hartree–Fock (HF) calculations, in post Hartree–Fock models (MP2, CCSD, Jastrow factors, etc.), and in the core Hamiltonian appearing in FCI-DMRG calculations [6, 298, 241, 128].

Given the finite basis set $\{g_\mu\}_{1\le\mu\le N_b}$, $g_\mu \in H^1(\mathbb{R}^3)$, the associated fourth-order two-electron integrals tensor $\mathbf{B} = [b_{\mu\nu\kappa\lambda}] \in \mathbb{R}^{N_b\times N_b\times N_b\times N_b}$ is defined entrywise by

$$b_{\mu\nu\kappa\lambda} = \int_{\mathbb{R}^3}\int_{\mathbb{R}^3} \frac{g_\mu(x)g_\nu(x)g_\kappa(y)g_\lambda(y)}{\|x-y\|}dxdy, \quad \mu,\nu,\kappa,\lambda \in \{1,\dots,N_b\} =: I_b. \tag{10.1}$$

The fast and accurate evaluation and effective storage of the fourth-order TEI tensor $\mathbf{B}$ of size N_b^4 is the challenging computational problem since it includes multiple 3D convolutions of the Newton kernel $1/\|x-y\|$, $x, y \in \mathbb{R}^3$, with strongly varying product-basis functions. Hence, in the limit of large N_b, the efficient numerical treatment and storage of the TEI tensor is considered as one of the central tasks in electronic structure calculations [247].

The traditional analytical integration using the representation of electronic orbitals in a Gaussian-type basis is the basement of most ab initio quantum chemical packages. Hence, the choice of a basis set $\{g_\mu\}_{1\le\mu\le N_b}$ is essentially restricted by the "analytic" integrability for efficient computations of the tensor entries represented by 6D integrals in (10.1). This approach possesses intrinsic limitations concerning the non-alternative constraint to the Gaussian-type basis functions, which may become

https://doi.org/10.1515/9783110365832-010

unstable and redundant for higher accuracy, larger molecules, or when considering heavy nuclei.

It is known in quantum chemistry simulations [17, 298, 303] that, in the case of compact molecules, the (pivoted) incomplete Cholesky factorization of the $N_b^2 \times N_b^2$ TEI matrix unfolding

$$B = [b_{\mu\nu;\kappa\lambda}] := \mathrm{mat}(\mathbf{B}) \quad \text{over } (I_b \otimes I_b) \times (I_b \otimes I_b) \tag{10.2}$$

reduces the asymptotic storage of the resultant low-rank approximation to $O(N_b^3)$. It was observed in numerical experiments that the particular rank-bound in the Cholesky decomposition scales linearly in N_b, depending mildly (e. g., logarithmically) on the error in the rank truncation. We refer to [130, 99, 121, 15, 286, 255] for more detail on the algebraic aspects of matrix Cholesky decomposition and the related ACA techniques. The Cholesky decomposition is applicable since the TEI matrix B is, indeed, the symmetric Gramm matrix of the product basis set $\{g_\mu g_\nu\}$ in the Coulomb metric $\langle \cdot, \frac{1}{\|x-y\|} \cdot \rangle$, ensuring its positive semidefiniteness. In some cases it is possible to reduce the storage, even to $O(N_b^2 \log N_b)$, taking into account the pointwise sparsity of matrix B in calculation of rather large extended systems [298].

For the Cholesky decomposition of a matrix B, we constructed in [157] the algebraically optimized redundancy-free factorization to the TEI matrix B, based on the reduced higher-order SVD [174], to obtain the low-rank separable representation of the discretized basis functions $\{g_\mu g_\nu\}$. Numerical experiments show that this minimizes the dimension of dominating subspace in $\mathrm{span}\{g_\mu g_\nu\}$ to $R_G \le N_b$, which allows one to reduce the number of 3D convolutions (by the order of magnitude) from $O(N_b^2)$ to R_G. Combined with the quantized-canonical tensor decompositions of long spatial n-vectors, this leads to the logarithmic scaling in n for storage $O(R_G \log n + N_b^2 R_G)$. An essential compression rate via the QTT approximation is observed in numerical experiments even for compact molecules, becoming stronger for more stretched compounds.

Computation of the rank-R_B, Cholesky decomposition employs only $R_B = O(N_b)$ selected columns in the TEI matrix B, calculated from precomputed factorizations of this matrix. We show by numerical experiments that each long N_b^2-vector of the L-factor in the Cholesky LL^T-decomposition can be further compressed using the quantized-TT (QTT) approximation, reducing the total storage from $O(N_b^3)$ to $O(N_b N_{\mathrm{orb}}^2)$, where the number of electron orbitals N_{orb} usually satisfies $N_b \sim 10 N_{\mathrm{orb}}$.

The presented grid-based approach benefits from the fast $O(n \log n)$ tensor-product convolution with the 6D Newton kernel over a large $n^3 \times n^3$ grid [166], which has already proved the numerical efficiency in the evaluation of the Hartree and exchange integrals [174, 145, 187]. However, in these papers, both the Coulomb and exchange operators are calculated directly on the fly at each DIIS iteration, and thus the use of TEI was avoided at the expense of time loops.

Recall that the beneficial feature of the grid-based tensor-structured methods is that it substitutes the 3D numerical integration by multilinear algebraic procedures like the scalar, Hadamard, and convolution products with linear 1D complexity $O(n)$. On the one hand, this weak dependence on the grid-size is the ultimate payoff for generality, in the sense that rather general approximating basis sets may be equally used instead of analytically integrable Gaussians. On the other hand, the approach also serves for structural simplicity of implementation, since the topology of the molecule is caught without any physical insight, and only by the algebraically determined rank parameters of the fully grid-based numerical scheme.

Due to $O(n \log n)$ complexity of the algorithms, there are rather weak practical restrictions on the grid-size n, allowing calculations on really large $n \times n \times n$ 3D Cartesian grids in the range $n \sim 10^3$–10^5, thereby avoiding grid refinement. The corresponding mesh sizes enable high resolution of the order of the size of atomic nuclei. For storage consuming operations, the numerical expense can be reduced to logarithmic level $O(\log n)$ by using the QTT representation of the discretized $3D$ basis functions and their convolutions.

In [157] it is shown that the rank-$O(N_b)$ Cholesky decomposition of the TEI matrix B, combined with the canonical-QTT data compression of long vectors, allows the reduction of the asymptotic complexity of grid-based tensor calculations in HF and some post-HF models. Alternative approaches to optimization of the HF, MPx, CCSD, and other post-HF models can be based on using physical insight to sparsify the TEI tensor $\mathbf{B}$ by zeroing-out all "small" elements [298, 241, 6, 268, 311].

10.2 Grid-based tensor representation of TEI in the full product basis

We assume that all basis functions $\{g_\mu\}_{1\le\mu\le N_b}$ have a support in a finite box $\Omega = [-b,b]^3 \subset \mathbb{R}^3$, and, for ease of presentation, we consider the case with $\operatorname{rank}(g_\mu) = 1$ (see Remark 10.2). The size of the computational box is chosen in such a way that the truncated part of the most slowly decaying basis function does not exceed the given tolerance $\varepsilon > 0$. Taking into account the exponential decay in molecular orbitals, the parameter $b > 0$ is chosen to be only few times larger than the molecular size.

Introduce the uniform $n \times n \times n$ rectangular grid in $[-b,b]^3$. Then each basis function $g_\mu(x)$ can be discretized by a three-dimensional tensor

$$\mathbf{G}_\mu = [g_\mu(x_1(i), x_2(j), x_3(k))]_{i,j,k=1}^{n} \in \mathbb{R}^{n\times n\times n}, \quad \mu = 1,\ldots,N_b,$$

obtained by sampling of $g_\mu(x)$ over the midpoints $(x_1(i), x_2(j), x_3(k))$ of the grid-cells with index (i,j,k). Given the discretized basis function $\mathbf{G}_\mu$, $(\mu = 1,\ldots,N_b)$, we assume (without loss of generality) that it is a rank-1 tensor, $\operatorname{rank}(\mathbf{G}_\mu) = 1$, i. e.,

$$\mathbf{G}_\mu = \mathbf{g}_\mu^{(1)} \otimes \mathbf{g}_\mu^{(2)} \otimes \mathbf{g}_\mu^{(3)} \in \mathbb{R}^{n\times n\times n} \tag{10.3}$$

with the skeleton vectors $\mathbf{g}_\mu^{(\ell)} \in \mathbb{R}^n$, $\ell = 1,2,3$, obtained as projections of the basis functions $g_\mu(x)$ on the uniform grid. Then the entries of $\mathbf{B}$ can be represented by using the tensor scalar product over the "grid" indices

$$b_{\mu\nu\kappa\lambda} = \langle \mathbf{G}_{\mu\nu}, \mathbf{H}_{\kappa\lambda}\rangle_{n^{\otimes 3}}, \tag{10.4}$$

where

$$\mathbf{G}_{\mu\nu} = \mathbf{G}_\mu \odot \mathbf{G}_\nu \in \mathbb{R}^{n^{\otimes 3}}, \quad \mathbf{H}_{\kappa\lambda} = \mathbf{P}_N * \mathbf{G}_{\kappa\lambda} \in \mathbb{R}^{n^{\otimes 3}}, \tag{10.5}$$

$\mu,\nu,\kappa,\lambda \in \{1,\dots,N_b\}$, with the rank-$R_N$ canonical tensor $\mathbf{P}_N \in \mathbb{R}^{n^{\otimes 3}}$ approximating the Newton potential $\frac{1}{\|x\|}$ (see Section 6.1). We recall that $*$ stands for the 3D tensor convolution (5.11) and $\odot$ denotes the 3D Hadamard product (2.37).

The element-wise accuracy of the tensor representation (10.4) is estimated by $O(h^2)$, where $h = 2b/n$ is the step-size of the Cartesian grid [166]. The Richardson extrapolation reduces the error to $O(h^3)$.

It is worth to emphasize that in our scheme the $n^{\otimes 3}$ tensor Cartesian grid does not depend on the positions of nuclei in a molecule. Consequently, the simultaneous rotation and translation of the nuclei positions still preserve the asymptotic approximation error on the level of $O(h^2)$.

Remark 10.1. The TEI tensor $\mathbf{B}$ has multiple symmetries

$$b_{\mu\nu\kappa\lambda} = b_{\nu\mu\kappa\lambda} = b_{\mu\nu\lambda\kappa} = b_{\kappa\lambda\mu\nu}, \quad \mu,\nu,\kappa,\lambda \in \{1,\dots,N_b\}.$$

The result is a direct consequence of definition (10.1) and symmetry of the convolution product. The above symmetry relation allows reducing the number of precomputed entries in the full TEI tensor to $N_b^4/8$. This property is also mentioned in [291].

Let us introduce the 5th-order tensors

$$\mathbf{G} = [\mathbf{G}_{\mu\nu}] \in \mathbb{R}^{N_b\times N_b\times n^{\otimes 3}} \quad \text{and} \quad \mathbf{H} = [\mathbf{H}_{\kappa\lambda}] \in \mathbb{R}^{N_b\times N_b\times n^{\otimes 3}}.$$

Then (10.4) is equivalent to the contracted product representation over $n^{\otimes 3}$-grid indexes

$$\mathbf{B} = \mathbf{G} \times_{n^{\otimes 3}} (\mathbf{P}_N *_{n^{\otimes 3}} \mathbf{G}) = \langle \mathbf{G}, \mathbf{P}_N *_{n^{\otimes 3}} \mathbf{G}\rangle_{n^{\otimes 3}} = \langle \mathbf{G}, \mathbf{H}\rangle_{n^{\otimes 3}}, \tag{10.6}$$

where the right-hand part is recognized as the discrete counterpart of the Galerkin representation (10.1) in the full product basis. When using the full grid calculations, the total storage cost for the $n \times n \times n$ product-basis tensor $\mathbf{G}$ and its convolution $\mathbf{H}$ amounts to $3\frac{N_b(N_b+1)}{2}n$ and $3R_N\frac{N_b(N_b+1)}{2}n$, respectively. The numerical cost of N_b^2 tensor-product convolutions to compute $\mathbf{H}$ is estimated by $O(R_N N_b^2 n \log n)$ [166]. Based on representation (10.6), each entry in the TEI tensor $\mathbf{B}$ of size N_b^4 can be calculated with the cost $O(R_N n)$, which might be too expensive for the large grid-size n. Thus a direct tensor calculation of TEI seems to be unfeasible except for small molecules, even when using the QTT tensor representation of the basis functions, as it was shown in [157].

Remark 10.2. If the separation rank of a basis set is larger than 1, then the complexity of scalar products in (10.6) increases quadratically in the rank parameter. However, the use of basis functions with the greater than one rank parameter (say, Slater-type functions) can be motivated by the reduction of the basis size N_b, which has a fourth-order effect on the complexity.

10.3 Redundancy-free factorization of the TEI matrix B

The efficient solution of the TEI problem introduced in [157] is based on the construction of the redundancy-free modified product basis by an algebraic "1D density fitting" and consequent Cholesky factorization to B. This approach minimizes the number of required convolution products in (10.6) by using the reduced HOSVD (RHOSVD), introduced in [174], for tensor-rank optimization in the canonical format. The RHOSVD-type factorization applied to the 3D canonical tensor $\mathbf{G}$ allows us to represent it in a "squeezed" form, in an optimized basis, obtained in a "black box" algebraic way.

10.3.1 Grid-based 1D density fitting scheme

For every space variable $\ell = 1, 2, 3$, we construct the side matrices corresponding to products of basis functions

$$G^{(\ell)} = [\mathbf{g}_\mu^{(\ell)} \odot \mathbf{g}_\nu^{(\ell)}]_{1 \le \mu,\nu \le N_b} \in \mathbb{R}^{n \times N_b^2}, \quad \mathbf{g}_\mu^{(\ell)}, \mathbf{g}_\nu^{(\ell)} \in \mathbb{R}^n, \tag{10.7}$$

which are associated with a product-basis tensor

$$\mathbf{G} = [\mathbf{G}_{\mu\nu}] := [\mathbf{G}_\mu \odot \mathbf{G}_\nu]_{1 \le \mu,\nu \le N_b} \in \mathbb{R}^{n \times n \times n \times N_b^2}. \tag{10.8}$$

The matrix $G^{(\ell)}$ is composed by concatenation of Hadamard products of the skeleton vectors $\mathbf{g}_\mu^{(\ell)} \odot \mathbf{g}_\nu^{(\ell)}$ of $\mathbf{G}$ in mode ℓ.

This representation serves to minimize the large number of convolution products in (10.1), that is, $\frac{N_b(N_b+1)}{2}$. The approach in [157] is based on the truncated SVD for finding the minimal set of dominating columns in the large site matrix $G^{(\ell)}$, $\ell = 1, 2, 3$ of size $n \times N_b^2$, representing the full (and highly redundant) set of product basis functions sampled on a grid. Given a tolerance $\varepsilon > 0$, we compute the ε-truncated SVD-based left-orthogonal decomposition of $G^{(\ell)}$ (1D density fitting),

$$G^{(\ell)} \cong U^{(\ell)} V^{(\ell)^T} \quad \text{such that} \quad \|G^{(\ell)} - U^{(\ell)} V^{(\ell)^T}\|_F \le \varepsilon, \quad \ell = 1, 2, 3, \tag{10.9}$$

with an orthogonal matrix $U^{(\ell)} \in \mathbb{R}^{n \times R_\ell}$ and a matrix $V^{(\ell)} \in \mathbb{R}^{N_b^2 \times R_\ell}$, where R_ℓ is the corresponding matrix ε-rank. Here, $U^{(\ell)}$, $V^{(\ell)}$ represent the so-called left and right redundancy-free basis sets, where only the grid-depending part $U^{(\ell)}$ is to be used in the convolution products.

Since the direct SVD of large rectangular matrices $G^{(\ell)} \in \mathbb{R}^{n\times N_b^2}$ can be prohibitively expensive, even for the moderate size molecules ($n \geq 2^{13}$, $N_b \geq 200$), the five-step algorithm was introduced in [157, 150], which reduces computational and storage costs to compute the low-rank approximation $G^{(\ell)} \cong U^{(\ell)} V^{(\ell)^T}$ with the guaranteed tolerance $\varepsilon > 0$, see Algorithm 1.

Algorithm 1 Fast low-rank ε-approximation of $G^{(\ell)}$.

Input: rectangular matrices $G^{(\ell)} \in \mathbb{R}^{n\times N_b^2}$, $\ell = 1, 2, 3$, tolerance $\varepsilon > 0$.
(1) Find the factor $\widetilde{U}^{(\ell)} \in \mathbb{R}^{n\times \widetilde{R}_\ell}$ of the truncated Cholesky decomposition to the Gram-matrix $G^{(\ell)} G^{(\ell)^T} \approx \widetilde{U}^{(\ell)} (\widetilde{U}^{(\ell)})^T$ by ε-thresholding the diagonal elements.
(2) Orthogonalize the column space of $\widetilde{U}^{(\ell)}$ by QR decomposition $\widetilde{U}^{(\ell)} := U^{(\ell)} R_U$.
(3) Project the initial matrix onto $U^{(\ell)}$, $\widetilde{V}^{(\ell)} := G^{(\ell)^T} U^{(\ell)}$ (can be executed in the data-sparse formats, e. g., in QTT).
(4) QR decomposition $\widetilde{V}^{(\ell)} := V^{(\ell)} R_V$ to obtain the orthogonal Q-factor $V^{(\ell)}$.
(5) Rank reduction ($\widetilde{R}_\ell$ to R_ℓ) by SVD of $R_V \in \mathbb{R}^{\widetilde{R}_\ell\times \widetilde{R}_\ell}$; update $U^{(\ell)}$ and $V^{(\ell)}$.
Output: Rank-R_ℓ decomposition of $G^{(\ell)} \approx U^{(\ell)} V^{(\ell)^T}$ with the orthogonal matrix $U^{(\ell)}$.

Numerical experiments show that the Frobenius error of these rank decompositions decays exponentially in the rank parameter R_ℓ:

$$\|G^{(\ell)} - U^{(\ell)} V^{(\ell)^T}\|_F \leq Ce^{-\gamma_\ell R_\ell}, \quad \ell = 1, 2, 3, \; \gamma_\ell > 0.$$

Figure 10.1 illustrates the exponential decay in singular values of $G^{(\ell)}$ for several moderate size molecules.

Step (3) in Algorithm 1 requires an access to the full matrix $G^{(\ell)}$. However, when this matrix allows data-sparse representation, the respective matrix–vector multipli-

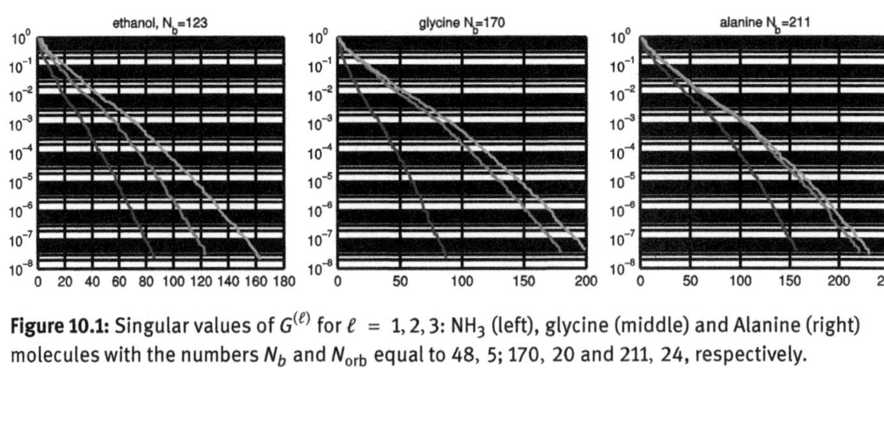

Figure 10.1: Singular values of $G^{(\ell)}$ for $\ell = 1, 2, 3$: NH_3 (left), glycine (middle) and Alanine (right) molecules with the numbers N_b and N_{orb} equal to 48, 5; 170, 20 and 211, 24, respectively.

cations can be implemented with reduced cost. For example, given the low-rank QTT representation of the column vectors in $G^{(\ell)}$, the matrix–matrix product at Step (3) can be implemented in $O(N_b^2 R_\ell \log n)$ operations. Notice that the QTT ranks of the column vectors are estimated in numerical experiments by $O(1)$ for all molecular systems considered so far, see also [68] concerning the QTT rank estimate of the Gaussian.

Another advantageous feature is due to a perfect parallel structure of the matrix–vector multiplication procedure at Step (3). Here, the algebraically optimized separation ranks R_ℓ are mostly determined by the geometry of a molecule, whereas the number $N_b^2 - R_\ell$ indicates the measure of redundancy in the product basis set. In numerical experiments we observe $R_\ell \leq N_b$ and $R_\ell \ll n$ for large n.

Figure 10.2, left, represents the ε-rank R_ℓ, $\ell = 1, 2, 3$, and R_B, computed on the examples of some compact molecules with $\varepsilon = 10^{-6}$. We observe that the Cholesky rank of B, R_B (see Section 10.3.2) is a multiple of N_b with a factor ~6 (see also Figure 10.3). Remarkably, the RHOSVD separation ranks $R_\ell \leq N_b$ remain to be very weakly dependent on N_b, but primarily depend on the topology of a molecule.

Figure 10.2 (right) provides average QTT ranks of column-vectors in $U^{(1)} \in \mathbb{R}^{n \times R_1}$ for NH_3, H_2O_2, N_2H_4, and C_2H_5OH molecules. Again, surprisingly, the rank portraits

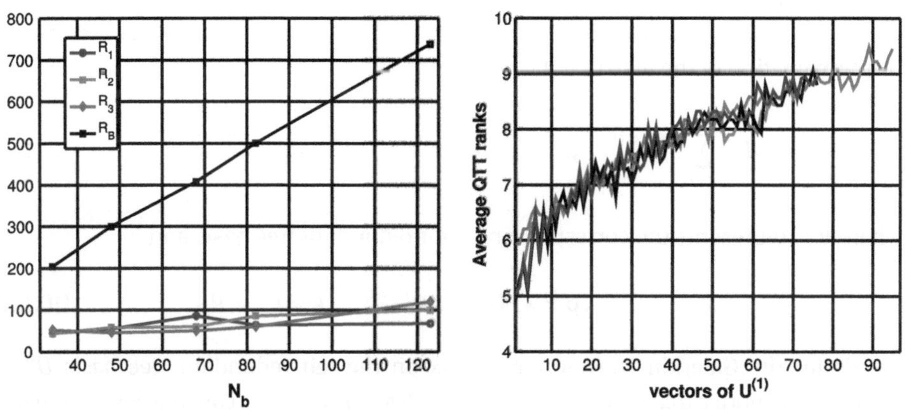

Figure 10.2: Left: ε-ranks R_ℓ and R_B for HF, NH_3, H_2O_2, N_2H_4, and C_2H_5OH molecules versus the number of basis functions N_b = 34, 48, 68, 82, and 123, respectively. Right: Average QTT ε-ranks of column-vectors in $U^{(1)} \in \mathbb{R}^{n \times R_\ell}$ for NH_3, H_2O_2, N_2H_4, and C_2H_5OH molecules, $\varepsilon = 10^{-6}$.

Table 10.1: Average QTT ε-ranks of $U^{(1)}$ and $V^{(1)}$ in $G^{(1)}$-factorization, $\varepsilon = 10^{-6}$.

Molecules	NH_3	H_2O_2	N_2H_4	C_2H_5OH
N_b; N_{orb}	48; 5	68; 9	82; 9	123; 13
Av. QTT rank of $U^{(1)}$	7.3	7.9	7.5	7.6
Av. QTT rank of $V^{(1)}$	15	21	24	37
(Av. QTT rank of $V^{(1)}$)/N_{orb}	3	2.3	2.6	2.85

appear to be nearly the same for different molecules, and the average rank over all indexes $m = 1, \dots, R_1$ is a small constant, about $r_0 \doteq 7$. The more detailed results are listed in Table 10.1.

10.3.2 Redundancy-free factorization of the TEI matrix *B*

Now we are in a position to represent the TEI matrix B in the factorized form using a reduced set of convolving functions. First, we recall that using the scalar product representation of $n \times n \times n$ arrays, we can rewrite the discretized integrals (10.1) in terms of tensor operations as in (10.4), (10.5). Then using representations (10.7) and (10.8) for each fixed multiindex $\mu\nu\kappa\lambda$, we arrive at the following *tensor factorization* of B [157]:

$$B = \sum_{k=1}^{R_N} \odot_{\ell=1}^{3} G^{(\ell)^T} (\mathbf{p}_k^{(\ell)} *_n G^{(\ell)}), \tag{10.10}$$

where $\mathbf{p}_k^{(\ell)}$, $\ell = 1, 2, 3$, are the column vectors in the side matrices of the rank-R_N canonical tensor representation $\mathbf{P}_N$ of the Newton kernel $\frac{1}{\|x\|}$ [166]. Substitution of the side matrix decomposition (10.9) to (10.10) leads to the redundancy-free factorized ε-approximation of the matrix B [157]:

$$B = \sum_{k=1}^{R_N} \odot_{\ell=1}^{3} G^{(\ell)^T} (\mathbf{p}_k^{(\ell)} *_n G^{(\ell)}) \cong \sum_{k=1}^{R_N} \odot_{\ell=1}^{3} V^{(\ell)} M_k^{(\ell)} V^{(\ell)^T} =: B_\varepsilon, \tag{10.11}$$

where $V^{(\ell)}$ represents the corresponding right redundant free basis and

$$M_k^{(\ell)} = U^{(\ell)^T} (\mathbf{p}_k^{(\ell)} *_n U^{(\ell)}) \in \mathbb{R}^{R_\ell \times R_\ell}, \quad k = 1, \dots, R_N, \tag{10.12}$$

stands for the Galerkin convolution matrix on the left redundant free basis $U^{(\ell)}$, $\ell = 1, 2, 3$. We notice that equation (10.12) includes only $R_\ell \ll N_b^2$ convolution products. The computational scheme for convolution matrices $M_k^{(\ell)}$ is described in Algorithm 2. Inspection of Algorithm 2 shows that the storage demand for representations (10.11)–(10.12) can be estimated by $R_N \sum_{\ell=1}^{3} R_\ell^2 + N_b^2 \sum_{\ell=1}^{3} R_\ell$ and $O((R_G + R_N)n)$, respectively.

Remark 10.3. The redundancy-free factorization (10.11) is completely parametrized by a set of thin matrices $V^{(\ell)}$ and small convolution factor matrices $M_k^{(\ell)}$, $\ell = 1, 2\,3$, precomputed by the 1D density fitting scheme. With this parametrization one can easily compute only the set of selected (required) columns in the matrix B by simple matrix-vector multiplications, thus completely avoiding calculation of 3D convolution products. In this concern, we notice that the standard Cholesky decomposition algorithm of TEI matrix B would require selected columns of this matrix of size N_b^2, where each entry needs calculation of the 3D convolution as in (10.1).

Algorithm 2 Computation of "convolution matrices" $M_k^{(\ell)}$.

Input: Rank-R_ℓ approximate decompositions $G^{(\ell)} \approx U^{(\ell)} V^{(\ell)^T}$, factor matrices $P^{(\ell)} = [\mathbf{p}_1^{(\ell)}, \dots, \mathbf{p}_{R_N}^{(\ell)}] \in \mathbb{R}^{n \times R_N}$, $\ell = 1, 2, 3$, in the rank-R_N canonical tensor $\mathbf{P}_N \in \mathbb{R}^{n \times n \times n}$.
(1) For $\ell = 1, 2, 3$, compute convolution products $\mathbf{p}_k^{(\ell)} *_n U^{(\ell)} \in \mathbb{R}^{n \times R_\ell}$, $k = 1, \dots, R_N$.
(2) For $\ell = 1, 2, 3$, compute and store Galerkin projections onto the left redundant free directional basis: $M_k^{(\ell)} = U^{(\ell)^T} (\mathbf{p}_k^{(\ell)} *_n U^{(\ell)}) \in \mathbb{R}^{R_\ell \times R_\ell}$.
Output: Right redundant free basis $V^{(\ell)}$; set of $R_\ell \times R_\ell$ matrices $M_k^{(\ell)}$ for $\ell = 1, 2, 3$, $k = 1, \dots, R_N$.

The following lemma proves the complexity and error estimates for tensor representations (10.11)–(10.12). Given the ε-truncated SVD-based left-orthogonal decomposition of $G^{(\ell)}$, $G^{(\ell)} \cong U^{(\ell)} V^{(\ell)^T}$, $\ell = 1, 2, 3$, with $n \times R_\ell$ and $N_b^2 \times R_\ell$ matrices, $U^{(\ell)}$ (orthogonal) and $V^{(\ell)}$, respectively, we denote $R_G = \max R_\ell$.

Lemma 10.4 ([157, 150]). *Given $\varepsilon > 0$, the redundancy-free factorized ε-approximations to the matrix B* (10.11) *and to the convolution matrix* (10.12) *exhibit the following properties:*

(A) *The storage demand for factorizations* (10.11) *and* (10.12) *is estimated by*

$$R_N \sum_{\ell=1}^{3} R_\ell^2 + N_b^2 \sum_{\ell=1}^{3} R_\ell, \quad \text{and} \quad O((R_G + R_N)n),$$

respectively. The numerical complexity of the ε-truncated representation (10.12) *is bounded by $O(R_N R_G^2 n + R_G R_N n \log n)$, where the second term includes the cost of tensor convolutions in the canonical format.*

(B) *The ε-rank of the matrix B_ε admits the following upper bound*

$$\operatorname{rank}(B_\varepsilon) \leq \min\left\{ N_b^2, R_N \prod_{\ell=1}^{3} R_\ell \right\}. \tag{10.13}$$

(C) *Denote $A_\ell(k) = G^{(\ell)^T} (\mathbf{p}_k^{(\ell)} *_n G^{(\ell)})$. Then we have the following error estimate in the Frobenius norm:*

$$\|B - B_\varepsilon\|_F \leq 6\varepsilon \max_\ell \|G^{(\ell)}\|_F \sum_{k=1}^{R_N} \max_\ell \|A_\ell(k)\|_F^2 \|\mathbf{p}_k^{(\ell)}\|_F. \tag{10.14}$$

Proof. (A) Using the Galerkin-type representation of the TEI tensor $\mathbf{B}$ as in (10.6), we obtain

$$B = \operatorname{mat}(\mathbf{B}) = \sum_{k=1}^{R_N} \odot_{\ell=1}^{3} G^{(\ell)^T} [\mathbf{p}_k^{(\ell)} *_n G^{(\ell)}].$$

Plugging the truncated SVD factorization of $G^{(\ell)}$ into the right-hand side leads to the desired representation

$$
\begin{aligned}
B_\varepsilon &= \sum_{k=1}^{R_N} \odot_{\ell=1}^{3} V^{(\ell)} U^{(\ell)^T} [\mathbf{p}_k^{(\ell)} *_n (U^{(\ell)} V^{(\ell)^T})] \\
&= \sum_{k=1}^{R_N} \odot_{\ell=1}^{3} V^{(\ell)} [U^{(\ell)^T} (\mathbf{p}_k^{(\ell)} *_n U^{(\ell)})] V^{(\ell)^T} \\
&= \sum_{k=1}^{R_N} \odot_{\ell=1}^{3} V^{(\ell)} M_k^{(\ell)} V^{(\ell)^T}.
\end{aligned} \tag{10.15}
$$

The storage cost for the RHOSVD-type factorization (10.15) to the $N_b^2 \times N_b^2$ matrix B is bounded by $R_N \sum_{\ell=1}^{3} R_\ell^2 + N_b^2 \sum_{\ell=1}^{3} R_\ell$ independently on the grid-size n.

The computational complexity at this step is dominated by the cost of the reduced Cholesky algorithm applied to the matrix $G^{(\ell)} G^{(\ell)^T}$ that computes truncated SVD of the side matrices $G^{(\ell)}$ at the cost $O(R_G(N_b^2+n))$ and by the total cost of convolution products in (10.12), $O(R_N R_G n \log n)$.

(B) Using the rank properties of Hadamard product of matrices, it is easy to see that (10.15) implies the direct ε-rank estimate for the matrix B_ε as in (10.13), where R_ℓ, $\ell = 1, 2, 3$ characterizes the effective rank in "1D density fitting".

(C) The error bound can be derived along the line of [174], Theorem 2.5(d), related to the RHOSVD error analysis. Indeed, the approximation error can be represented explicitly by

$$
B - B_\varepsilon = \sum_{k=1}^{R_N} (\odot_{\ell=1}^{3} G^{(\ell)} \mathbf{p}_k^{(\ell)} *_n G^{(\ell)} - \odot_{\ell=1}^{3} V^{(\ell)} U^{(\ell)^T} \mathbf{p}_k^{(\ell)} *_n U^{(\ell)} V^{(\ell)^T}).
$$

Denote $\widetilde{A}_\ell(k) = V^{(\ell)} U^{(\ell)^T} (\mathbf{p}_k^{(\ell)} *_n U^{(\ell)} V^{(\ell)^T})$. Then for each fixed $k = 1, \ldots R_N$, we have

$$
\|A_\ell - \widetilde{A}_\ell\| \le 2\varepsilon \|\mathbf{p}_k^{(\ell)}\| \|G^{(\ell)}\| \tag{10.16}
$$

because of the stability in the Frobenius norm $\|U^{(\ell)} V^{(\ell)^T}\| \le \|G^{(\ell)}\|$. Now, for fixed k, we obtain

$$
\begin{aligned}
A_1 \odot A_2 \odot A_3 - \widetilde{A}_1 \odot \widetilde{A}_2 \odot \widetilde{A}_3 = {} & A_1 \odot A_2 \odot A_3 - \widetilde{A}_1 \odot A_2 \odot A_3 \\
& + \widetilde{A}_1 \odot A_2 \odot A_3 - \widetilde{A}_1 \odot \widetilde{A}_2 \odot A_3 \\
& + \widetilde{A}_1 \odot \widetilde{A}_2 \odot A_3 - \widetilde{A}_1 \odot \widetilde{A}_2 \odot \widetilde{A}_3.
\end{aligned}
$$

Summing up the above representation over $k = 1, \ldots, R_N$ and taking into account (10.16), we arrive at the bound

$$
\|B - B_\varepsilon\|_F \le 6\varepsilon \max_\ell \|G^{(\ell)}\|_F \sum_{k=1}^{R_N} \max_\ell \|A_\ell(k)\|_F^2 \|\mathbf{p}_k^{(\ell)}\|_F, \tag{10.17}
$$

which proves the result. □

Proof of Lemma 10.4 is constructive and outlines the way to an efficient implementation of (10.11), (10.12). Some numerical results on the performance of the corresponding black-box algorithm are shown in Sections 10.3.3 and 11.4.

The RHOSVD factorization (10.11), (10.12) is reminiscent of the exact Galerkin representation (10.6) in the right redundancy free basis, whereas matrices $M_k^{(\ell)}$ play the role of "directional" Galerkin projections of the Newton kernel onto the left redundancy free basis. This factorization can be applied directly to fast calculation of the reduced Cholesky decomposition of the matrix B considered in the next section.

Finally, we point out that our RHOSVD-type factorization can be viewed as the algebraic tensor-structured counterpart of the *density fitting scheme* commonly used in quantum chemistry [3, 217, 237]. We notice that in our approach the "1D density fitting" is implemented independently for each space dimension, reducing the ε-ranks of the dominating directional bases to the lowest possible value. The robust error control in the proposed basis optimization approach is based on purely algebraic SVD-like procedure that allows eliminating the redundancy in the product basis set up to given precision $\varepsilon > 0$.

Further storage reduction can be achieved by the quantized-TT (QTT) approximation of the column vectors in $U^{(\ell)}$ and $V^{(\ell)}$ in (10.12). Specifically, the required storage amounts to $O((R_G + R_N)\log n)$ reals.

In some cases the representation (10.11) may provide the direct low-rank decomposition of the matrix B. In fact, suppose that $R_\ell \le C_\ell |\log \varepsilon| N_{\mathrm{orb}}$ with constants $C_\ell \le 1$, $\ell = 1, 2, 3$. Then the ε-rank of the matrix B is bounded by

$$\operatorname{rank}(B_\varepsilon) \le \min\left\{N_b^2, R_N |\log \varepsilon|^3 N_{\mathrm{orb}}^3 \prod_{\ell=1}^{3} C_\ell\right\}. \tag{10.18}$$

Indeed, in accordance to [157], we have the rank estimate $\operatorname{rank}(B_\varepsilon) \le \min\{N_b^2, R_N \prod_{\ell=1}^3 R_\ell\}$, which proves the statement.

Rank estimate (10.13) outlines the way to efficient implementation of (10.11), (10.12). Here, the algebraically optimized directional separation ranks R_ℓ, $\ell = 1, 2, 3$, are only determined by the entanglement properties of a molecule, whereas the numbers $N_b^2 - R_\ell$ indicate the measure of redundancy in the product basis set. Normally, we have $R_\ell \ll n$ and $R_\ell \le N_b$, $\ell = 1, 2, 3$. The asymptotic bound $R_\ell \le C_\ell |\log \varepsilon| N_{\mathrm{orb}}$ can be seen in Figure 10.1. One can observe that in the case of glycine molecule, the first mode-rank is much smaller than others, indicating the flattened shape of the molecule. However, the a priori rank estimate (10.13) looks too pessimistic compared to the results of numerical experiments, though in the case of flattened or extended molecules (some of directional ranks are small), this estimate provides much lower bound.

10.3.3 Low-rank Cholesky decomposition of the TEI matrix *B*

The Hartree–Fock calculations for the moderate size molecules are usually based on the incomplete Cholesky decomposition [303, 130, 17] applied to the symmetric and positive definite TEI matrix B,

$$B \approx LL^T, \quad L \in \mathbb{R}^{N_b^2 \times R_B}, \tag{10.19}$$

where the separation rank $R_B \ll N_b^2$ is of order $O(N_b)$. This decomposition can be efficiently computed by using the precomputed (off-line step) factorization of B as in (10.11), which requires only a small number of adaptively chosen column vectors in B, [157]. The detailed computational scheme is presented in Algorithm 3.

In this section, we describe the economical computational scheme introduced in [157, 150], providing the $O(N_b)$-rank truncated Cholesky factorization of the TEI matrix B with complexity $O(N_b^3)$. This approach requires only computation of the selected columns in B, without the need to compute the whole TEI matrix. The Cholesky scheme requires only $O(N_b)$ adaptively chosen columns in B, calculated on-line using the results of redundancy-free factorization (10.11).

Further the complexity can be reduced to $O(N_{\text{orb}}^2 N_b)$ using the quantized representation of the Cholesky vectors.

We denote the long indexes in the $N \times N$ $(N = N_b^2)$ matrix unfolding B by

$$i = \operatorname{vec}(\mu, \nu) := (\mu - 1)N_b + \nu, \quad j = \operatorname{vec}(\kappa, \lambda), \quad i, j \in I_N := \{1, \dots, N\}.$$

Lemma 10.5 ([157]). *The unfolding matrix B is symmetric and positive semidefinite.*

Proof. The symmetry is enforced by the definition (see Lemma 10.1). The positive semi-definiteness follows from the observation that the matrix B can be viewed as the Galerkin matrix $\langle -\Delta^{-1} u_i, u_j \rangle$, $i, j \in I_N$, in the finite product basis set $\{u_i\} = \{g_\mu g_\nu\}$, where Δ^{-1} is the inverse of the self-adjoint and positive definite in $H^1(\mathbb{R}^3)$ Laplacian operator subject to the homogeneous Dirichlet boundary conditions as $x \to \infty$. □

We consider the ε-truncated Cholesky factorization of $B \approx B_\varepsilon = LL^T$, where

$$\|B - LL^T\| \le C\varepsilon, \quad L \in \mathbb{R}^{N \times R_B}.$$

Based on the previous observation, we will postulate rather general ε-rank estimate (in electronic structure calculations this conventional fact traces back to [17]); see numerics on Figure 10.3.

Remark 10.6. Given a fixed truncation error $\varepsilon > 0$, for the Gaussian-type AO basis functions, we have $R_B = \operatorname{rank}(LL^T) \le CN_b$, where the constant $C > 0$ is independent of N_b.

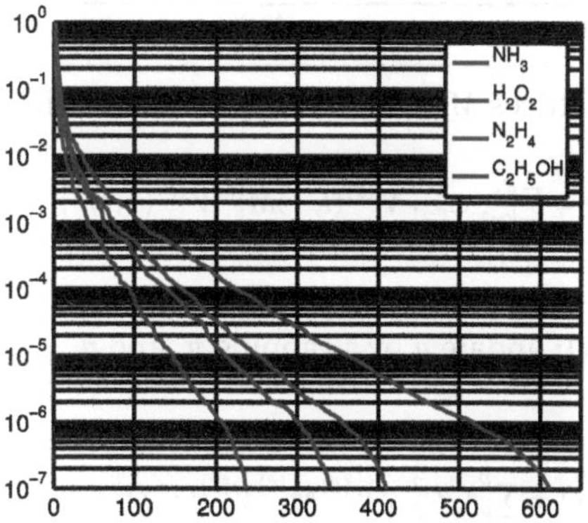

Figure 10.3: Singular values of $B_\varepsilon = LL^T$ for NH_3, H_2O_2, N_2H_4, and C_2H_5OH molecules with the number N_b of basis functions 48, 68, 82, and 123, respectively.

Clearly, the fastest version of the numerical Cholesky decomposition is possible in the case of precomputed full TEI tensor **B**. In this case the CPU time for the Cholesky decomposition becomes negligible compared to those to compute the TEI tensor **B**. However, the practical use of the algorithm is limited to the small basis sets because of the large storage requirements N_b^4.

The following approach was introduced in [157] to compute the truncated Cholesky decomposition with reduced storage demands by using the redundancy-free RHOSVD-type factorization of B in form (10.11), see Remark 10.3. Using this representation, one can calculate the truncated Cholesky decomposition of B, calculating on the fly a few columns and also diagonal elements in the TEI matrix B by the following cheap tensor operations:

$$B(:,j^*) = \sum_{k=1}^{R_N} \odot_{\ell=1}^{3} V^{(\ell)} M_k^{(\ell)} V^{(\ell)}(:,j^*)^T$$

and

$$B(i,i) = \sum_{k=1}^{R_N} \odot_{\ell=1}^{3} V^{(\ell)}(i,:) M_k^{(\ell)} V^{(\ell)}(:,i)^T,$$

respectively, as shown in Algorithm 3, [150].

The results of our numerical experiments using Matlab implementation of Algorithm 3 indicate that the truncated Cholesky decomposition with the separation rank $O(N_b)$ ensures the satisfactory numerical precision $\varepsilon > 0$ of order 10^{-5}–10^{-6}. The refined rank estimate $O(N_b|\log\varepsilon|)$ was observed in numerical experiments for every molecular system we calculated so far.

The factorization (10.11) essentially reduces the amount of work on the "preprocessing" stage in the limit of large N_b (see Lemma 10.4) since the number of convolutions is now estimated by $O(N_b)$ instead of N_b^2.

Other methods of tensor contraction in TEI calculations have been discussed in [134, 233].

Algorithm 3 Truncated Cholesky factorization of the matrix $B \in \mathbb{R}^{N\times N}$, $N = N_b^2$.

Input: Right RF basis $V^{(\ell)}$; set of $R_\ell \times R_\ell$ matrices $M_k^{(\ell)}$ for $\ell = 1,2,3$, $k = 1,\dots,R_N$, error tolerance $\varepsilon > 0$;
(1) Compute the diagonal $b = \operatorname{diag}(B)$: $B(i,i) = \sum_{k=1}^{R_N} \odot_{\ell=1}^{3} V^{(\ell)}(i,:)M_k^{(\ell)} V^{(\ell)}(:,i)^T$;
(2) Set $r = 1$, err $= \|b\|_1$ and initialize $\pi = \{1,\dots,N\}$;
While err $> \varepsilon$ perform (3)–(9)
(3) Find $m = \arg\max\{b(\pi_j) : j = r, r+1,\dots,N\}$; update π by swapping π_r and π_m;
(4) Set $\ell_{r,\pi_r} = \sqrt{b(\pi_r)}$;
For $r + 1 \le m \le N$ perform (5)–(7)
(5) Compute the entire column of B via $B(:,r) = \sum_{k=1}^{R_N} \odot_{\ell=1}^{3} V^{(\ell)} M_k^{(\ell)} V^{(\ell)}(:,r)^T$;
(6) Compute the L-column $\ell_{r,\pi_m} = (B(r,\pi_m) - \sum_{j=1}^{r-1} \ell_{j,\pi_r}\ell_{j,\pi_m})$;
(7) Update the stored diagonal $b(\pi_m) = b(\pi_m) - \ell_{r,\pi_m}^2$;
(8) Compute err $= \sum_{j=r+1}^{N} b(\pi_m)$;
(9) Increase $r = r + 1$;
Output: Low-rank decomposition of B, $B_\varepsilon = LL^T$ such that $\operatorname{tr}(B - B_\varepsilon) \le \varepsilon$.

10.4 On QTT compression to the Cholesky factor *L*

This section collects the important observations obtained in numerical experiments. In the QTT analysis of the TEI matrix B for several moderate size compact molecules, we revealed that, for fixed approximation error $\varepsilon > 0$, the average QTT ranks of the Cholesky vectors indicate the behavior $r_{QTT} \sim k_{chol}N_{orb}$ with $k_{chol} \le 3$. From this numerics we make a conclusion that the factor $k_{chol} = 3$ is due to the spatial dimensionality of the considered molecular system (or problem) observed for compact compounds, and it becomes closer to 2 for more stretched molecules; see Table 10.2.

Table 10.2: Average QTT ranks of the Cholesky vectors vs. N_{orb} for some molecules.

Molecule	HF	H_2O	NH_3	H_2O_2	N_2H_4	C_2H_5OH
N_{orb}	5	5	5	9	9	13
r_{QTT}	12	13.6	15	21	24	37
$k_{chol} = r_{QTT}/N_{orb}$	2.4	2.7	3	2.3	2.6	2.85

Based on this numerical experiments, we formulate our hypothesis [157]:

Hypothesis 10.7. The structural complexity of the Cholesky factor L of the matrix B in the QTT representation is characterized by the rank parameter

$$r_{QTT}(L) \cong 3N_{orb}.$$

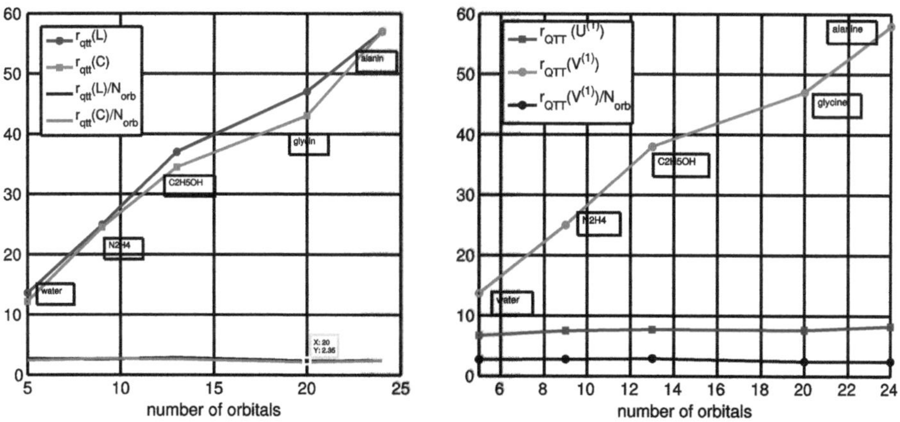

Figure 10.4: (Left): Average QTT ranks of the column vectors in L, $r_{QTT}(L)$, and in the vectorized coefficient matrix, $r_{QTT}(C)$, for several compact molecules. The "constant" lines at the level 2.35–2.85 indicate the corresponding ratios $r_{QTT}(L)/N_{orb}$ and $r_{QTT}(C)/N_{orb}$ for the respective molecule. (Right): QTT ranks of skeleton vectors in factorization (10.11)–(10.12) for H_2O, N_2H_4, C_2H_5OH, $C_2H_5NO_2$ (glycine), $C_3H_7NO_2$ (alanine) calculations, with N_{orb} equal to 5, 9, 13, 20, and 24, respectively.

The effective representation complexity of the Cholesky factor $L \in \mathbb{R}^{N \times R_B}$ is estimated by

$$9 R_B N_{orb}^2 \ll R_B N_b^2.$$

Assuming that the conventional relation $N_b \approx 10 N_{orb}$ is fulfilled, we conclude that the reduction factor in the storage size with QTT representation of L is about 10^{-1} (for QTT representation, we used the TT-Toolbox 2.2[1]).

Similar rank characterizations have been observed by the QTT analysis of $U^{(\ell)}$ and $V^{(\ell)}$ factors in the rank factorization of the initial product bases tensors $\mathbf{G}^{(\ell)}$, $\ell = 1, 2, 3$ (see Table 10.1). In particular, the average QTT ranks of the reduced higher-order SVD factors $V^{(\ell)} \in \mathbb{R}^{N_b^2 \times R_\ell}$ in the rank factorization of the initial product bases tensors $\mathbf{G}^{(\ell)}$, $\ell = 1, 2, 3$, have almost the same rank scaling, $r_{QTT}(V^{(\ell)}) \leq 3 N_{orb}$, as a factor $k_{chol} \approx 3$ in the Cholesky decomposition of the matrix B (see Table 10.1). Hence, the QTT representation complexity for the factor $V^{(\ell)}$ in (10.11) can be reduced to

$$10 N_{orb}^2 R_G \approx \frac{1}{10} N_b^2 R_G.$$

Figure 10.4 illustrates QTT-ranks behavior versus N_{orb} for skeleton vectors in factorization (10.11) for some compact molecules with different numbers of electron orbitals N_{orb}.

1 Free download from http://github.com/oseledets/TT-Toolbox, Skolkovo, Moscow.

11 Fast grid-based Hartree–Fock solver by factorized TEI

In this section, we describe the fast black-box Hartree–Fock solver[1] by the rank-structured tensor numerical methods introduced in [147]. It follows the conventional HF computational scheme, which relies on the precomputed grid-based TEI in a form of low-rank factorization. The DIIS- type iteration [238] for solving the nonlinear eigenvalue problem runs by updating the density matrix that gainfully employes the factorized representation of TEI tensor. The computational scheme is performed in a "black box" way; one has to specify only the size of a computational domain and the related $n \times n \times n$ 3D Cartesian grid, the skeleton vectors of the Galerkin basis functions discretized on that grid, and the coordinates and charges of nuclei in a molecule. The iterative solution process is terminated by chosen ε-threshold defining the accuracy of the rank-truncation operations.

In this approach, the grid-based tensor-structured calculation of factorized TEI tensor and of the core Hamiltonian are employed [157, 150, 156]. The routine size of the 3D grid for TEI calculations in MATLAB on a terminal is of order $n^3 \simeq 10^{14}$ (with $n = 32768$), yielding fine mesh resolution of order $h \simeq 10^{-4}$ Å.

The performance of this Hartree-Fock solver in MATLAB implementation [147] is comparable with the standard quantum chemical packages both in computation time and accuracy. Ab initio Hartree–Fock iteration for large compact molecules, up to amino acids glycine ($C_2H_5NO_2$) and alanine ($C_3H_7NO_2$), can run on a laptop.

The discretized Gaussians are used as the "global" Galerkin basis due to simplicity of comparison with the MOLPRO package [299] calculations for the same basis sets. Here, the primitive Gaussians of the basis sets of type cc-pVDZ are used for all considered molecules. Due to the grid representation of the basis set, this Hartree–Fock solver can be considered as a "laboratory" for development and testing of the optimized bases of general type. In particular, in this framework, the substitution of the the set of steepest core electron Gaussians by a Slater-type functions for every non-hydrogen nuclei, may be employed, essentially reducing the number of basis functions.

11.1 Grid representation of the global basis functions

The initial eigenvalue problem is posed in the finite volume box $\Omega = [-b, b]^3 \in \mathbb{R}^3$ subject to the homogeneous Dirichlet boundary conditions on $\partial\Omega$. For a given discretization parameter $n \in \mathbb{N}$, we use the equidistant $n \times n \times n$ tensor grid $\omega_{3,n} = \{x_{\mathbf{i}}\}$,

1 Also abbreviated as fast TESC Hartree-Fock solver, see Section 7.4.

https://doi.org/10.1515/9783110365832-011

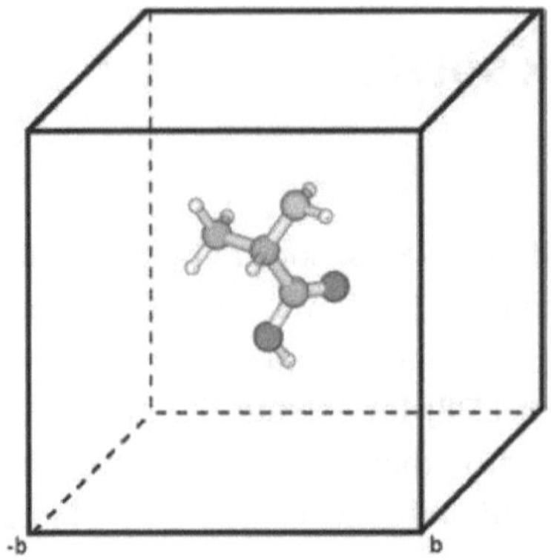

Figure 11.1: The computational box $[-b,b]^3$ routine size is $b = 20$ au (~10.5 Å).

$\mathbf{i} \in \mathcal{I} := \{1,\ldots,n\}^3$, with the mesh-size $h = 2b/(n+1)$,, see Figure 11.1. For the set of "global" separable Galerkin basis functions $\{g_k\}_{1\le k\le N_b}$, $k = 1,2,\ldots,N_b$, we define approximating functions $\overline{g}_k := \mathbf{I}_1 g_k$, $k = 1,\ldots,N_b$, by linear tensor-product interpolation via the set of product "local" basis functions $\{\xi_{\mathbf{i}}\} = \xi_{i_1}(x_1)\xi_{i_2}(x_2)\xi_{i_3}(x_3)$, $\mathbf{i} \in \mathcal{I}$, associated with the respective grid-cells in $\omega_{\mathbf{3},n}$. The local basis functions are chosen as piecewise linear (hat functions) for tensor calculation of the Laplace operator [156] or piecewise constant for factorized calculations of two-electron integrals [157] and the direct tensor calculation of the nuclear potential operator V_c [156]. Recall that the linear interpolant $\mathbf{I}_1 = \mathrm{I}_1 \times \mathrm{I}_1 \times \mathrm{I}_1$ is a product of 1D interpolation operators $\overline{g}_k^{(\ell)} = \mathrm{I}_1 g_k^{(\ell)}$, $\ell = 1,2,3$, where $\mathrm{I}_1 : C^0([-b,b]) \to W_h := \mathrm{span}\{\xi_i\}_{i=1}^n$ is defined over the set of (piecewise linear or piecewise constant) local basis functions $(\mathrm{I}_1 w)(x_\ell) := \sum_{i=1}^N w(x_{\ell,i})\xi_i(x_\ell)$, $x_{\mathbf{i}} \in \omega_{\mathbf{3},N}$. This leads to the separable grid-based approximation of the initial basis functions $g_k(x)$,

$$g_k(x) \approx \overline{g}_k(x) = \prod_{\ell=1}^{3} \overline{g}_k^{(\ell)}(x_\ell) = \prod_{\ell=1}^{3}\sum_{i=1}^{N} g_k^{(\ell)}(x_{\ell,i})\xi_i(x_\ell) \tag{11.1}$$

such that the rank-1 coefficient tensor $\mathbf{G}_k$ is given by

$$\mathbf{G}_k = \mathbf{g}_k^{(1)} \otimes \mathbf{g}_k^{(2)} \otimes \mathbf{g}_k^{(3)}, \quad k = 1,\ldots,N_b, \tag{11.2}$$

with the canonical vectors $\mathbf{g}_k^{(\ell)} = \{g_{k_i}^{(\ell)}\} \equiv \{g_k^{(\ell)}(x_i^{(\ell)})\}$. The discretized Galerkin basis is then represented by the set of rank-1 canonical tensors $\mathbf{G}_k$, $k = 1,\ldots,N_b$.

Since the tensor-structured calculation of the operators in the Hartree–Fock equation is reduced to one-dimensional rank-structured algebraic operations, the size n of the tensor-product grid $\omega_{\mathbf{3},n}$ can be chosen differently for different parts of the Fock operator. For example, the entries of the matrices in A_g and V_g in (see Section 9), corresponding to kinetic and nuclear energy parts in the core Hamiltonian, can be computed using different grid sizes n for discretizing the "global" Gaussian basis functions. The same concerns with the grid size n in rank-structured calculations of the two-electron integrals by using piecewise constant basis functions, which can be much smaller than the grid-size n required for calculation of both A_g and V_g, since J and K are integral operators. Thus, the discretization step-size for the grid representation of the Galerkin basis is specified only by accuracy needs for the particular part of the Fock operator of interest.

11.2 3D Laplace operator in $O(n)$ and $O(\log n)$ complexity

Recall that the grid-based calculation of the core Hamiltonian part (7.5), $H_c = -\frac{1}{2}\Delta + V_c$, has been discussed in Section 9. In particular, given the Gaussian-type Galerkin basis $\{g_k(x)\}_{1\leq k\leq N_b}$, $x \in \mathbb{R}^3$, the Laplace operator takes the matrix form $A_g = [a_{km}] \in \mathbb{R}^{N_b\times N_b}$ with the entries

$$a_{km} = \langle -\Delta g_k(x), g_m(x)\rangle, \quad k,m = 1,\dots N_b,$$

which can be computed by using simple multilinear algebra with rank-1 tensor $\mathbf{G}_k$. The exact Galerkin matrix A_g is approximated using (11.2) as in [156], $A_g \approx A_G = \{\overline{a}_{km}\}$, $k,m = 1,\dots N_b$, with

$$\overline{a}_{km} = -\langle A_\Delta \mathbf{G}_k, \mathbf{G}_m\rangle, \tag{11.3}$$

which should be calculated with large grid-size n that resolve the sharp Gaussian basis functions.

To overcome the limitations caused by the large mode size n of the target tensors, the QTT tensor format [167, 165] can be used for calculation of the Laplace part in the Fock operator [147]. This allows calculation of the multidimensional functions and operators in logarithmic complexity $O(\log n)$. For the Laplace operator

$$A_\Delta = \Delta_1^{(1)} \otimes I^{(2)} \otimes I^{(3)} + I^{(1)} \otimes \Delta_1^{(2)} \otimes I^{(3)} + I^{(1)} \otimes I^{(2)} \otimes \Delta_1^{(3)}, \tag{11.4}$$

the exact rank-2 tensor train representation was introduced in [142]:

$$\Delta_{\mathrm{TT}} = \begin{bmatrix}\Delta_1 & I\end{bmatrix} \otimes_b \begin{bmatrix} I & 0 \\ \Delta_1 & I\end{bmatrix} \otimes_b \begin{bmatrix} I \\ \Delta_1\end{bmatrix}, \tag{11.5}$$

where the sign $\otimes_b$ (sometimes also denoted by $\bowtie$) means the matrix product of block core matrices with blocks being multiplied by means of the tensor product. Suppose that $n = 2^L$. Then the quantized representation of Δ_1 takes the form [142, 170]

$$\Delta_{1_Q} = \begin{bmatrix} I & J' & J\end{bmatrix} \otimes_b \begin{bmatrix} I & J' & J \\ & J & \\ & & J'\end{bmatrix}^{\otimes_b (L-2)} \otimes_b \begin{bmatrix} 2I - J - J' \\ -J \\ -nJ' \end{bmatrix}, \tag{11.6}$$

where L is equal to the number of the virtual dimensions in the quantized format, and

$$I = \begin{pmatrix} 1 & 0 \\ 0 & 1 \end{pmatrix}, \quad J = \begin{pmatrix} 0 & 1 \\ 0 & 0 \end{pmatrix}.$$

For the discretized representation (11.2) of basis functions, the entries of the matrix $A_G = \{\overline{a}_{km}\}$, $k, m = 1, \ldots N_b$, are calculated as

$$\overline{a}_{km} = -\langle A_\Delta \mathbf{G}_k, \mathbf{G}_m \rangle \approx -\langle \Delta_{\mathrm{QTT}} \mathbf{Q}_k^{(1)} \otimes \mathbf{Q}_k^{(2)} \otimes \mathbf{Q}_k^{(3)}, \mathbf{Q}_m^{(1)} \otimes \mathbf{Q}_m^{(2)} \otimes \mathbf{Q}_m^{(3)} \rangle, \tag{11.7}$$

where the matrix Δ_{QTT} is obtained by plugging the QTT Laplace representation (11.6) into (11.5), and a tensor $\mathbf{Q}_k^{(\ell)}$, $\ell = 1, 2, 3$, is the quantized representation of a vector $\mathbf{g}_k^{(\ell)} \in \mathbb{R}^n$.

Table 11.1: QTT calculations of the Laplacian matrix for H_2O molecule.

p	15	16	17	18	19	20
$n^3 = 2^{3p}$	$32\,767^3$	$65\,535^3$	$131\,071^3$	$262\,143^3$	$524\,287^3$	$1\,048\,575^3$
err(A_G)	0.0027	$6.8 \cdot 10^{-4}$	$1.7 \cdot 10^{-4}$	$4.2 \cdot 10^{-5}$	$1.0 \cdot 10^{-5}$	$2.6 \cdot 10^{-6}$
RE	–	$1.0 \cdot 10^{-5}$	$8.3 \cdot 10^{-8}$	$2.6 \cdot 10^{-9}$	$3.3 \cdot 10^{-10}$	0
time (sec)	12.8	17.4	25.7	42.6	77	135

Table 11.1 demonstrates weak dependence of the calculation time on the size of the 3D Cartesian grid. In the case of water molecule, it shows the approximation error for the Laplacian matrix $\mathrm{err}(A_G) = \|A_{\mathrm{Molpro}} - A_G\|$ represented in the discretized basis of $N_b = 41$ Cartesian Gaussians, where A_{Molpro} is the result of analytical computations with the same Gaussian basis from MOLPRO program [299]. Time is given for MATLAB implementation. The line "RE" in Table 11.1 represents the approximation error for the discrete Laplacian A_G obtained by the Richardson extrapolation on two adjacent grids, where the grid-size is given by $n = 2^p$, $p = 1, \ldots, 20$. The QTT ranks of the canonical vectors $\mathbf{g}_k^{(\ell)}$ are bounded by several ones. The approximation order $O(h^2)$ can be observed.

11.3 Nuclear potential operator in $O(n)$ complexity

Now we recall shortly computation of the nuclear potential operator by direct tensor summation of electrostatic potentials described in detail in Chapter 9:

$$V_c(x) = \sum_{\nu=1}^{M_0} \frac{Z_\nu}{\|x - a_\nu\|}, \quad Z_\nu > 0, \ x, a_\nu \in \Omega \subset \mathbb{R}^3, \tag{11.8}$$

where M_0 is the number of nuclei in Ω. Using the canonical tensor representation of a reference 3D Newton kernel $\frac{1}{\|x\|}$ described in Section 6.1

$$\widetilde{\mathbf{P}}_R = \sum_{q=1}^{R} \mathbf{p}_q^{(1)} \otimes \mathbf{p}_q^{(2)} \otimes \mathbf{p}_q^{(3)} \in \mathbb{R}^{2n\times 2n\times 2n} \tag{11.9}$$

and the rank-1 shifting-windowing operator (see Section 9.2)

$$\mathcal{W}_\nu = \mathcal{W}_\nu^{(1)} \otimes \mathcal{W}_\nu^{(2)} \otimes \mathcal{W}_\nu^{(3)} \quad \text{for } \nu = 1,\dots,M_0,$$

the total electrostatic potential $V_c(x)$ in the computational box Ω is approximated by the direct canonical tensor sum (see also (9.11))

$$\mathbf{P}_c = \sum_{\nu=1}^{M_0} Z_\nu \sum_{q=1}^{R} \mathcal{W}_\nu^{(1)} \mathbf{p}_q^{(1)} \otimes \mathcal{W}_\nu^{(2)} \mathbf{p}_q^{(2)} \otimes \mathcal{W}_\nu^{(3)} \mathbf{p}_q^{(3)} \in \mathbb{R}^{n\times n\times n}. \tag{11.10}$$

Then for a given tensor representation of the basis function as a rank-1 canonical tensor (11.2), the sum $V_c(x)$ of potentials in a box as in (11.8) is represented in a given basis set by a matrix $V_g \approx V_G = \{\overline{v}_{km}\} \in \mathbb{R}^{N_b\times N_b}$ whose entries are calculated by simple tensor operations [156, 147]:

$$v_{km} = \int_{\mathbb{R}^3} V_c(x) g_k(x) g_m(x) dx \approx \{\overline{v}_{km}\} := \langle \mathbf{G}_k \odot \mathbf{G}_m, \mathbf{P}_c \rangle, \quad 1 \le k, m \le N_b. \tag{11.11}$$

Note that for the grid-based representation of the core potential, $V_c(x)$, $\mathbf{P}_c$, the univariate grid size n can be noticeably smaller than the size of the grid used for the piecewise linear discretization for the Laplace operator.

11.4 Coulomb and exchange operators by factorized TEI

Here we recall the multilinear algebraic calculation of the Coulomb and exchange matrices in the Fock operator discussed in full detail in [157, 147]. For precomputed two-electron integrals, in view of (7.14), the Coulomb matrix is given by

$$J(D)_{\mu\nu} = \sum_{\kappa,\lambda=1}^{N_b} b_{\mu\nu,\kappa\lambda} D_{\kappa\lambda}. \tag{11.12}$$

Vectorizing matrices $\overline{J} = \mathrm{vec}(J)$ and $\overline{D} = \mathrm{vec}(D)$ and taking into account the rank structure in TEI matrix B, we arrive at the simple matrix representation for the Coulomb matrix

$$\overline{J} = B\overline{D} \approx L(L^T \overline{D}). \tag{11.13}$$

The straightforward calculation by (11.13) amounts to $O(R_B N_b^2)$ operations, where R_B is the ε-rank of B.

For the exchange operator K, tensor evaluation is more involved due to summation over permuted indices:

$$K(D)_{\mu\nu} = -\frac{1}{2} \sum_{\kappa,\lambda=1}^{N_b} b_{\mu\lambda,\nu\kappa} D_{\kappa\lambda}, \tag{11.14}$$

which diminishes the advantages of the low-rank structure in the matrix B. Introducing the permuted tensor $\widetilde{\mathbf{B}} = \mathrm{permute}(\mathbf{B}, [2,3,1,4])$ and the respective unfolding matrix $\widetilde{B} = \mathrm{mat}(\widetilde{\mathbf{B}})$, we then obtain

$$\mathrm{vec}(K) = \overline{K} = \widetilde{B}\overline{D}. \tag{11.15}$$

The direct calculation by (11.15) amounts to $O(R_B N_b^3)$ operations. However, using the rank-N_{orb} decomposition of the density matrix $D = 2CC^T$ reduces the cost to $O(R_B N_{\mathrm{orb}} N_b^2)$ via the representation

$$K(D)_{\mu\nu} = -\sum_{i=1}^{N_{\mathrm{orb}}} \Big(\sum L_{\mu\lambda} C_{\lambda i}\Big)\Big(\sum L_{\kappa\nu} C_{\kappa i}\Big)^T,$$

where $L_{\mu\nu} = \mathrm{reshape}(L, [N_b, N_b, R_B]) \in \mathbb{R}^{N_b \times N_b \times R_B}$ is the $N_b \times N_b \times R_B$-folding of the Cholesky factor L.

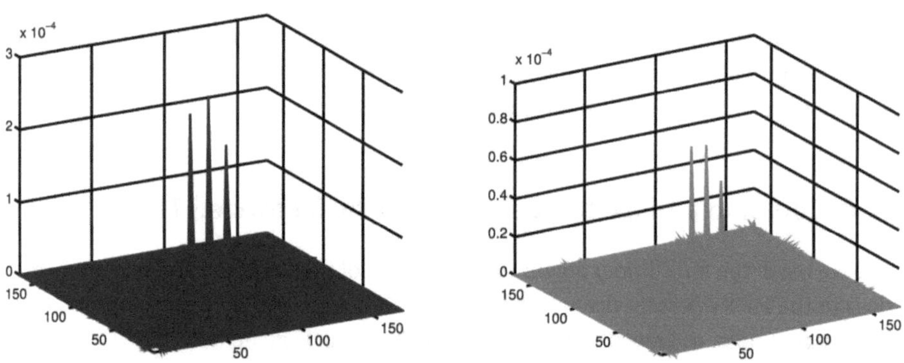

Figure 11.2: Approximation accuracy for the Coulomb matrix of glycine molecule using TEI computed on the grid with $n^3 = 32\,768^3$ (left) and $n^3 = 65\,536^3$ (right).

Figure 11.2 presents the error in computation of the Coulomb matrix for glycine amino acid ($N_b = 170$) using TEI computed on the grids $n^3 = 32\,768^3$ (left) and $n^3 = 65\,536^3$ (right). The numerical error scales quadratically in the grid size $O(h^2)$ and can be improved to $O(h^3)$ by the Richardson extrapolation. The observed decay ratio $1 : 4$ indicates the applicability of the Richardson extrapolation to the results on a pair of

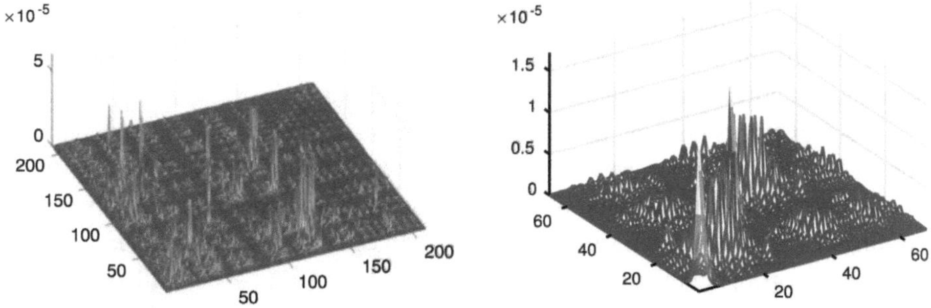

Figure 11.3: Left: the error in density matrix for the amino acid alanine ($N_b = 210$) for the TEI computed with $n^3 = 131\,072^3$. Right: the error in exchange matrix for H_2O_2 ($N_b = 68$) computed by TEI using the grid of size $n^3 = 131\,072^3$.

diadically refined grids. Figure 11.3 (left) demonstrates the error in computation of the density matrix of alanine molecule ($N_b = 210$) using TEI computed on the grid with $n^3 = 131\,072^3$. Figure 11.3 (right) displays the error in exchange matrix computation for the H_2O_2 molecule ($N_b = 68$) using TEI with $n^3 = 131\,072^3$.

11.5 Algorithm of the black-box HF solver

Our grid-based TESC HF solver operates in a black-box way; the input only includes setting up the (x, y, z)-coordinates and charges of nuclei in Z_ν, $\nu = 1, \ldots, M_0$, in the molecule and the Galerkin basis functions discretized on a tensor grid. For lattice systems, it is necessary to give the coordinates and Galerkin basis of a “reference atom”, the interval between atoms in a lattice, and its length.

Recalling the discussion in Section 7.3, we have to solve the eigenvalue problem for the coefficients matrix $C = \{c_{i\mu}\} \in \mathbb{R}^{N_{\mathrm{orb}} \times N_b}$,

$$F(C)C = SC\Lambda, \quad \Lambda = \operatorname{diag}(\lambda_1, \ldots, \lambda_{N_b}) \tag{11.16}$$

with the overlap matrix S for the chosen Galerkin basis (7.10) and the Fock operator

$$F(C) = H + J(C) + K(C), \tag{11.17}$$

where the matrices $J(C)$ and $K(C)$ depend on the solution matrix C. To solve the eigenvalue problem (11.16), we start self-consistent field iteration with $F(C) = H$ and with zero matrices for both Coulomb J and exchange K operators. In the course of SCF iteration, we control the residual, computing the maximum-norm of the difference in the virtual part of the eigenvectors from two consequent iterations

$$\|C(1, N_{\mathrm{orb}} : N_b)_{\mathrm{it}-1} - C(1, N_{\mathrm{orb}} : N_b)_{\mathrm{it}}\|_\infty \leq \varepsilon. \tag{11.18}$$

Iteration may be terminated when this value becomes smaller than a given ε-threshold or the number of iterations may be predefined. Since iteration times are negligibly small, we usually use a predefined number of iterations.

The first step is defining the global Galerkin basis. In what follows, for comparison with MOLPRO output, we discretize the rank-1 basis functions given as a product of polynomials with Gaussians. We choose in advance the appropriate grid sizes according to the desired accuracy of calculations. In general, one can set an $n_x \times n_y \times n_z$ 3D Cartesian grid, but in our current calculations, we use a cubic box with equal sizes n in every space variable. As it was already noted, the univariate grid-size n of the $n \times n \times n$ 3D Cartesian grid can be chosen differently in calculation of the discretized Laplacian, the nuclear potential operator, and the two-electron integrals tensor. Using finer (larger) grids need more CPU time. Therefore, there is a playoff between the required accuracy and computational cost.

Given the coordinates of nuclei and the Galerkin basis, the black-box HF solver performs the following computation steps.

(1) Choose the grid size n and the ε-threshold for rank truncation. Set up the grid representation of the basis functions.
(2) Compute the nuclear energy shift E_{nuc}, by (7.17).
(3) Compute the core Hamiltonian H by the three-dimensional grid-based calculation of the Galerkin matrix A_G for the Laplacian by (11.3) or (11.7) and for the nuclear potential operator V_G by (11.11).
(4) Using grid-based "1D density fitting", compute the factorized TEI matrix in a form of low-rank Cholesky decomposition $B = LL^T$ by (10.11), (10.12).
(5) Set up the input data for SCF iteration:
   - threshold ε for the residual (alternatively, a maximal number of iterations);
   - number *Mopt* specifying the design of DIIS scheme [238];
   - define initial Coulomb and exchange matrices as $J = 0$ and $K = 0$.
(6) Start the SCF iteration for solving nonlinear eigenvalue problem:
   - solve the linear spectral problem (11.16) with the current Fock matrix

$$F = \frac{1}{2}A_G - V_G + J - K;$$

   - update the residual (11.18) (difference in the virtual parts of the eigenvectors);
   - update matrices $J(C)$ and $K(C)$ by computing (11.12) and (11.14);
   - compute the ground-state energy $E_{0,\mathrm{it}}$ at current iteration.

   When the residual arrives at the given ε (or when the maximal iteration number is reached), iteration is terminated.
(7) Compute the ground-state energy $E_{0,n}$.
(8) Calculate the MP2 corrections by factorizations introduced in [150]; see Section 11.8.

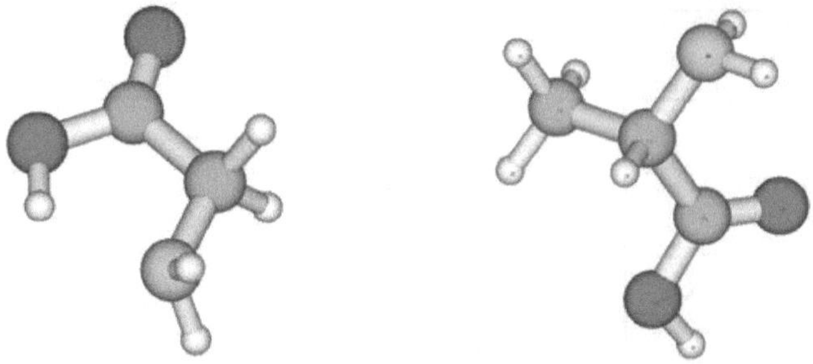

Figure 11.4: The largest molecules considered for numerical examples (below): amino acids glycine $C_2H_5NO_2$ (left) and alanine $C_3H_7NO_2$ (right). The ball-stick picture of molecules is generated by the MOLDEN program [258].

Table 11.2: Times for one SCF iteration in the tensor-based Hartree–Fock solver (step 6) in MATLAB implementation.

Molecule	NH_3	H_2O_2	N_2H_4	C_2H_5OH	glycine	alanine
N_b	48	68	82	123	170	210
Time (sec)	0.2	0.3	0.4	0.6	1.6	3.3

For small and moderate size molecules the solver in MATLAB works in one run from the first step to the end of SCF iteration using 3D Cartesian grids for TEI calculations up to $n^3 = 131\,072^3$. The total computation time usually does not exceed several minutes, see Table 11.2 illustrating times for one SCF iteration by fast TESC Hartree-Fock solver in MATLAB implementation.

For larger molecules (amino acids, see Figure 11.4), accurate calculations with the grids exceeding $n^3 = 65\,536^3$ need an off-line precomputing of TEI, which requires less than one hour of Matlab calculations. CPU time for TEI calculations depends mostly on the number of basis functions rather than on the size of the grid. The grid size is mainly limited by the available storage of the computer: storage demand for the first step in TEI calculations (factorization of the side matrices $G^{(\ell)} \in \mathbb{R}^{n \times N_b^2}$, $\ell = 1, 2, 3$), is estimated by $O(3nN_b^2)$, whereas for the second step of TEI calculations (Cholesky decomposition of the TEI matrix B), it is bounded by $O(N_b^3)$.

11.6 Ab initio ground state energy calculations for compact molecules

Numerical simulations are performed in MATLAB on an 8 AMD Opteron Dual-Core/2800 computer cluster. The molecule is considered in a computational box $[-b, b]^3$ with $b = 20$ au ($\approx$10.6 Å). In TEI calculations, we use the uniform mesh sizes up to finest level with $h = 2.5 \cdot 10^{-4}$ corresponding to approximately $1.3 \cdot 10^{-4}$ Å. For

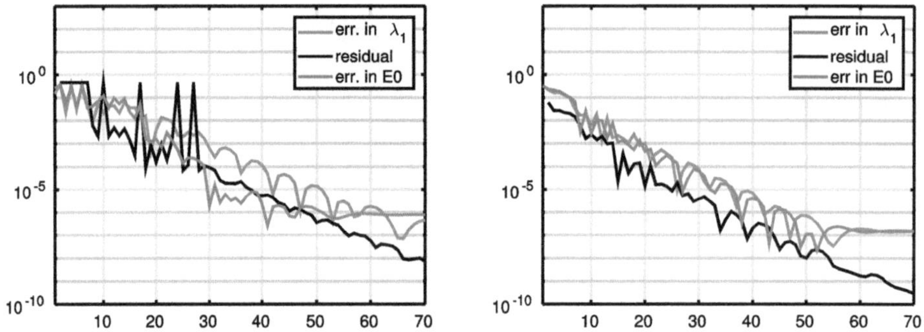

Figure 11.5: SCF iterations history for glycine and H_2O molecules.

the core Hamiltonian calculations finer grids are required with the mesh size about $h = 3.5 \cdot 10^{-5}$ au (~$1.8 \cdot 10^{-5}$ Å). These corresponds to large 3D Cartesian grids of size $n^3 = 65\,535^3$ and $n^3 = 1\,048\,576^3$ entries, correspondingly. In the following examples, we present calculations of the ground-state energy for several compact molecules.

Figure 11.5 shows convergence of the SCF iterations for glycine ($N_b = 170$) amino acid (left) and water ($N_b = 41$) molecule (right) using the factorized representation of TEI precomputed with $n^3 = 131\,072^3$. The black line shows convergence of the residual computed as the maximum-norm of the difference of the eigenvectors from two consequent iterations $\|C(1,:)_{\mathrm{it}-1} - C(1,:)_{\mathrm{it}}\|_\infty$. The green line presents the difference between the lowest eigenvalue computed by the grid-based solver and the respective eigenvalue from MOLPRO calculations, $\Delta\lambda_{1,\mathrm{it}} = |\lambda_{1,\mathrm{Molpro}} - \lambda_{1,\mathrm{it}}|$. The red line is the difference in ground-state energy with the MOLPRO results, $\Delta E_{0,\mathrm{it}} = |E_{0,\mathrm{Molpro}} - E_{0,\mathrm{it}}|$.

Figures 11.6–11.8 demonstrate the convergence of the ground-state energy versus self-consistent field iteration for glycine amino acids ($N_b = 170$), NH_3 ($N_b = 48$) and water ($N_b = 41$) molecules. Left figures show convergence history over 70 iterations; right figures show the zoom of last 30 iterations. The black line corresponds to $E_{0,\mathrm{Molpro}}$ computed by MOLPRO for the same Gaussian basis.

Figure 11.9 presents the output of the solver for alanine molecule. Figure 11.10 presents the last $30 + k$ iterations on convergence of the ground-state energy for H_2O_2 molecule. The red, green and blue lines correspond to grid sizes $n^3 = 32768^3, 65536^3$, and $131\,072^3$, correspondingly.

Table 11.3: Glycine, basis of 170 Gaussians (cc-pVDZ): error in ground-state energy versus the mesh size h. MOLPRO result is $E_{0,\mathrm{Molpro}} = -282.8651$.

p	13	15	16	17
$n^3 = 2^{3p}$	8192^3	$32\,767^3$	$65\,535^3$	$131\,072^3$
h	0.0039	$9.7 \cdot 10^{-4}$	$4.9 \cdot 10^{-4}$	$2.5 \cdot 10^{-4}$
$E_{0,n}$	−282.8679	−282.8655	−282.8654	−282.8653
$\mathrm{er}(E_0)$	0.0024	$3.5 \cdot 10^{-4}$	$2.2 \cdot 10^{-4}$	$2.2 \cdot 10^{-4}$

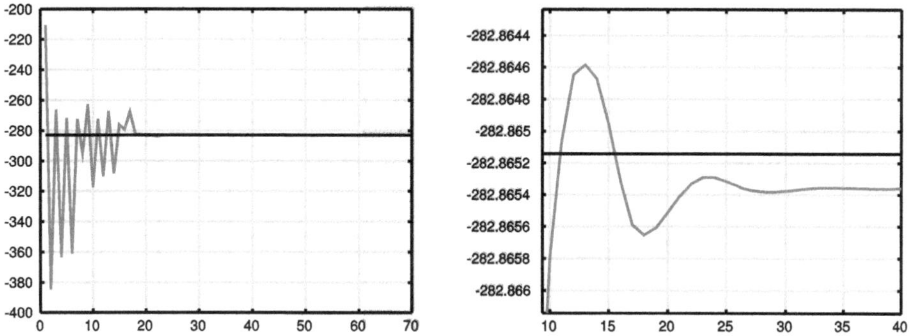

Figure 11.6: Convergence of the ground-state energy for the glycine molecule (left), with the grid size for TEI calculation, $n^{\otimes 3} = 131\,072^3$; zoom for last 30 iterations (right).

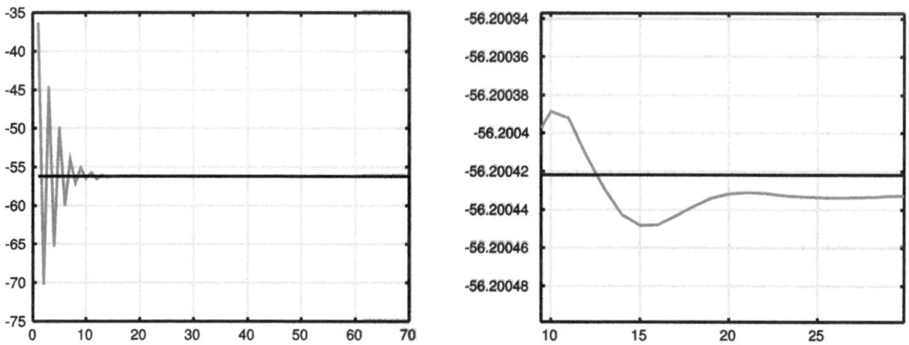

Figure 11.7: Convergence of the ground-state energy for the NH_3 molecule (left), with TEI grid size $n^{\otimes 3} = 131\,072^3$; zoom of the last 30 iterations (right).

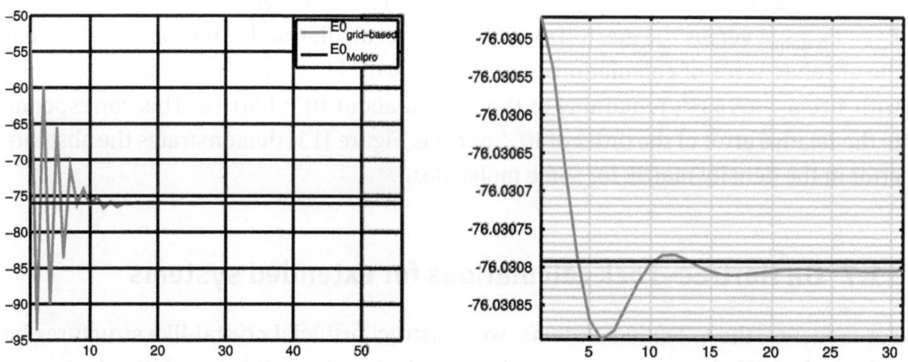

Figure 11.8: Convergence of the ground-state energy for the H_2O molecule (left), with the TEI grid size $n^{\otimes 3} = 131\,072^3$; zoom of the last 30 iterations (right).

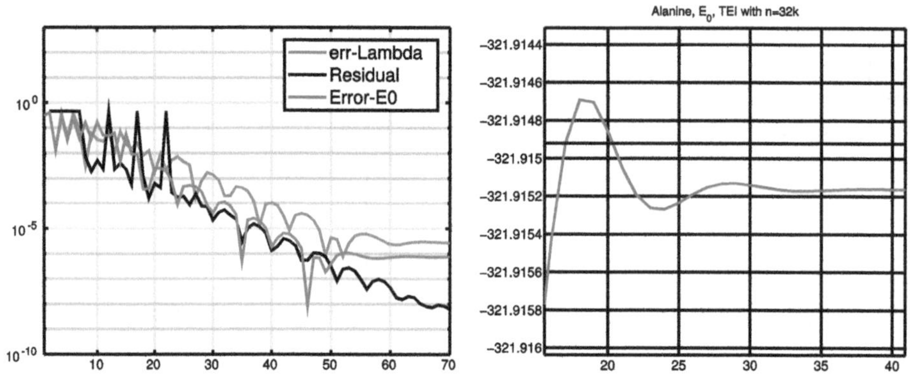

Figure 11.9: Left: SCF iteration for alanine molecule ($N_b = 211$) with TEI computed on the grid $n^{\otimes 3} = 32\,768^3$. Right: convergence of $E_{0,\mathrm{it}}$ at last 30 iterations.

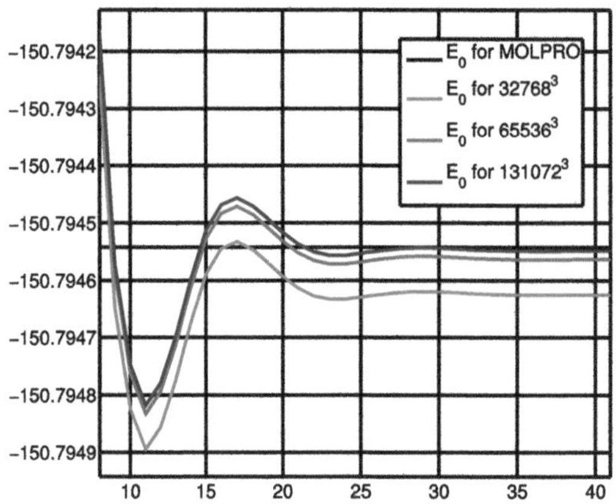

Figure 11.10: Molecule H_2O_2, convergence of $E_{0,n}$ after $30 + k$ iterations; with TEI calculated on a sequence of grids.

Table 11.3 presents the error in the ground-state energy for glycine molecule $\mathrm{er}(E_0) = E_{0,n} - E_{0,\mathrm{Molpro}}$ versus the mesh size of the grid for calculating TEI tensor, n. Notice that the absolute error of calculations with grid-based TEI changes only mildly for grids with size $n \geq 65\,535^3$, remaining at the level of about 10^{-4} hartree. This corresponds to the relative error of the order of 10^{-7} hartree. Figure 11.11 demonstrates the absolute error in the density matrix for some molecules.

11.7 On Hartree–Fock calculations for extended systems

For modeling the extended systems, we construct artificial crystal-like structures by using a single Hydrogen atom as an initiating block and multiply translating it at equal intervals d_1, d_2, d_3 in every of three spacial directions x, y, and z, respectively. Thus, a 3D lattice cluster of size $m_1 \times m_2 \times m_3$ is assembled, where m_1, m_2, m_3 are the numbers of atoms in the spatial directions x, y, and z, see an example in Figure 11.12.

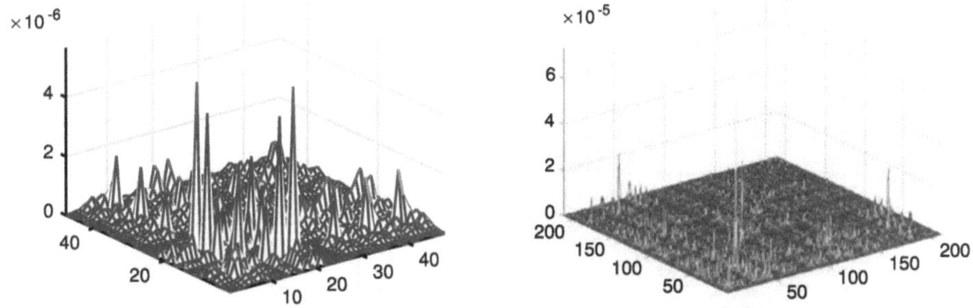

Figure 11.11: Absolute error of the density matrix for NH_3 molecule (left) and alanine amino acid (right) compared with MOLPRO output.

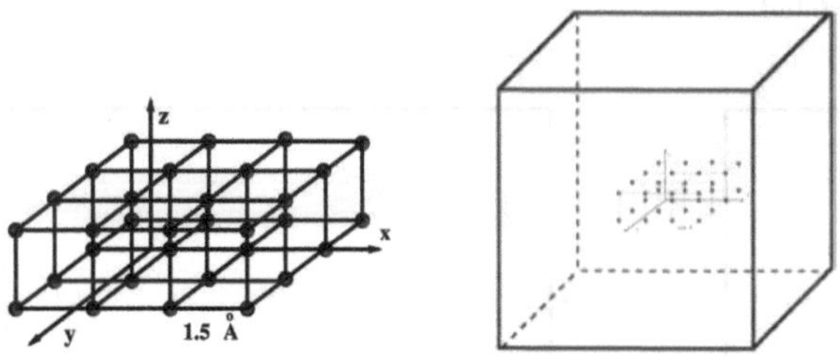

Figure 11.12: The lattice structure of size $4.5 \times 4.5 \times 1.5$ Å^3 in the computational box $[-b, b]^3$ with $b = 16$ au (~8.5 Å).

Several basis functions (e. g., Gaussians) taken for a single atom as the "initialization basis" are duplicated for the lattice atoms, thus, creating the basis set for the whole molecular system. For model problems, we construct artificial structures using the Hydrogen atoms, for example, in a form of the $4 \times 4 \times 2$ lattice, using Hydrogen molecule H_2 as the "initiating" building block, with the distance between atoms 1.5 Å. Then for a lattice system as described above, one can apply the fast Hartree-Fock solver. Figure 11.13 shows the slice of the nuclear potential calculated for the slab with $4 \times 4 \times 2$ Hydrogen atoms. Figure 11.14 shows the output of the Hartree–Fock eigenvalue problem solver for a cluster of $4 \times 4 \times 2$ Hydrogen atoms. The left figure shows the convergence of the ground-state energy, and the right one demonstrates the lower part of the spectrum $\{\lambda_\mu\}$, $\mu = 1, \ldots, N_b$, where every line corresponds to one λ_μ.

Tensor Hartree–Fock calculations do not have special requirements on the positions of nuclei on the 3D grid; the nuclei in the investigated molecular systems may have an arbitrary position in (x, y, z)-coordinates in the computational box.

Solving the ab initio Hartree–Fock problem for larger clusters of Hydrogen-like atoms by using block circulant and Toeplitz structures in the framework of the linearized Fock operator is considered in [151, 154]. The reformulation of the nonlinear Hartree–Fock equation for periodic molecular systems, based on the Bloch theory [37], has been addressed in the literature for more than forty years ago, and nowadays there are several implementations mostly relying on the analytic treatment of arising integral operators [72, 235, 88]. Mathematical analysis of spectral problems for PDEs

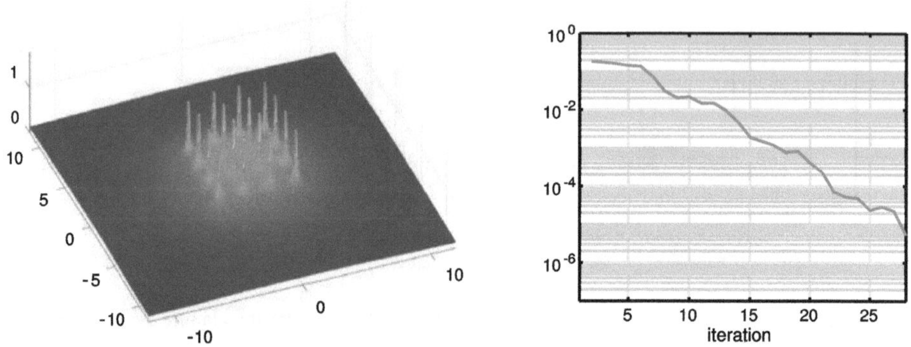

Figure 11.13: Left: cross-section of the nuclear potential for the $8 \times 4 \times 1$ cluster of H atoms. Right: convergence of the residual in SCF iteration.

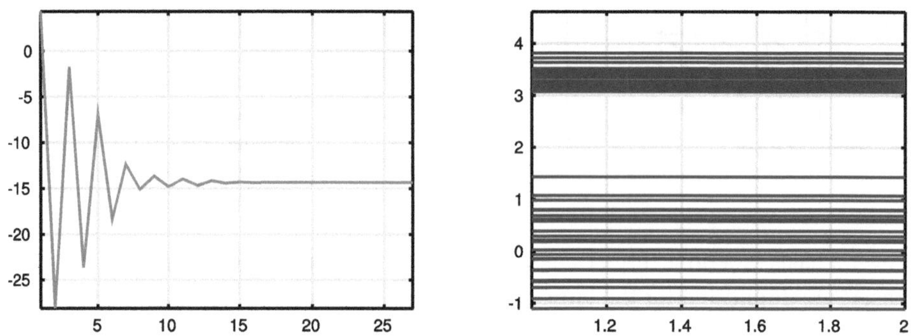

Figure 11.14: Convergence of the ground-state energy for the $4 \times 4 \times 2$ cluster of H atoms (left) and a part of its spectrum (right).

with the periodic-type coefficients was an attractive topic in the recent decade; see [46, 47, 45, 77] and the references therein.

In [154], the new grid-based tensor approach to approximate solution of the elliptic eigenvalue problem for the 3D lattice-structured systems is introduced and analyzed, where the linearized Hartree–Fock equation is considered over a spatial $L_1 \times L_2 \times L_3$ lattice for both periodic and non-periodic problem settings, was discretized in the localized Gaussian-type orbitals basis. In the periodic case, the Galerkin system matrix obeys a three-level block-circulant structure that allows the FFT-based diagonalization, whereas for the finite extended systems in a box (Dirichlet boundary conditions) this matrix allows the perturbed block-Toeplitz representation providing fast matrix-vector multiplication and low storage size.

The above mentioned grid-based tensor techniques manifest the twofold benefits: (a) the entries of the Fock matrix are computed by 1D operations using low-rank tensors represented on a 3D grid; (b) in the periodic case, the low-rank tensor structure in the diagonal blocks of the Fock matrix in the Fourier space reduces the conventional 3D FFT to the product of 1D FFTs.

Lattice-type systems in a box with Dirichlet boundary conditions are treated numerically by the tensor solver as for single molecules, which makes possible calculations on rather large $L_1 \times L_2 \times L_3$ lattices due to reduced numerical cost for 3D problems. The numerical simulations for both box-type and periodic $L \times 1 \times 1$ lattice chains in a 3D rectangular "tube" with L up to several hundred confirm the theoretical complexity bounds for the block-structured eigenvalue solvers at the limit of large L, see [154].

11.8 MP2 calculations by factorized TEI

The Møller–Plesset perturbation theory (MP2) provides an efficient tool for correcting the Hartree–Fock energy by relatively modest numerical efforts [222, 3, 128]. It facilitates the accurate calculation of the molecular gradient energy and other quantities [127]. Since the straightforward calculation of the MP2 correction scales as $O(N_b^5)$ flops with respect to the number of basis functions, efficient methods are consistently developed making the problem tractable for larger molecular systems. The direct method for evaluating the MP2 energy correction and the energy gradient, which reduces the storage needs to $O(N_b^2)$ at the expense of calculation time has been reported in [125].

The advantageous technique using the Cholesky factorization of the two-electron integrals introduced in [17] was efficiently applied for MP2 calculations [4]. A linear scaling MP2 scheme for extended systems is considered in [6]. Recently, the MP2 scheme attracted much interest due to efficient algorithms for the multi-electron integrals [304, 247], its density fitting approach exhibiting a low cost (considering extended molecular systems [216, 298, 241]), and owing to application of tensor factorization methods [306]. An efficient MP2 algorithm that is applicable to large extended molecular systems in the framework of the DFT model is based on the Laplace transform reformulation of the problem and usage of the multipole expansion [311].

Following [150], here we describe an approach to compute the Møller–Plesset correction to the Hartree–Fock energy with reduced numerical cost based on using the factorized tensor representation of TEI matrix. Notice that the auxiliary redundancy-free factorization of TEI is obtained in a "black-box" way, that is, without physical insight into the molecular configuration.

The TEI matrix is precomputed in a low-rank format obtained via truncated Cholesky factorization (approximation). This induces separability in the molecular orbitals transformed TEI matrix and in the doubles amplitude tensor. Such an approach reduces the asymptotic complexity of the MP2 calculations from $O(N_b^5)$ to $O(N_b^3 N_{\mathrm{orb}})$, where N_b is the total number of basis functions, whereas N_{orb} denotes the number of occupied orbitals. The rank parameters are estimated for both the orbital basis transformed TEI and for the doubles amplitude tensors. Notice that using the QTT tensor approximation [167] of long N_b^2-vectors in the Cholesky factor allows to reduce the storage consumption and CPU times by a factor of about ≃10 in both TEI and MP2 calculations.

The efficiency of MP2 energy correction algorithm was tested in [150] for some compact molecules, including glycine and alanine amino acids. Due to factorized tensor representations of the involved multidimensional data arrays, the MP2 calculation times turned out to be rather moderate compared to those for TEI tensor, ranging from one second for water molecule to approximately 4 minutes for glycine molecule. The numerical accuracy is controlled by the given threshold $\varepsilon > 0$ due to stable tensor-rank reduction algorithms.

11.8.1 Two-electron integrals in a molecular orbital basis

In what follows, we describe the main ingredients of the computational scheme introduced in [150], which reduces the cost by using low-rank tensor decompositions of arising multidimensional data arrays.

Let $C = \{C_{\mu i}\} \in \mathbb{R}^{N_b \times N_b}$ be the coefficient matrix representing the Hartree–Fock molecular orbitals (MO) in the atomic orbitals (AO) basis set $\{g_\mu\}_{1\le\mu\le N_b}$ (obtained in the Hartree–Fock calculations). First, one has to transform the TEI tensor $\mathbf{B} = [b_{\mu\nu\lambda\sigma}]$ computed in the initial AO basis set to that represented in the MO basis

$$\mathbf{B} \mapsto \mathbf{V} = [v_{iajb}] : v_{iajb} = \sum_{\mu,\nu,\lambda,\sigma=1}^{N_b} C_{\mu i} C_{\nu a} C_{\lambda j} C_{\sigma b} b_{\mu\nu\lambda\sigma}, \quad a, b \in I_{\text{vir}},\ i, j \in I_{\text{occ}}, \tag{11.19}$$

where $I_{\text{occ}} := \{1, \dots, N_{\text{orb}}\}$, $I_{\text{vir}} := \{N_{\text{orb}} + 1, \dots, N_b\}$, with N_{orb} denoting the number of occupied orbitals. In what follows, we shall use the notation

$$N_{\text{vir}} = N_b - N_{\text{orb}}, \qquad N_{ov} = N_{\text{orb}} N_{\text{vir}}.$$

Hence we have $\mathbf{V} \in \mathbb{R}^{\mathcal{I}}$, where

$$\mathcal{I} := (I_{\text{vir}} \times I_{\text{occ}}) \times (I_{\text{vir}} \times I_{\text{occ}}) \subset I_b^{\otimes 4}.$$

Straightforward computation of the tensor $\mathbf{V}$ in the above representation makes the dominating impact to the overall numerical cost of MP2 calculations $O(N_b^5)$.

Given the tensor $\mathbf{V} = [v_{iajb}]$, the second-order MP2 perturbation to the HF energy is calculated by

$$E_{\text{MP2}} = -\sum_{a,b\in I_{\text{vir}}} \sum_{i,j\in I_{\text{occ}}} \frac{v_{iajb}(2v_{iajb} - v_{ibja})}{\varepsilon_a + \varepsilon_b - \varepsilon_i - \varepsilon_j}, \tag{11.20}$$

where the real numbers ε_k, $k = 1, \dots, N_b$, represent the HF eigenvalues. Notice that the denominator in (11.20) remains strongly positive if $\varepsilon_a > 0$ for $a \in I_{\text{vir}}$ and $\varepsilon_i < 0$

for $i \in I_{occ}$. The latter conditions (nonzero homo lumo gap) will be assumed in the following.

Introduce the so-called doubles amplitude tensor **T**,

$$\mathbf{T} = [t_{iajb}] : \quad t_{iajb} = \frac{(2v_{iajb} - v_{ibja})}{\varepsilon_a + \varepsilon_b - \varepsilon_i - \varepsilon_j}, \quad a, b \in I_{vir};\ i, j \in I_{occ},$$

then the MP2 perturbation takes the form of a scalar product of rank-structured tensors:

$$E_{MP2} = -\langle \mathbf{V}, \mathbf{T} \rangle = -\langle \mathbf{V} \odot \mathbf{T}, \mathbf{1} \rangle,$$

where the summation is restricted to the subset of indices $\mathcal{I}$, and **1** denotes the rank-1 all-ones tensor. Define the reciprocal "energy" tensor

$$\mathbf{E} = [e_{abij}] := \left[\frac{1}{\varepsilon_a + \varepsilon_b - \varepsilon_i - \varepsilon_j} \right], \quad a, b \in I_{vir};\ i, j \in I_{occ}, \tag{11.21}$$

and the partly transposed tensor (transposition in indices a and b)

$$\mathbf{V}' = [v'_{iajb}] := [v_{ibja}].$$

Now the doubles amplitudes tensor **T** will be further decomposed into the sum

$$\mathbf{T} = \mathbf{T}^{(1)} + \mathbf{T}^{(2)} = 2\mathbf{V} \odot \mathbf{E} - \mathbf{V}' \odot \mathbf{E}, \tag{11.22}$$

where each term in the right-hand side above will be treated separately.

11.8.2 Separation rank estimates and numerical illustrations

In this section, we show that the rank $R_B = O(N_b)$ approximation to the symmetric TEI matrix $B \approx LL^T$ with the Cholesky factor $L \in \mathbb{R}^{N \times R_B}$ leads to the low-rank representation of the tensor **V** and the R_B-term decomposition of **T**. This reduces the asymptotic complexity of MP2 calculations to $O(N_b^3 N_{orb})$ and also provides certain computational benefits. In particular, it reduces the storage costs.

Lemma 11.1 ([150]). *Given the rank-R_B Cholesky decomposition of the matrix B, the matrix unfolding $V = [v_{ia;jb}]$ allows a rank decomposition with rank $\le R_B$. Moreover, the tensor $\mathbf{V}' = [v_{ibja}]$ enables an R_B-term decomposition of mixed form.*

Proof. Let us denote by $L_k = L_k(\mu; \nu)$, $k = 1, \ldots, R_B$, a matrix unfolding of the vector $L(:, k) \in \mathbb{R}^{N_b \times N_b}$ in the Cholesky factor $L \in \mathbb{R}^{N_b^2 \times R_B}$; notice that the Cholesky factorization can be written pointwise as follows:

$$b_{\mu\nu;\lambda\sigma} \approx \sum_{k=1}^{R_B} L_k(\mu; \nu) L_k(\sigma; \lambda).$$

Let $C_m = C(:,m)$, $m = 1,\dots,N_b$ be the mth column of the coefficient matrix $C = \{C_{\mu i}\} \in \mathbb{R}^{N_b \times N_b}$. Then, the rank-$R_B$ representation of the matrix unfolding $V = [v_{ia;jb}] \in \mathbb{R}^{N_{ov} \times N_{ov}}$ takes the form

$$V = L_V L_V^T, \quad L_V \in \mathbb{R}^{N_{ov} \times R_B},$$

where

$$L_V((i-1)N_{\mathrm{vir}} + a; k) = C_i^T L_k C_a, \quad k = 1,\dots,R_B,\ a = 1,\dots,N_{\mathrm{vir}},\ i = 1,\dots,N_{\mathrm{orb}}.$$

This is justified by the following transformations:

$$\begin{aligned} v_{iajb} &= \sum_{\mu,\nu,\lambda,\sigma=1}^{N_b} C_{\mu i} C_{\nu a} C_{\lambda j} C_{\sigma b} b_{\mu\nu\lambda\sigma} \\ &\approx \sum_{k=1}^{R_B} \sum_{\mu,\nu,\lambda,\sigma=1}^{N_b} C_{\mu i} C_{\nu a} C_{\lambda j} C_{\sigma b} L_k(\mu;\nu) L_k(\sigma;\lambda) \\ &= \sum_{k=1}^{R_B} \left(\sum_{\mu,\nu=1}^{N_b} C_{\mu i} C_{\nu a} L_k(\mu;\nu) \right) \left(\sum_{\lambda,\sigma=1}^{N_b} C_{\lambda j} C_{\sigma b} L_k(\sigma;\lambda) \right) \\ &= \sum_{k=1}^{R_B} (C_i^T L_k C_a)(C_b^T L_k^T C_j). \end{aligned} \tag{11.23}$$

This proves the first statement. Furthermore, the partly transposed tensor $\mathbf{V}' := [v_{ibja}]$ allows an R_B-term decomposition derived similarly to (11.23):

$$v'_{iajb} = v_{ibja} = \sum_{k=1}^{R_B} (C_i^T L_k C_b)(C_a^T L_k^T C_j). \tag{11.24}$$

This completes the proof. □

It is worth noting that one has to compute and store the only L_V factor in the above symmetric factorizations of V and V'. Hence, the storage cost of decompositions (11.23) and (11.24) restricted to the active index set $I_{\mathrm{vir}} \times I_{\mathrm{occ}}$ amounts to $R_B N_{\mathrm{vir}} N_{\mathrm{orb}}$ numbers. The complexity of straightforward computation can be estimated by $O(R_B N_b^2 N_{\mathrm{orb}})$.

Next, we consider separable representation of the tensor $\mathbf{T}$ in (11.22). To that end, we first apply low-rank canonical ε-approximation to the tensor $\mathbf{E}$. The following lemma describes the canonical approximation to the tensor $\mathbf{E}$ that converges exponentially fast in the rank parameter.

Lemma 11.2 ([150]). *Suppose that the so-called homo lumo gap is estimated by*

$$\min_{a \in I_{\mathrm{vir}}, i \in I_{\mathrm{occ}}} |\varepsilon_a - \varepsilon_i| \geq \frac{\delta}{2} > 0.$$

Then the rank-R_E canonical approximation to the tensor $\mathbf{E} \approx \mathbf{E}_{R_E}$, *with* $R_E = 2M + 1$,

$$e_{a,b,i,j} \approx \sum_{p=-M}^{M} c_p e^{-\alpha_p(\varepsilon_a+\varepsilon_b-\varepsilon_i-\varepsilon_j)}, \quad \alpha_p > 0, \tag{11.25}$$

with the particular choice

$$h = \pi/\sqrt{M}, \quad \alpha_p = e^{ph}, \quad c_p = h\alpha_p, \text{ and } M = O(|\log\varepsilon \log\delta|)$$

provides the error bound

$$\|\mathbf{E} - \mathbf{E}_{R_E}\|_F \leq O(\varepsilon).$$

Proof. Consider the sinc-quadrature approximation of the Laplace transform applied to the 4th-order Hilbert tensor

$$\frac{1}{x_1 + x_2 + x_3 + x_4} = \int_0^\infty e^{-t(x_1+x_2+x_3+x_4)} dt \approx \sum_{p=-M}^{M} c_p e^{-\alpha_p(x_1+x_2+x_3+x_4)}$$

for $x_i \geq 0$ such that $\sum x_i > \delta$, which converges exponentially in M (see [111, 93]). This proves the statement. □

Notice that the matrix V exhibits an exponential decay in the singular values (observed in numerical experiments; see Figure 11.15), which means that the approximation error $\varepsilon > 0$ can be achieved with the separation rank $R_V = O(|\log\varepsilon|)$. Figure 11.15 illustrates the exponential convergence in the rank parameter for the low-rank approximation of matrices V and $E = [e_{ab;ij}]$.

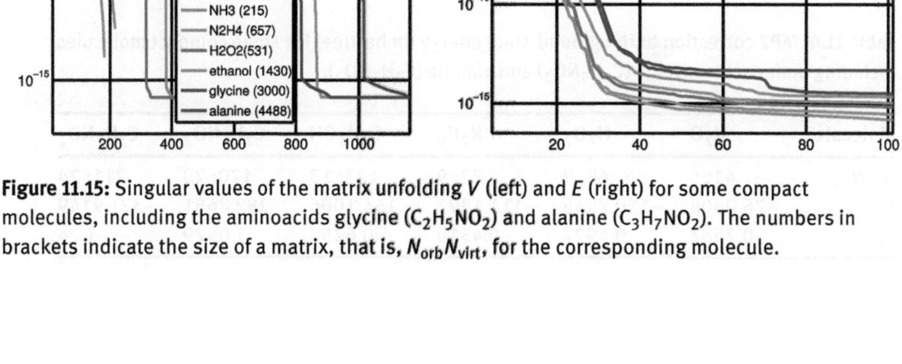

Figure 11.15: Singular values of the matrix unfolding V (left) and E (right) for some compact molecules, including the aminoacids glycine ($C_2H_5NO_2$) and alanine ($C_3H_7NO_2$). The numbers in brackets indicate the size of a matrix, that is, $N_{orb}N_{virt}$, for the corresponding molecule.

11.8.3 Complexity bounds, sketch of algorithm, QTT compression

Lemmas 11.1 and 11.2 result in the following complexity bound: The Hadamard product $\mathbf{V} \odot \mathbf{T}$ and the resultant functional E_{MP2} can be evaluated at the expense

$$O(R_E R_B^2 N_{\mathrm{occ}} N_{\mathrm{vir}}).$$

Indeed, the first term in the splitting $\mathbf{T} = \mathbf{T}^{(1)} + \mathbf{T}^{(2)}$ is represented by rank-structured tensor operations

$$\mathbf{T}^{(1)} = 2\mathbf{V} \odot \mathbf{E} = 2[t^{(1)}_{iajb}],$$

where

$$t^{(1)}_{iajb} = \sum_{p=1}^{R_E} c_p \sum_{k=1}^{R_B} (e^{\alpha_p \varepsilon_i} C_i^T L_k e^{-\alpha_p \varepsilon_a} C_a)(e^{-\alpha_p \varepsilon_b} C_b^T L_k^T e^{\alpha_p \varepsilon_j} C_j) \tag{11.26}$$

and $L_k = L_k(:,:)$ stands for the $N_b \times N_b$ matrix unfolding of the Cholesky vector $L(:,k)$. Then the numerical complexity of this rank-$(R_E R_B)$ separable approximation is estimated via the multiple of R_E with the corresponding cost for the treatment of the tensor $\mathbf{V}$, that is, $O(R_E R_B N_{\mathrm{occ}} N_{\mathrm{vir}})$. Furthermore, the R_B-term decomposition of $\mathbf{V}' := [v_{ibja}]$ (see (11.24)) again leads to the summation over $(R_E R_B)$-term representation of the second term in the splitting of $\mathbf{T}$,

$$\mathbf{T}^{(2)} = [t^{(2)}_{iajb}] = \mathbf{V}' \odot \mathbf{E},$$

where

$$t^{(2)}_{iajb} = \sum_{p=1}^{R_E} c_p \sum_{k=1}^{R_B} (e^{\alpha_p \varepsilon_i} C_i^T L_k e^{-\alpha_p \varepsilon_a} C_b)(e^{-\alpha_p \varepsilon_b} C_a^T L_k^T e^{\alpha_p \varepsilon_j} C_j). \tag{11.27}$$

This makes the main contribution to the overall cost.

Based on the rank decompositions of the matrix B, the energy tensor $\mathbf{E}$, and the doubles amplitude tensor $\mathbf{T}$, we utilize the final Algorithm 4 to compute the MP2 energy correction (see [150]).

Table 11.4: MP2 correction to the ground-state energy (in hartree) for some compact molecules, including aminoacids glycine ($C_2H_5NO_2$) and alanine ($C_3H_7NO_2$).

Molecules	H_2O	H_2O_2	N_2H_4	C_2H_5OH	$C_2H_5NO_2$	$C_3H_7NO_2$
$N_b; N_{\mathrm{orb}}$	41; 5	68; 9	82; 9	123; 13	170; 20	211; 24
E_0	-76.0308	−150.7945	−111.1897	−154.1006	−282.8651	−321.9149
E_{MP2}	−0.2587	−0.4927	−0.4510	−0.6257	−1.0529	−1.24

Algorithm 4 Fast tensor-structured computation of the MP2 energy correction.

Input: Rank-R_B factorization LL^T of B, coefficient matrix C, and Hartree–Fock eigenvalues $\varepsilon_1, \dots, \varepsilon_{N_b}$, error tolerance $\varepsilon > 0$.
(1) Compute the column vectors in the rank-R_B decomposition of matrix $V = [v_{ia;jb}]$, $C_i^T L_k C_a$, $k = 1, \dots, R_B$ $(i, a = 1, \dots, N_b)$ as in (11.23).
(2) Precompute the matrix factors in R_B-term decomposition of $V' = [v_{ib;ja}]$ as in (11.24).
(3) Construct the canonical decomposition of "energy"tensor $\mathbf{E} = [e_{a,b,i,j}]$ by the sinc-quadrature $e_{a,b,i,j} \approx \sum_{p=-M}^{M} c_p e^{-\alpha_p(\varepsilon_a+\varepsilon_b-\varepsilon_i-\varepsilon_j)}$ as in (11.25) (Lemma 11.2).
(4) Compute the tensor $\mathbf{T}^{(1)} = 2\mathbf{V} \odot \mathbf{E}$ as in (11.26) using rank decompositions of V and $\mathbf{E}$.
(5) Compute the tensor $\mathbf{T}^{(2)} = \mathbf{V}' \odot \mathbf{E}$ as in (11.27) using R_B-term decompositions of V' and $\mathbf{E}$.
(6) Compute the MP2 correction by the "formatted"scalar product operation:

$$E_{\mathrm{MP2}} = -\langle \mathbf{V}, \mathbf{T}^{(1)} + \mathbf{T}^{(2)} \rangle.$$

Output: MP2 energy correction E_{MP2}.

Table 11.4 presents the effect of MP2 correction for several compact molecules. In most cases, this correction amounts to about 0.4 % of the total energy.

The tensor-structured factorization of the TEI matrix B makes it possible to reduce the overall cost of MP2 calculations to $O(N_b^2 N_{\mathrm{vir}} N_{\mathrm{orb}})$ by using the QTT approximation of the long column vectors in the Cholesky factor L. Figure 10.4 (left) indicates that the average QTT ranks of columns vectors in the Cholesky factor and of the vectorized density matrix $C \in \mathbb{R}^{N_b \times N_b}$ remains almost the same (they depend only on the entanglement properties of a molecule), and they can be estimated by

$$\mathrm{rank}_{\mathrm{QTT}}(L(:,k)) \approx \mathrm{rank}_{\mathrm{QTT}}(C_k) \le 3N_{\mathrm{orb}}, \quad k = 1, \dots, N_B.$$

This hidden structural property implies that the computation and storage cost for the matrix $V = L_V L_V^T$ involved in Algorithm 4 (the most expensive part in the MP2 calculation) can be reduced to $O(N_{\mathrm{orb}}^2)$ at the main step in (11.23), that is, computing $C_i^T L_k C_a$ instead of N_b^2, thus indicating the reduced redundancy in the AO basis in the case of compact molecules. Since the QTT rank enters quadratically the storage cost for QTT vectors, we conclude that

$$(3N_{\mathrm{orb}})^2 \le C N_b^2,$$

where the constant C is estimated by $C \approx 0.1$, taking into account that the typical relation $N_b \approx 10 \cdot N_{\mathrm{orb}}$ holds in the case of Gaussian-type basis sets.

Further reduction of the numerical complexity can be based on taking into account the more specific properties of the matrix unfolding V when using a physical insight to the problem (say, flat or extended molecules, multiple symmetries, lattice type or periodic structures, accounting data sparsity, etc.).

Other methods for high-accuracy energy calculations are based on coupled clusters technique, which requires much larger computations resources; see, for example, [260, 13, 249].

12 Calculation of excitation energies of molecules

12.1 Numerical solution of the Bethe–Salpeter equation

Recently, computation of excitation energies and absorption spectra for molecules and surfaces of solids attracted much interest due to the related promising applications, in particular, in the development of *sustainable energy* technologies. The traditional methods for computer simulation of excitation energies for molecular systems require large computational facilities. Therefore, there is a steady need for new algorithmic approaches for calculating the absorption spectra of molecules with less computational cost and having a good potential for application to larger systems. The tensor-based approach seems to present a good alternative to conventional methods.

One of the well established ab initio methods for computation of excited states is based on the solution of the Bethe–Salpeter equation (BSE) [252, 126], which in turn is based on the Green's function formalism and many-body perturbation theory, providing calculation of the excitation energies in a self-consistent way [224, 259, 194, 245]. The BSE method leads to a challenging computational task of solving a large eigenvalue problem for a fully populated (dense) matrix, which, in general, is non-symmetric. Another commonly used approach for computation of the excitation energies is based on time-dependent density functional theory (TDDFT) [251, 107, 51, 274, 56, 248].

The size of the BSE matrix scales quadratically $\mathcal{O}(N_b^2)$ in the size N_b of the atomic orbitals basis sets commonly used in ab initio electronic structure calculations. The direct diagonalization of $\mathcal{O}(N_b^6)$-complexity becomes prohibitive even for moderate size molecules with size of the atomic orbitals basis set $N_b \approx 100$. Therefore, an approximation that relies entirely on multiplications of the governing BSE matrix, or its approximation with vectors in the framework of some iterative procedure, is the only feasible strategy. In turn, fast matrix–vector computations can be based on the use of low-rank matrix representations since such data structures allow efficient storage and basic linear algebra operations with linear complexity scaling in the matrix size.

An efficient method was introduced in [23] for approximate numerical solution of the BSE eigenvalue problem by using the low-rank approximation, which leads to relaxation of the numerical costs from $O(N^6)$ down to $O(N^2)$. It is based on the construction of a simplified problem by a diagonal plus rank-structured representation of a system matrix so that the related spectral problem can be solved iteratively. Then model reduction via the projection onto a reduced basis is constructed by using the representative set of eigenvectors of a simplified system matrix. The further enhancement based on the block-diagonal plus low-rank approximation to BSE matrix for accuracy improvement was presented in [25].

The particular construction of the BSE system matrix in [23] is based on the non-interacting Green's function in terms of eigenfunctions and eigenvalues of the

https://doi.org/10.1515/9783110365832-012

Hartree–Fock operator introduced in [243, 244], where it was applied to the simple H_2 molecule in the minimal basis of two Slater functions, and where the system matrix entries are evaluated analytically. In [23] it was shown that this computational scheme for solving the BSE becomes practically applicable to moderate size compact molecules when using the tensor-structured Hartree–Fock calculations [147, 152] yielding efficient representation of the two-electron integrals (TEI) in the molecular orbitals basis in a form of a low-rank Cholesky factorization [157, 150].

The low-rank representation of TEI tensor stipulates the beneficial structure of the BSE matrix blocks, thus enabling efficient numerical algorithms for solution of large structured eigenvalue problems. The simplified block decomposition in the BSE system matrix is characterized by the separation rank of order $O(N_b)$, which enables compact storage and fast matrix–vector multiplications in the framework of iterations on a subspace for computation of a few (lowest/largest) eigenvalues. To reduce the error of the diagonal plus low-rank approximation, it was proposed in [25] to represent the static screen interaction part in the BSE matrix by a small fully populated subblock with adaptively chosen size.

In [25] efficient iterative schemes are introduced for computing several tens of the smallest in modulo eigenvalues for both the BSE problem and its Tamm–Dancoff approximation (TDA). The most efficient subspace iteration is based on application of the matrix inverse, which for the considered matrix formats can be evaluated in a structural form by using the Sherman–Morrison–Woodbury formula [269]. The numerical experiments show that our method is economical (at least up to small amino-acids), where the numerical cost for computing several hundreds eigenvalues decreases by the orders of magnitude. Usually, the smallest in modulo singular values of BSE problem are of most interest in applications.

12.2 Prerequisites from Hartree–Fock calculations

As the prerequisites for constructing the generating matrices for the BSE eigenvalue problem, we use the results of ab initio tensor-based Hartree–Fock and MP2 calculations; see Sections 11.5 and 11.8. They provide a full set of necessary quantities for fast and economical computation of a set of lowest in magnitude part of the BSE spectrum (here, we follow notations from [23]):

- full set of eigenvalues of the Hartree–Fock EVP

$$\varepsilon_1, \dots, \varepsilon_{N_b};$$

- the full set of Galerkin coefficients of the expansion of molecular orbitals in a Gaussian basis

$$C = \{c_{\mu i}\} \in \mathbb{R}^{N_b \times N_b};$$

- the two-electron integral matrix $B = [b_{\mu\nu,\kappa\lambda}] \in \mathbb{R}^{N_b^2 \times N_b^2}$ computed in a form of low-rank Cholesky factorization

$$B \approx LL^T, \quad L \in \mathbb{R}^{N_b^2 \times R_B}, \quad R_B = O(N_b), \tag{12.1}$$

and presented in the molecular orbitals basis

$$B \mapsto \mathbf{V} = [v_{iajb}], \quad \text{where } v_{iajb} = \sum_{\mu,\nu,\kappa,\lambda=1}^{N_b} C_{\mu i} C_{\nu a} C_{\kappa j} C_{\lambda b} b_{\mu\nu,\kappa\lambda}. \tag{12.2}$$

The indices $i,j \in \mathcal{I}_o := \{1,\dots,N_{\text{orb}}\}$ correspond to occupied orbitals, and $a,b \in \mathcal{I}_v$ to virtual ones, $\mathcal{I}_v := \{N_{\text{orb}},\dots,N_b\}$. Denote $N_v = N_b - N_{\text{orb}}$ and $N_{ov} = N_{\text{orb}} N_v$ and further use the short notation $N_o = N_{\text{orb}}$.

The BSE calculations utilize the two subtensors of $\mathbf{V}$ specified by the index sets $\mathcal{I}_o$ and $\mathcal{I}_v$. The first subtensor is defined as in the MP2 calculations [150]

$$\mathbf{V} = [v_{iajb}] : \quad a,b \in \mathcal{I}_v,\ i,j \in \mathcal{I}_o, \tag{12.3}$$

whereas the second one lives on the extended index set

$$\widehat{\mathbf{V}} = [\widehat{v}_{turs}] : \quad r,s \in \mathcal{I}_v,\ t,u \in \mathcal{I}_o. \tag{12.4}$$

In what follows, $\{C_i\}$ and $\{C_a\}$ denote the sets of occupied and virtual orbitals, respectively. We shall also use the notation.

Denote the associated matrix by $V = [v_{ia,jb}] \in \mathbb{R}^{N_{ov} \times N_{ov}}$ in case (12.3), and similarly by $\widehat{V} = [\widehat{v}_{tu,rs}] \subset \mathbb{R}^{N_o^2 \times N_v^2}$ in case (12.4). The straightforward computation of the matrix V by the above representations accounts for the dominating impact on the overall numerical cost of order $O(N_b^5)$ in the evaluation of the block entries in the BSE matrix.

Recall that the rank $R_B = O(N_b)$ approximation to matrix $B \approx LL^T$ with the $N \times R_B$ Cholesky factor L allows introducing the low-rank representation of the tensor $\mathbf{V}$, and then to reduce the asymptotic complexity of calculations to $O(N_b^4)$, [150]; see Section 11.8, Lemma 11.1. A similar factorization can be derived in the case of (12.4).

The following statement is a slight modification of Lemma 11.1.

Lemma 12.1. *Let the rank-R_B Cholesky decomposition of the matrix B be given by* (12.1). *Then the R_B-term representation of the matrix $V = [v_{ia;jb}]$ takes the form*

$$V = L_V L_V^T, \quad L_V \in \mathbb{R}^{N_{ov} \times R_B}, \tag{12.5}$$

where the columns of L_V are given by

$$L_V((i-1)N_{\text{vir}} + a - N_o; k) = C_i^T L_k C_a, \quad k = 1,\dots,R_B,\ a \in \mathcal{I}_v,\ i \in \mathcal{I}_o.$$

On the index set (12.4), *we have*

$$\widehat{V} = U_{\widehat{V}} W_{\widehat{V}}^T \in \mathbb{R}^{N_o^2 \times N_v^2}$$

with $U_{\widehat{V}} \in \mathbb{R}^{N_o^2 \times R_B}$, $W_{\widehat{V}} \in \mathbb{R}^{N_v^2 \times R_B}$.

The numerical cost is determined by the computation complexity and storage size for the factors L_V, $U_{\widehat{V}}$, and $W_{\widehat{V}}$ in the above rank-structured factorizations.

Lemma 12.1 provides the upper bounds on rank(V) in the representation (12.5), which might be reduced by the SVD based ε-rank truncation. It can be shown that the ε-rank of the matrix V remains of the same magnitude as that for the TEI matrix B obtained by its ε-rank truncated Cholesky factorization (see the numerical illustration in Section 12.4).

Numerical tests in [150] (see also Sections 10 and 11.8) indicate that the singular values of the TEI matrix B decay exponentially as

$$\sigma_k \leq Ce^{-\frac{\gamma}{N_b}k}, \tag{12.6}$$

where the constant $\gamma > 0$ in the exponential depends weakly on the molecule configuration. If we define $R_B(\varepsilon)$ as the minimal number satisfying the condition

$$\sum_{k=R_B(\varepsilon)+1}^{R_B} \sigma_k^2 \leq \varepsilon^2, \tag{12.7}$$

then estimate (12.6) leads to the ε-rank bound $R_B(\varepsilon) \leq CN_b|\log \varepsilon|$, which will be postulated in the following.

Note that the matrix rank $R_V(\varepsilon)$ increases only logarithmically in ε, similarly to the bound for $R_B(\varepsilon)$. This can be formulated as the following lemma (see [23]),

Lemma 12.2. *For given $\varepsilon > 0$, there exist a rank-r approximation V_r of the matrix V and a constant $C > 0$ not depending on ε such that $rR_V(\varepsilon) \leq R_B(\varepsilon)$ and*

$$\|V_r - V\| \leq CN_b\varepsilon|\log \varepsilon|.$$

12.3 Tensor factorization of the BSE matrix blocks

Here, we discuss the main ingredients for calculation of blocks in the BSE matrix and their reduced rank approximate representation. We compose the $2N_{ov} \times 2N_{ov}$ BSE matrix by following equations (46a) and (46b) in [243], though the construction of static screened interaction matrix $w(ij, ab)$ in equation (12.11) below may slightly differ; see also [23].

Construction of the BSE matrix includes computation of several auxiliary quantities. First, introduce a fourth-order diagonal "energy" matrix by

$$\mathbf{\Delta\varepsilon} = [\Delta\varepsilon_{ia,jb}] \in \mathbb{R}^{N_{ov} \times N_{ov}}: \quad \Delta\varepsilon_{ia,jb} = (\varepsilon_a - \varepsilon_i)\delta_{ij}\delta_{ab}$$

which can be represented in the Kronecker product form

$$\mathbf{\Delta\varepsilon} = I_o \otimes \text{diag}\{\varepsilon_a : a \in \mathcal{I}_v\} - \text{diag}\{\varepsilon_i : i \in \mathcal{I}_o\} \otimes I_v,$$

where I_o and I_v are the identity matrices on respective index sets. It is worth noting that if the so-called *homo lumo* gap of the system is positive, i. e.,

$$\varepsilon_a - \varepsilon_i > \delta > 0, \quad a \in \mathcal{I}_v, i \in \mathcal{I}_o,$$

then the matrix $\mathbf{\Delta\varepsilon}$ is invertible.

Using the matrix $\mathbf{\Delta\varepsilon}$ and the $N_{ov} \times N_{ov}$ TEI matrix $V = [v_{ia,jb}]$ represented in the MO basis as in (12.2), the *dielectric function* ($N_{ov} \times N_{ov}$ matrix) $Z = [z_{pq,rs}]$ is defined by

$$z_{pq,rs} := \delta_{pr}\delta_{qs} - v_{pq,rs}[\chi_0(\omega = 0)]_{rs,rs},$$

where $\chi_0(\omega)$ is the matrix form of the so-called Lehmann representation to the *response function*. In turn, the representation of the inverse matrix of $\chi_0(\omega)$ is known to have a form

$$\chi_0^{-1}(\omega) = -\begin{pmatrix} \mathbf{\Delta\varepsilon} & \mathbf{0} \\ \mathbf{0} & \mathbf{\Delta\varepsilon} \end{pmatrix} + \omega \begin{pmatrix} \mathbf{1} & \mathbf{0} \\ \mathbf{0} & \mathbf{-1} \end{pmatrix},$$

implying

$$\chi_0(0) = -\begin{pmatrix} \mathbf{\Delta\varepsilon}^{-1} & \mathbf{0} \\ \mathbf{0} & \mathbf{\Delta\varepsilon}^{-1} \end{pmatrix}.$$

Define the rank-1 matrix $\mathbf{1} \otimes \mathbf{d}_\varepsilon$, where $\mathbf{1} \in \mathbb{R}^{N_{ov}}$ is the all-ones vector, and $\mathbf{d}_\varepsilon = \operatorname{diag}\{\mathbf{\Delta\varepsilon}^{-1}\} \in \mathbb{R}^{N_{ov}}$ is the diagonal vector of $\mathbf{\Delta\varepsilon}^{-1}$. In this notation, the matrix $Z = [z_{pq,rs}]$ takes a compact form

$$Z = I_o \otimes I_v + V \odot (\mathbf{1} \cdot \mathbf{d}_\varepsilon^T). \tag{12.8}$$

Introducing the inverse matrix Z^{-1}, we finally define the so-called *static screened interaction* matrix by

$$W = [w_{pq,rs}]: \quad w_{pq,rs} := \sum_{t\in\mathcal{I}_v, u\in\mathcal{I}_o} z^{-1}_{pq,tu} v_{tu,rs}. \tag{12.9}$$

In the forthcoming calculations, this equation is considered on the conventional and extended index sets $\{p, s \in \mathcal{I}_o\} \times \{q, r \in \mathcal{I}_v\}$ and $\{p, q \in \mathcal{I}_o\} \times \{r, s \in \mathcal{I}_v\}$, respectively, such that $v_{tu,rs}$ corresponds either to sub-tensor in (12.3) or in (12.4).

On the conventional index set, we obtain the following matrix factorization of $W := [w_{ia,jb}]$,

$$W = Z^{-1}V, \quad \text{provided that } a, b \in \mathcal{I}_v,\ i, j \in \mathcal{I}_o,$$

where V is calculated by (12.3). Lemma 12.1 suggests the existence of a low-rank factorization for the matrix W defined above.

Lemma 12.3 ([23]). *Let the matrix Z defined by (12.8) over the index set $a, b \in \mathcal{I}_v$, $i, j \in \mathcal{I}_o$ be invertible. Then the rank of the respective matrix $W = Z^{-1}V$ is bounded by*

$$\mathrm{rank}(W) \le \mathrm{rank}(V) \le R_B.$$

Furthermore, equation (46a) in [243] includes matrix entries $w_{ij,ab}$ for $a, b \in \mathcal{I}_v$, $i,j \in \mathcal{I}_o$. To this end, the modified matrix $\overline{W} = [\overline{w}_{pq,rs}]$ is computed by (12.9) on the extended index set $\{p, q \in \mathcal{I}_o\} \times \{r, s \in \mathcal{I}_v\}$ by using entries $\widehat{v}_{ij,ab}$ in the matrix unfolding of the tensor $\widehat{\mathbf{V}}$ in (12.4) multiplied from the left with the $N_o^2 \times N_o^2$ sub-matrix of Z^{-1}.

Now the matrix representation of the Bethe–Salpeter equation in the (ov, vo) subspace reads as the following eigenvalue problem:

$$F\begin{pmatrix} \mathbf{x}_n \\ \mathbf{y}_n \end{pmatrix} \equiv \begin{pmatrix} A & B \\ B^* & A^* \end{pmatrix}\begin{pmatrix} \mathbf{x}_n \\ \mathbf{y}_n \end{pmatrix} = \omega_n \begin{pmatrix} I & 0 \\ 0 & -I \end{pmatrix}\begin{pmatrix} \mathbf{x}_n \\ \mathbf{y}_n \end{pmatrix}, \tag{12.10}$$

determining the excitation energies ω_n and the respective excited states. Here, the matrix blocks are defined in the index notation by (see (46a) and (46b) in [243] for more detail)

$$a_{ia,jb} := \Delta\varepsilon_{ia,jb} + v_{ia,jb} - \overline{w}_{ij,ab}, \tag{12.11}$$

$$b_{ia,jb} := v_{ia,bj} - w_{ib,aj}, \quad a, b \in \mathcal{I}_v,\ i, j \in \mathcal{I}_o. \tag{12.12}$$

In the matrix form, we obtain

$$A = \boldsymbol{\Delta\varepsilon} + V - \widehat{W},$$

where the matrix elements in $\widehat{W} = [\widehat{w}_{ia,jb}]$ are defined by $\widehat{w}_{ia,jb} = \overline{w}_{ij,ab}$, computed by (12.9). Here, the diagonal plus low-rank sparsity structure in $\boldsymbol{\Delta\varepsilon} + V$ can be recognized in view of Lemma 12.1. For the matrix block B, we have

$$B = \widetilde{V} - \widetilde{W} = V - \widetilde{W},$$

where the matrix $\widetilde{V}$, which is an unfolding of the partly transposed tensor, is defined entrywise by

$$\widetilde{\mathbf{V}} = [\widetilde{v}_{iajb}] := [v_{iabj}] = [v_{iajb}],$$

and hence it coincides with V in (12.3) due to the symmetry properties. Here, $\widetilde{W} = [\widetilde{w}_{ia,jb}] = [w_{ib,aj}]$ is defined by permutation. The ε-rank structure in the matrix blocks A and B, resulting from the corresponding factorizations of V, has been analyzed in [23].

Solutions of equation (12.10) can be grouped in pairs: excitation energies ω_n with eigenvectors $(\mathbf{x}_n, \mathbf{y}_n)$ and de-excitation energies $-\omega_n$ with eigenvectors $(\mathbf{x}_n^*, \mathbf{y}_n^*)$.

The block structure in the matrices A and B is inherited from the symmetry of the TEI matrix V, $v_{ia,jb} = v_{ai,bj}^*$ and the matrix W, $w_{ia,jb} = w_{bj,ai}^*$. In particular, it is known from the literature that the matrix A is Hermitian, and the matrix B is (complex) symmetric (since $v_{ia,bj} = v_{jb,ai}$ and $w_{ib,aj} = w_{ja,bi}$), which we presuppose in the matrix construction. The literature concerning the discussion of skew-symmetric (Hamiltonian) block structure in BSE matrix can be found in [23].

In the following discussion, we confine ourselves to the case of real spin orbitals; that is, the matrices A and B remain real. The dimension of the matrix in (12.10) is $2N_o N_v \times 2N_o N_v$, where N_o and N_v denote the numbers of occupied and virtual orbitals, respectively. In general, $N_o N_v$ is asymptotically of size $O(N_b^2)$. That is, the spectral problem (12.10) may be computationally extensive. Indeed, the direct eigenvalue solver for (12.10) via diagonalization becomes infeasible due to $O(N_b^6)$ complexity scaling. Furthermore, the numerical cost for calculation of the matrix elements based on the precomputed TEI integrals from the Hartree–Fock equation scales as $O(N_{ov}^2) = O(N_b^4)$, where the low-rank structure in the matrix V can be adapted.

The challenging computational tasks arise in the case of lattice-structured compounds, where the number of basis functions increases proportionally to the lattice size $L \times L \times L$, that is $N_b \approx N_{b,0} L^3$, which quickly leads to intractable problems even for small lattices.

12.4 The reduced basis approach using low-rank approximations

Notice that in realistic quantum chemical simulations of excitation energies, the calculation of several tens of eigenpairs may be sufficient.

As we have already seen, the part $\mathbf{\Delta\varepsilon} + V$ in the matrix block A allows the accurate diagonal plus low-rank (DPLR) structured approximation. Moreover, the sub-matrix $\widetilde{V} = V$ in the block B also inherits the low-rank approximation. Taking into account these structures, a special solver was proposed in [23] for partial eigenvalue problem based on the use of a reduced basis obtained from the eigenvectors of the reduced matrix that picks up only the essential part of the initial BSE matrix with the DPLR structure. The iterative solver is based on fast matrix–vector multiplication and efficient storage of all data involved in the computational scheme. Using the reduced basis approach, the initial problem is then approximated by its Galerkin projection onto a reduced basis of moderate size.

We summarize that the low-rank decomposition of the matrix V,

$$V \approx L_V L_V^T, \quad L_V \in \mathbb{R}^{N_{ov} \times R_V}, \quad R_V = R_V(\varepsilon) = O(N_b |\log \varepsilon|) \le R_B \qquad (12.13)$$

can be optimized depending on the truncation error $\varepsilon > 0$; see also Section 11.8.

In the construction of simplified matrix, we represent matrix blocks A and B included in the BSE matrix by using the rank-structured decompositions.

The properties of the Hadamard product imply that the matrix Z exhibits the representation

$$Z = I_o \otimes I_v + L_V L_V^T \odot (\mathbf{1} \cdot \mathbf{d}_\varepsilon^T) = I_{N_{ov}} + L_V (L_V \odot \mathbf{d}_\varepsilon)^T,$$

where the rank of the second summand does not exceed R_V. Hence, the linear system solver $W = Z^{-1}V$ can be implemented by algorithms tailored to the DPLR structure by adapting the Sherman–Morrison–Woodbury formula.

The computational cost for setting up the full BSE matrix F in (12.10) can be estimated by $O(N_{ov}^2)$, which includes the cost $O(N_{ov}R_B)$ for generating the matrix V and the dominating cost $O(N_{ov}^2)$ for setting up $\widehat{W}$.

We further rewrite the spectral problem (12.10) in the equivalent form

$$F_1\begin{pmatrix}\mathbf{x}_n\\ \mathbf{y}_n\end{pmatrix} \equiv \begin{pmatrix}A & B\\ -B^* & -A^*\end{pmatrix}\begin{pmatrix}\mathbf{x}_n\\ \mathbf{y}_n\end{pmatrix} = \omega_n\begin{pmatrix}\mathbf{x}_n\\ \mathbf{y}_n\end{pmatrix}. \tag{12.14}$$

It is worth noting that the so-called Tamm–Dancoff approximation (TDA) simplifies equation (12.14) to the standard Hermitian eigenvalue problem

$$A\mathbf{x}_n = \mu_n\mathbf{x}_n, \quad \mathbf{x}_n \in \mathbb{R}^{N_{ov}},\ A \in \mathbb{R}^{N_{ov}\times N_{ov}}, \tag{12.15}$$

with the factor two smaller matrix size N_{ov}.

The main idea of the *reduced basis approach* presented here is as follows: Instead of solving the partial eigenvalue problem for finding m_0 eigenpairs thereby satisfying equation (12.14), we first solve the simplified auxiliary spectral problem with a modified matrix F_0. The approximation F_0 is obtained from F_1 by using low-rank approximations of the parts $\widehat{W}$ and $\widetilde{W}$ of the matrix blocks A and B, respectively; that is, A and B are replaced by

$$A \mapsto A_0 := \boldsymbol{\Delta\varepsilon} + V - \widehat{W}_r \quad\text{and}\quad B \mapsto B_0 := V - \widetilde{W}_r, \tag{12.16}$$

respectively. Here, we assume that the matrix V is already represented in the low-rank format in the form (12.13).

The modified auxiliary problem reads

$$F_0\begin{pmatrix}\mathbf{u}_n\\ \mathbf{v}_n\end{pmatrix} \equiv \begin{pmatrix}A_0 & B_0\\ -B_0^* & -A_0^*\end{pmatrix}\begin{pmatrix}\mathbf{u}_n\\ \mathbf{v}_n\end{pmatrix} = \lambda_n\begin{pmatrix}\mathbf{u}_n\\ \mathbf{v}_n\end{pmatrix}. \tag{12.17}$$

This structured eigenvalue problem is much simpler than (12.10) since the matrix blocks A_0 and B_0, defined in (12.16), are composed of diagonal and low-rank matrices. Figures 12.1 and 12.2 illustrate the structure of A_0 and B_0 submatrices in a BSE system matrix.

Given the set of m_0 eigenpairs

$$\{(\lambda_n, \psi_n) = (\lambda_n, (\mathbf{u}_n, \mathbf{v}_n)^T)\},$$

computed for the modified (simplified) problem (12.17), we solve the full eigenvalue problem for the reduced matrix obtained by the Galerkin projection of the initial equation onto the problem-adapted small basis set $\{\psi_n\}$ of size m_0, $\{\psi_n\} \in \mathbb{R}^{2N_{ov}\times 1}$, $n = 1,\dots,m_0$. Here, the quantities λ_n represent the closest to zero eigenvalues of F_0.

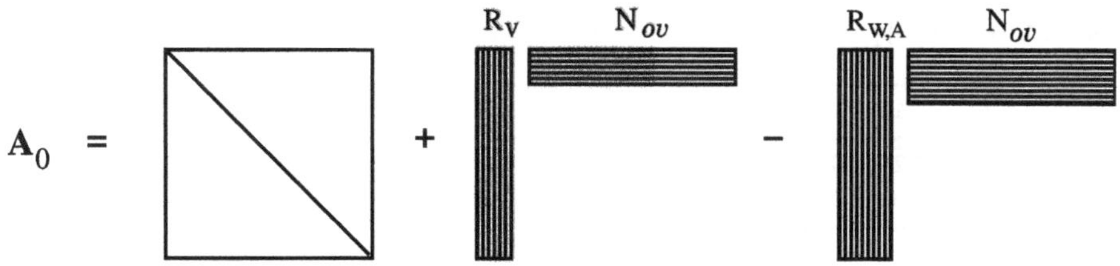

Figure 12.1: The diagonal plus low-rank structure of A_0 block in the modified BSE system matrix.

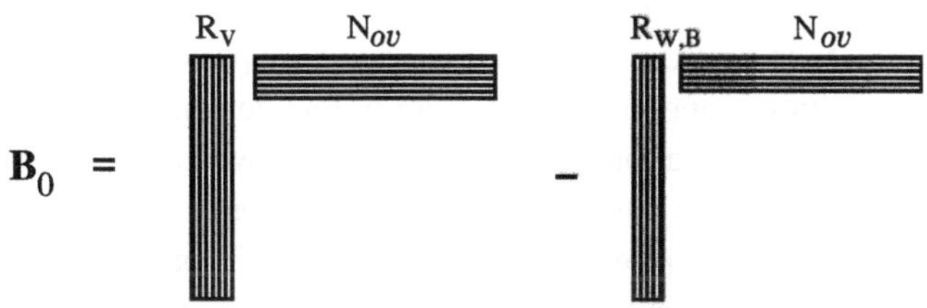

Figure 12.2: The low-rank structure of the block B_0 in the modified BSE matrix.

Define a matrix

$$G_1 = [\psi_1, \psi_2, \ldots, \psi_{m_0}] \in \mathbb{R}^{2N_{ov} \times m_0}$$

whose columns are computed by the vectors in the reduced basis, and then compute the stiffness and mass matrices by projection of the initial BSE matrix F_1 onto the reduced basis specified by the columns in G_1,

$$M_1 = G_1^T F_1 G_1, \quad S_1 = G_1^T G_1 \in \mathbb{R}^{m_0 \times m_0}.$$

The projected generalized eigenvalue problem of small size $m_0 \times m_0$ reads

$$M_1 \mathbf{y} = \gamma_n S_1 \mathbf{y}, \quad \mathbf{y} \in \mathbb{R}^{m_0}. \tag{12.18}$$

The portion of eigenvalues γ_n, $n = 1, \ldots, m_0$, computed by the direct diagonalization is thought to be very close to the corresponding excitation energies ω_n ($n = 1, \ldots, m_0$) in the initial spectral problem (12.10).

The reduced basis approach via low-rank approximation can be applied directly to the TDA equation, such that the simplified auxiliary problem reads

$$A_0 \mathbf{u} = \lambda_n \mathbf{u}, \tag{12.19}$$

where we are interested in finding the m_0 smallest eigenvalues.

Table 12.1 illustrates that the larger the size m_0 of the reduced basis, the better the accuracy of the lowest excitation energy γ_1, as expected [23].

Table 12.1: The error $|\gamma_1 - \omega_1|$ vs. the size of reduced basis, m_0.

m_0	5	10	20	30	40	50
H_2O	0.025	0.025	0.014	0.01	0.01	0.005
N_2H_4	0.02	0.02	0.015	0.015	0.015	0.005

Notice that the matrix $\widehat{W}$ might have rather large ε-rank for small values of ε, which increases the cost of high accuracy solutions. Numerical tests show (see Table 12.2) that the ε-rank approximation to the matrix $\widehat{W}$ with a moderate rank parameter leads to a numerical error in the excitation energies of the order of few percents. For this reason, the paper [23] studies another approximation strategy in which the rank-approximation of the matrix $\widehat{W}$ remains fixed, whereas the matrices V and $\widetilde{W}$ are substituted by their adaptive ε-rank approximations. This approach only slightly improves the numerical efficiency of the method.

Table 12.2: Accuracy (in eV) for the first eigenvalue, $|\gamma_1 - \omega_1|$, vs. ε-ranks for V, $\widehat{W}$, and $\widetilde{W}$.

	ε	$4 \cdot 10^{-1}$	$2 \cdot 10^{-1}$	10^{-1}	10^{-2}
H_2O	$|\gamma_1 - \omega_1|$	0.27	0.27	0.21	$2.1 \cdot 10^{-4}$
	ranks $V,\widehat{W},\widetilde{W}$	6, 9, 6	13, 13, 10	25, 72, 36	60, 180, 92
N_2H_4	$|\gamma_1 - \omega_1|$	0.38	0.38	0.27	$1.6 \cdot 10^{-4}$
	ranks $V,\widehat{W},\widetilde{W}$	11, 17, 11	26, 25, 15	49, 144, 54	117, 657, 196
C_2H_5OH	$|\gamma_1 - \omega_1|$	0.81	0.81	0.4	$1.6 \cdot 10^{-4}$
	ranks $V,\widehat{W},\widetilde{W}$	16, 17, 14	39, 29, 20	71, 105, 74	171, 1430, 296

Matrix blocks in the auxiliary equation (12.17) are obtained by rather rough ε-rank approximation to the initial system matrix. However, we observe much smaller approximations error $\gamma_n - \omega_n$ for solving the projected reduced basis system (12.18) compared with that for auxiliary equation (12.17); see Figures 12.3 and 12.4.

Numerical tests indicate that the difference $\gamma_n - \omega_n$ behaves merely quadratically in the rank truncation parameter ε; see [23] for a more detailed discussion.

In the case of a symmetric matrix, the above-mentioned effect of "quadratic" convergence rate can be justified by a well-known property of the quadratic error behavior in the approximate eigenvalue, computed by the Rayleigh quotient with respect to the perturbed eigenvector (vectors of the reduced basis ψ_n in our construction), compared with the perturbation error in the eigenvector, which is of order $O(\varepsilon)$. This beneficial property may explain the efficiency of the reduced basis approach in this particular application.

In the BSE formulation based on the Hartree–Fock molecular orbitals basis, we may have a slight perturbation of the symmetry in the matrix block $\widehat{W}$; that is, the

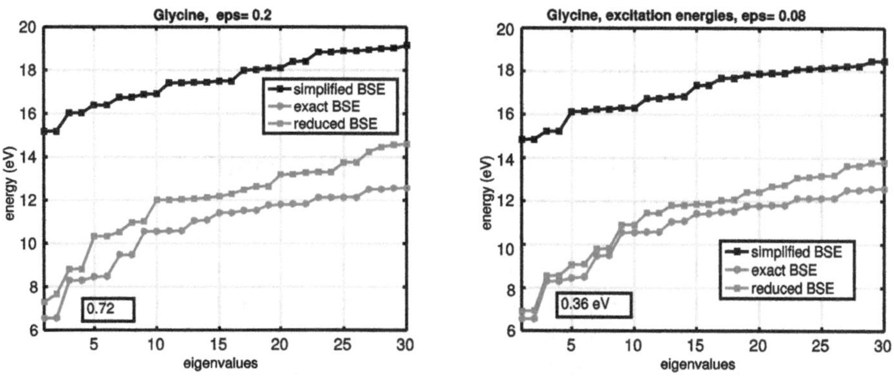

Figure 12.3: Comparison of $m_0 = 30$ lower eigenvalues for the reduced and exact BSE systems vs. ε in the case of Glycine amino acid.

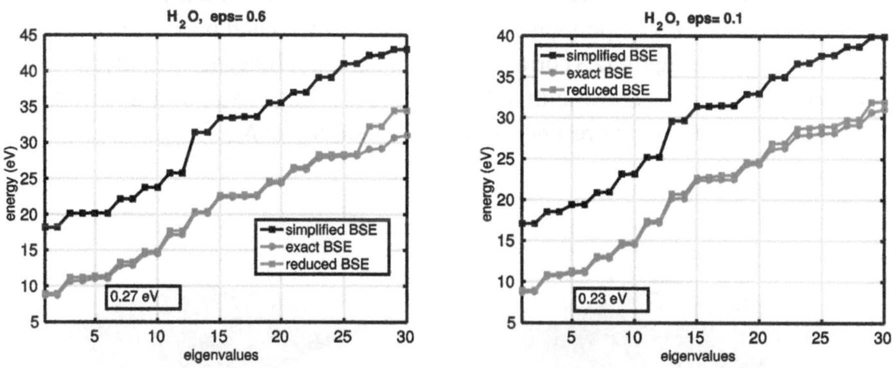

Figure 12.4: Comparison of $m_0 = 30$ lower eigenvalues for the reduced and exact BSE systems for H_2O molecule: $\varepsilon = 0.6$, left; $\varepsilon = 0.1$, right.

above argument does not apply directly. However, we observe the same quadratic error decay in all numerical experiments implemented so far. It is also worth noting that due to the symmetry features of the eigenproblem, the approximation computed by the reduced basis approach is always an *upper bound* of the true excitation energies obtained from the full BSE model. Again, this is a simple consequence of the variational properties of the Ritz values being upper bounds on the smaller eigenvalues for symmetric matrices. The "upper bound" character is also clearly visible in Figures 12.3 and 12.4.

Table 12.2 shows numerics for molecular systems H_2O (360 × 360), N_2H_4 (1430 × 1430), and C_2H_5OH (2860 × 2860), where the BSE matrix size is given in brackets. It demonstrates the quadratic decay of the error $|\gamma_1 - \omega_1|$ in the lowest excitation energy with respect to the approximation error $|\lambda_1 - \omega_1|$ for the modified auxiliary BSE problem

(12.17). The error is controlled by the tolerance $\varepsilon > 0$ in the rank truncation procedure applied to the BSE submatrices V, $\widehat{W}$, and $\widetilde{W}$; see [23] for the detailed discussion.

12.5 Approximating the screened interaction matrix in a reduced-block format

Numerical results in [23] (see Table 12.2) show that using simple diagonal plus low-rank structures for accurate approximation of the BSE system matrix we arrive at large ranks for the representation of the screened matrix $\widehat{W}$, thus deteriorating computational efficiency. The remedy to this problem was found in [25]. It was proposed to substitute the low-rank representation of this part of matrix A_0 by a smaller-size active sub-matrix of corresponding size.

This approach was motivated by the numerical consideration (observed for all molecular systems considered so far) that eigenvectors in the central part of the spectrum have dominating components supported by a rather small part of the full index set of size $2N_{ov}$; see Figure 12.5 corresponding to $m_0 = 30$. Indeed, their effective support is compactly located at the first "active" indexes $\{1, \ldots, N_W\}$ and in the cluster $\{N_{ov}+1, \ldots, N_{ov}+N_W\}$ in the respective blocks, where $N_W \ll N_{ov}$.

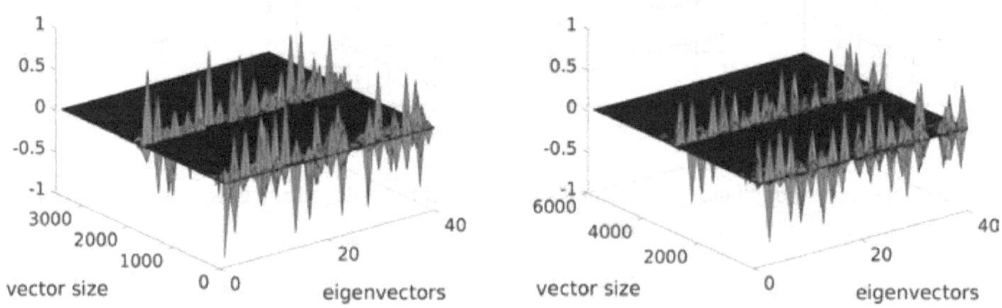

Figure 12.5: Visualizing the first m_0 BSE eigenvectors for the H_{32} chain with $N_W = 554$ (left) and Glycine amino acid molecule with $N_W = 880$ (right).

Following [25], we define the selected sub-matrix $\widehat{W}_b$ in $\widehat{W}$ by keeping the balance between the storage size for the active sub-block $\widehat{W}_b$ and the storage for the matrix V. Since the storage and numerical complexity of the rank-R_V matrix V is bounded by $2R_V N_{ov}$, we control the size of the restricted $N_W \times N_W$ block $\widehat{W}_b$ by the balancing relation

$$N_W \approx C_W \sqrt{2 R_V N_{ov}}, \tag{12.20}$$

where the constant C_W is close to 1. The approximation error introduced due to the corresponding matrix truncation can be controlled by the choice of the constant C_W.

Keeping the diagonal in the matrix $\widehat{W}$ unchanged, we define the simplified matrix by $\widehat{W} \mapsto \widehat{W}_{N_W} \in \mathbb{R}^{N_{ov} \times N_{ov}}$, where

$$\widehat{W}_{N_W}(i,j) = \begin{cases} \widehat{W}(i,j), & i,j \le N_W \text{ or } i = j, \\ 0 & \text{otherwise.} \end{cases} \tag{12.21}$$

The simplified matrix $\widehat{A}$ is then given by

$$A \mapsto \widehat{A} := \mathbf{\Delta\varepsilon} + V - \widehat{W}_{N_W}, \tag{12.22}$$

whereas the modified block B_0 remains the same as in (12.16). The corresponding structure of the simplified matrix $\widehat{A}$ is illustrated in Figure 12.6.

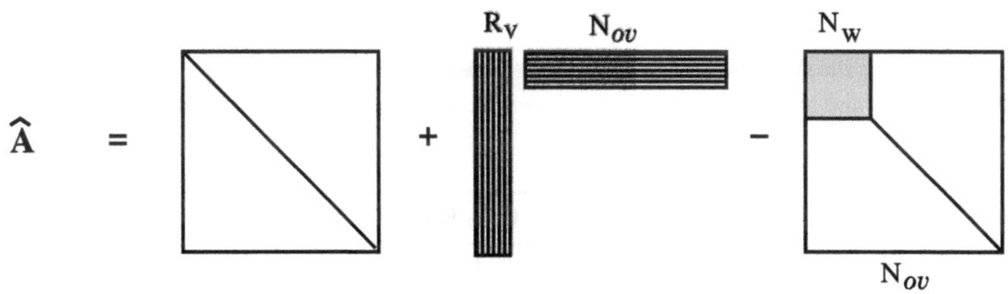

Figure 12.6: Diagonal plus low-rank plus reduced-block structure of the matrix $\widehat{A}$.

This construction guarantees that the storage and matrix–vector multiplication complexity for the simplified matrix block $\widehat{A}$ remains of the same order as that for the matrix V, characterized by a low ε-rank. Table 12.3 demonstrates how the ratio N_W/N_{ov} decreases with the increasing problem size.

Table 12.3: The ratio N_W/N_{ov} for some molecules.

Molecule	H_2O	H_2O_2	N_2H_4	C_2H_5OH	H_{32}	$C_2H_5NO_2$	$C_3H_7NO_2$
N_{ov}	180	531	657	1430	1792	3000	4488
N_W/N_{ov}	0.63	0.5	0.4	0.3	0.32	0.29	0.25

We modify the simplified matrix

$$F_0 \mapsto \widehat{F} \quad \text{by replacing} \quad A_0 \mapsto \widehat{A} \quad \text{in (12.17),}$$

which leads to the corrections in the eigenvalues $\lambda_n \mapsto \widehat{\lambda}_n$ and eigenvectors $G_0 \mapsto \widehat{G} = [\psi_1, \ldots, \psi_{m_0}] \in \mathbb{R}^{2N_{ov} \times m_0}$ by solving the simplified problem

$$\widehat{F}\psi_n = \widehat{\lambda}_n \psi_n \tag{12.23}$$

defined by the low-rank plus block-diagonal approximation $\widehat{F}$ to the initial BSE matrix F. The corresponding eigenvalues $\widehat{\gamma}_n$ of the modified reduced system (12.23) are computed by direct solution of the small size reduced eigenvalue problem

$$\widehat{M}\mathbf{q}_n = \widehat{\gamma}_n \widehat{S}\mathbf{q}_n, \quad \mathbf{q}_n \in \mathbb{R}^{m_0}, \tag{12.24}$$

where the Galerkin and stiffness matrices are specified by

$$\widehat{M} = \widehat{G}^T F \widehat{G}, \quad \widehat{S} = \widehat{G}^T \widehat{G} \in \mathbb{R}^{m_0 \times m_0}.$$

Table 12.4 illustrates the decrease of the approximation error of the simplified and reduced BSE problems by the order of magnitude.

Table 12.4: Accuracies (in eV) of eigenvalues for the reduced BSE problem via simple low-rank approximation $|\omega_1 - \gamma_1|$ and for block diagonal plus low-rank approximation to BSE matrices $|\omega_1 - \widehat{\gamma}_1|$ with the $\epsilon = 0.1$.

Molecule	H_2O	N_2H_4	C_2H_5OH	$C_2H_5NO_2$	$C_3H_7NO_2$
BSE size	360^2	1314^2	2660^2	6000^2	8976^2
$\lvert\omega_1 - \gamma_1\rvert$	0.2	0.27	0.4	0.38	0.53
$\lvert\omega_1 - \widehat{\gamma}_1\rvert$	0.02	0.03	0.08	0.05	0.1

Proposition 12.4 ([25]). *The numerical results indicate the important property observed for all molecular systems tested so far: the close to zero eigenvalues $\widehat{\lambda}_k$ and $\widehat{\gamma}_k$ provide lower and upper bounds for the exact BSE eigenvalues ω_k; that is,*

$$\widehat{\lambda}_k \le \omega_k \le \widehat{\gamma}_k, \quad k = 1, 2, \ldots, \underline{m}_0 \le m_0.$$

The upper bound via the eigenvalues $\widehat{\gamma}_k$ can be explained by the variational form of the reduced problem setting. However, the understanding of the lower bound property, when using the output $\widehat{\lambda}_k$ from the simplified system addresses an interesting open problem.

Figure 12.7 demonstrates the two-sided error estimates declared in Proposition 12.4. Here, the "black" line represents the eigenvalues for the auxiliary problem (12.17), but with the modified matrix $\widehat{F}$, whereas the blue line represents the eigenvalues of the reduced equation (12.24) of type (12.18) with the Galerkin matrices $\widehat{M}$ and $\widehat{S}$. We observe a considerable decrease of the approximation error for both simplified and reduced problems with the diagonal plus low rank plus small block approach for submatrix A as compared with the error of the straightforward diagonal plus low-rank approach presented in Figures 12.3 and 12.4.

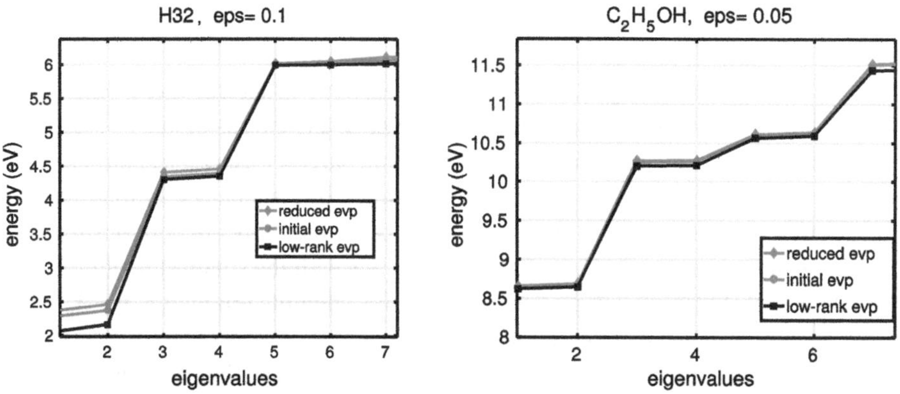

Figure 12.7: Two-sided bounds for the BSE excitation energies for the H_{32} chain (left) and $C_2H_5NO_2$ molecule (right).

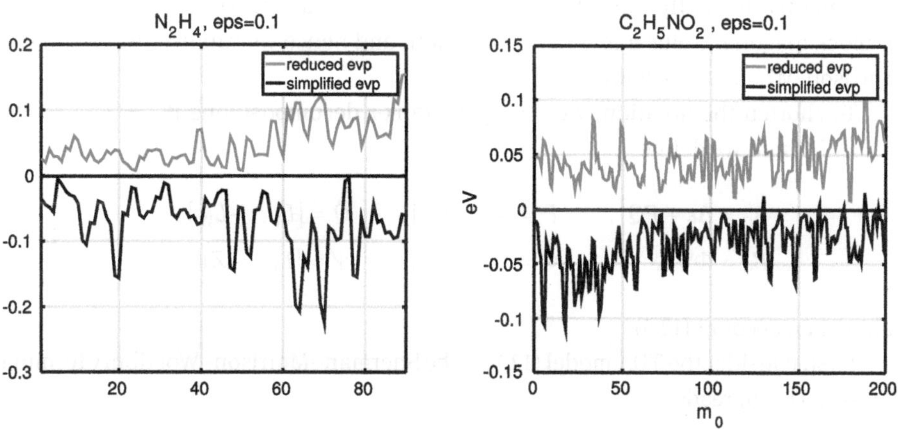

Figure 12.8: Two-sided error bounds: The errors (in eV) in m_0 smallest eigenvalues for simplified and reduced schemes; N_2H_4 molecule (left) and Glycine amino acid $C_2H_5NO_2$ (right).

Figure 12.8 represents examples of upper and lower bounds, i. e., $\widehat{\lambda}_k - \omega_k$ and $\omega_k - \widehat{\gamma}_k$, for the whole sets of $m_0 \leq 250$ eigenvalues for larger molecules. We observe that the lower bound is violated only by few larger excitation energies at the level below the truncation error ϵ.

We conclude that the reduced basis approach, based on the modified auxiliary matrix $\widehat{M}$ via reduced-block approximation (12.22), provides considerably better accuracies $\omega_k - \widehat{\gamma}_k$ than that for γ_k corresponding to matrix M_0. Table 12.4 compares the accuracies $|\omega_1 - \gamma_1|$ for the first eigenvalues of the reduced BSE problem based on the straightforward low-rank approximation from equation (12.18) with accuracies $|\omega_1 - \widehat{\gamma}_1|$ resulting from combined block plus low-rank approximation all computed for several molecules.

12.6 Inverse iteration for diagonal plus low-rank matrix

In this section, following [25], we discuss the efficient structural eigenvalue solver for problem (12.23). Iterative eigenvalue solvers, such as Lanczos or Jacobi–Davidson methods, are quite efficient in approximation of the largest eigenvalues, but may suffer from slow convergence if applied to computation of the smallest or intermediate eigenvalues. We are interested in both of these scenarios. There are both positive and negative eigenvalues in (12.17), and we need the few with the smallest magnitude. In the TDA model (12.15), we solve a symmetric positive definite problem $A_0\mathbf{u} = \lambda_n\mathbf{u}$, but again, the smallest eigenvalues are required.

In both cases, the remedy is to invert the system matrix so that the eigenvalues of interest become largest. The MATLAB interface to ARPACK (procedure `eigs`) assumes by default that the user-defined function solves a linear system with the matrix instead of multiplying it when the smallest eigenvalues are requested. In our case, we can implement this efficiently since the matrix consists of an easily invertible part (diagonal or block-diagonal), plus a low-rank correction, and hence we can use the Sherman–Morrison–Woodbury formula [269].

To shorten the notation, we set up the rank-r decompositions $\widehat{W}_r = L_W L_W^T$ and $\widetilde{W}_r = YZ^T$, and we define

$$\begin{aligned} A_0 &= \mathbf{\Delta\varepsilon} + PQ^T, & P &= [L_V \quad L_W], & Q &= [L_V \quad -L_W], \\ B_0 &= \Phi\Psi^T, & \Phi &= [L_V \quad Y], & \Psi &= [L_V \quad -Z], \end{aligned} \tag{12.25}$$

taking into account (12.5).

First, consider the TDA model (12.15). The Sherman–Morrison–Woodbury formula for A_0 in (12.25) reads

$$A_0^{-1} = \mathbf{\Delta\varepsilon}^{-1} - \mathbf{\Delta\varepsilon}^{-1}P(I + Q^T\mathbf{\Delta\varepsilon}^{-1}P)^{-1}Q^T\mathbf{\Delta\varepsilon}^{-1}. \tag{12.26}$$

Here, the $2r\times 2r$ core matrix $K = (I+Q^T\mathbf{\Delta\varepsilon}^{-1}P)^{-1}$ is small and can be computed explicitly at the expense $\mathcal{O}(r^3+r^2N_{ov})$. Hence, the matrix–vector product $A_0^{-1}\mathbf{u}_n$ requires multiplication by the diagonal matrix $\mathbf{\Delta\varepsilon}^{-1}$ and the low-rank matrix in the second summand. This amounts to the overall cost $\mathcal{O}(N_{ov}r)$. To invert the matrix F_0 in the simplified BSE, we first derive its LU decomposition,

$$F_0 = \begin{bmatrix} A_0 & B_0 \\ -B_0^T & -A_0^T \end{bmatrix} = \begin{bmatrix} A_0 & 0 \\ -B_0^T & I \end{bmatrix} \begin{bmatrix} I & A_0^{-1}B_0 \\ 0 & S \end{bmatrix}, \qquad S = -A_0^T + B_0^T A_0^{-1} B_0. \tag{12.27}$$

To solve a system

$$F_0 \begin{bmatrix} \mathbf{z} \\ \mathbf{y} \end{bmatrix} = \begin{bmatrix} \mathbf{u} \\ \mathbf{v} \end{bmatrix},$$

we need one action of A_0^{-1} and of the inverse of the Schur complement S^{-1}. Indeed,

$$\begin{aligned} \tilde{\mathbf{z}} &= A_0^{-1}\mathbf{u}, & \tilde{\mathbf{y}} &= \mathbf{v} + B_0^T\tilde{\mathbf{z}}, \\ \mathbf{y} &= S^{-1}\tilde{\mathbf{y}}, & \mathbf{z} &= \tilde{\mathbf{z}} - A_0^{-1}B_0\mathbf{y}. \end{aligned} \tag{12.28}$$

Note that $A_0^{-1}B_0$ is a low-rank matrix and can be precomputed in advance. The action of A_0^{-1} is given by (12.26), so we address now the inversion of the Schur complement.

Plugging (12.26) into S, we obtain

$$S = -\boldsymbol{\Delta\varepsilon} - QP^T + \Psi\Phi^T A_0^{-1}\Phi\Psi^T = -(\boldsymbol{\Delta\varepsilon} + Q_S P_S^T),$$

where

$$Q_S = \begin{bmatrix} Q & \Psi(\Phi^T\boldsymbol{\Delta\varepsilon}^{-1}PKQ^T\boldsymbol{\Delta\varepsilon}^{-1}\Phi - \Phi^T\boldsymbol{\Delta\varepsilon}^{-1}\Phi) \end{bmatrix}, \quad P_S = \begin{bmatrix} P & \Psi \end{bmatrix}. \tag{12.29}$$

Therefore,

$$S^{-1} = -(\boldsymbol{\Delta\varepsilon}^{-1} - \boldsymbol{\Delta\varepsilon}^{-1}Q_S K_S P_S^T\boldsymbol{\Delta\varepsilon}^{-1}), \quad K_S = (I + P_S^T\boldsymbol{\Delta\varepsilon}^{-1}Q_S)^{-1}. \tag{12.30}$$

Keeping intermediate results in these calculations, we can trade off the memory against the CPU time. The computational cost of (12.29), and then (12.30), is again bounded by $\mathcal{O}(r^2 N_{ov})$, whereas the implementation of (12.28) takes $\mathcal{O}(rN_{ov})$ operations.

Table 12.5: Times (s) for eigenvalue problem solvers applied to simplified TDA matrix A_0 ("–" means that iterations did not converge).

Molecular syst.	H_2O	N_2H_4	C_2H_5OH	H_{32}	$C_2H_5NO_2$	H_{48}	$C_3H_7NO_2$
TDA matrix size	180^2	657^2	1430^2	1792^2	3000^2	4032^2	4488^2
eig(A_0)	0.02	0.5	4.3	9.8	37.6	91	127.4
lobpcg(A_0)	0.22	0.6	5.4	2.77	18.2	5.6	34.2
lobpcg(inv(A_0))	0.03	0.06	0.15	0.5	0.53	0.5	1.4
eigs(A_0)	0.07	0.29	1.7	0.49	–	–	–
eigs(inv(A_0))	0.05	0.08	0.17	0.11	0.32	0.34	0.5

Precomputation of intermediate matrices and their use in the structured matrix inversion are shown in Algorithms 1 and 2 in [25]. Table 12.5 compares CPU times (sec) for full `eig` and the rank-structured iteration for TDA problem (12.15) in Matlab implementation [25]. The rank-truncation threshold is $\varepsilon = 0.1$; the number of computed eigenvalues is $m_0 = 30$. The bottom line shows the CPU times (sec) of the `eigs` procedure applied with the inverse matrix–vector product $A_0^{-1}\mathbf{u}$ marked by "inv". The other lines show results of the corresponding algorithms, which used the traditional product $A_0\mathbf{u}$ (A_0 in the diagonal plus low-rank form). Notice that the results for Matlab version of

LOBPCG by [190] are presented for comparison. We see that the inverse-based method is superior in all tests.

Notice that the initial guess for the subspace iteration applied to the full BSE can be constructed, thereby replicating the eigenvectors computed in the TDA model. It provides rather accurate approximation to the exact eigenvectors for the initial BSE system (12.14). In [23] it was shown numerically that the TDA approximation error $|\mu_n - \omega_n|$ of order 10^{-2} eV is achieved for the compact and extended molecules presented in Table 12.5.

Table 12.6 compares CPU times (sec) for the full `eig`-solver, and the rank-structured `eigs`-iteration applied to the inverse of simplified rank-structured BSE system (12.17); see [25] for more detail.

Table 12.6: Times (s) for the simplified rank-structured BSE matrix F_0.

Molecule	H_2O	N_2H_4	C_2H_5OH	H_{32}	$C_2H_5\ NO_2$	H_{48}	$C_3H_7\ NO_2$
N_o, N_b	5, 41	9, 82	13, 123	16, 128	20, 170	24, 192	24, 211
BSE matrix size	360^2	1314^2	2860^2	3584^2	6000^2	8064^2	8976^2
`eig`(F_0)	0.08	4.2	33.7	68.1	274	649	903
`eigs`(inv(F_0))	0.13	0.28	0.7	0.77	2.2	2.3	3.9

12.7 Inversion of the block-sparse matrices

If the matrix $\widehat{W}_{N_W}$ is kept in the block-diagonal form as in (12.21)–(12.22), inverting of $\widehat{A} = \mathbf{\Delta\varepsilon} + V - \widehat{W}_{N_W}$ is also easy; the same applies to the case (12.16). The same Sherman–Morrison–Woodbury scheme can be used as in Algorithms 1 and 2 in [25]. To that end, we aggregate $\mathbf{\Delta\varepsilon}_W = \mathbf{\Delta\varepsilon} - \widehat{W}_{N_W}$, whereas in the low-rank factors only $P = Q = L_V$ remains. After that, all calculations as in Subsection 12.6 are retained unchanged just by replacing all $\mathbf{\Delta\varepsilon}$ by $\mathbf{\Delta\varepsilon}_W$, where the latter is now a block-diagonal matrix.

The particular simple modifications for the enhanced algorithm are as follows (see [25]): Let us split $\mathbf{\Delta\varepsilon} = \operatorname{blockdiag}(\mathbf{\Delta\varepsilon}_1, \mathbf{\Delta\varepsilon}_2)$, where $\mathbf{\Delta\varepsilon}_1$ has the size N_W, and $\mathbf{\Delta\varepsilon}_2 \in \mathbb{R}^{N'_W \times N'_W}$ with $N'_W = N_{ov} - N_W$ representing the remaining values. The same applies to $\widehat{W}_{N_W} = \operatorname{blockdiag}(W_b, \operatorname{diag}(w_2))$, where w_2 contains the elements on the diagonal of $\widehat{W}$, which do not belong to W_b. Then the implementation of the matrix inverse

$$\mathbf{\Delta\varepsilon}_W^{-1} = \operatorname{blockdiag}((\mathbf{\Delta\varepsilon}_1 - W_b)^{-1}, (\mathbf{\Delta\varepsilon}_2 - \operatorname{diag}(w_2))^{-1}) \tag{12.31}$$

requires inversion of an $N_W \times N_W$ dense matrix, and a diagonal matrix of size $N'_W = N_{ov} - N_W$. Since N_W is chosen small, the complexity of this operation is moderate. Now all steps requiring multiplication with $\Delta\boldsymbol{\varepsilon}^{-1}$ in Algorithms 1, 2 in [25] can be substituted by (12.31). The numerical complexity of the new inversion scheme is estimated in the following lemma.

Lemma 12.5 ([25] [Complexity of the reduced-block algorithm]). *Suppose that the rank parameters in the decomposition of V and $\widetilde{W}$ do not exceed r and that the block-size N_W is chosen from equation* (12.20). *Then the rank structured plus reduced-block representations of the inverse matrices $\widehat{A}^{-1}$ and $\widehat{F}^{-1}$ can be set up with the overall cost $\mathcal{O}(N_{ov}^{3/2}r^{3/2}+N_{ov}r^2)$. The complexity of each inversion $\widehat{A}^{-1}\mathbf{u}$ or $\widehat{F}^{-1}\mathbf{w}$ is bounded by $\mathcal{O}(N_{ov}r)$.*

Proof. Inversion of the $N_W \times N_W$ dense block in (12.31) requires $\mathcal{O}(N_W^3)$ operations. Hence, condition (12.20) ensures that the cost of setting up the matrix (12.31) is bounded by $\mathcal{O}(N_{ov}^{3/2}r^{3/2})$. After that, multiplication of (12.31) by an $N_{ov} \times r$ matrix requires $\mathcal{O}(N_W^2 r + N_W' r) = \mathcal{O}(N_{ov}(r^2 + r))$ operations. Multiplication of (12.31) by a vector is performed with $\mathcal{O}(N_W^2 + N_W') = \mathcal{O}(N_{ov}r)$ cost. The complexity of the other steps is the same as for diagonal plus low-rank approach. □

Numerical illustrations for the enhanced data sparsity via block-diagonal plus low-rank approximation are presented in Table 12.7.

Table 12.7: Block-sparse matrices: times (s) for eigensolvers applied to TDA and BSE systems. The bottom line shows the error (eV) for the case of block-sparse approximation to the diagonal matrix block $\widehat{A}$, $\varepsilon = 0.1$.

Molecular syst.	H_2O	N_2H_4	C_2H_5OH	H_{32}	$C_2H_5NO_2$	H_{48}	$C_3H_7NO_2$
BSE matrix size	360^2	1314^2	2860^2	3584^2	6000^2	8064^2	8976^2
eigs(inv($\widehat{A}$))	0.07	0.09	0.25	0.77	0.54	3.0	1.0
eigs(inv($\widehat{F}$))	0.21	0.37	1.11	1.10	2.4	2.92	4.6
BSE vs. $\widehat{F}$: $\lvert\widehat{\gamma}_1 - \omega_1\rvert$	0.02	0.03	0.08	0.07	0.05	0.10	0.1

Notice that the performance of the diagonal plus low-rank and block-sparse plus low-rank solvers is comparable, but the second one provides better sparsity and higher accuracy in the computed eigenvalues (see Section 12.5). It is remarkable that the approach, based on the inverse iteration applied to the low-rank plus reduced-block approximation, outperforms the full eigenvalue solver by several orders of magnitude (see Tables 12.6 and 12.7).

The data in previous tables correspond to the choice $m_0 = 30$. Figure 12.9 indicates a merely linear increase in the computational time for the eigs(inv($\widehat{F}$)) solver with respect to the increasing value of m_0.

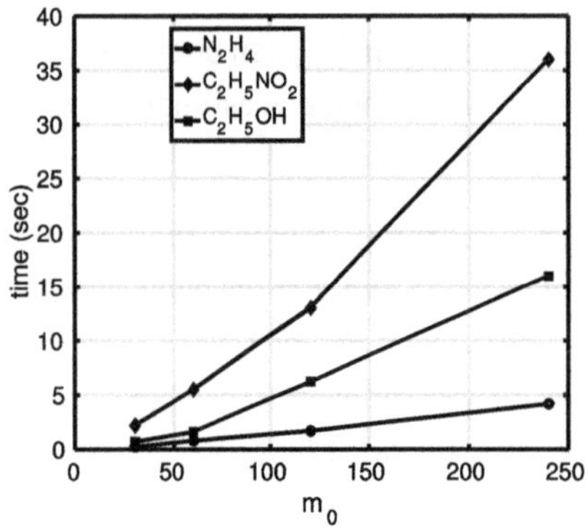

Figure 12.9: CPU times vs. m_0 for N_2H_4 (dashed line), $C_2H_5NO_2$ (solid line), and C_2H_5OH (doted line) molecules.

12.8 Solving BSE spectral problems in the QTT format

Solving the BSE problem in the QTT format [167] was introduced in [25]. In this approach, reduction of the numerical cost in the case of large system size is achieved by adapting the ALS-type iteration (in particular, the DMRG iteration) for computing the eigenvectors in the block-QTT tensor representation [70]. Application of the QTT-approximation is motivated by the observation, known from [150], (see also Section 11.8). Accordingly, the generating Cholesky factors in the TEI tensor exhibit the average QTT-ranks proportional only to the number of occupied orbitals in the molecular system N_o, but they do not depend on the total BSE matrix size $\mathcal{O}(N_b^2)$.

For eigenvectors in the block-QTT format, the QTT ranks are even smaller; typically they are proportional to the number of computed eigenvectors, which makes this approach to solving the BSE eigenvalue problem very competitive. Contrary to the conventional QTT matrix representations in [25], only the columns in the Cholesky factor of a low-rank part in the BSE matrix are approximated in the QTT format, thus keeping the low-rank form $V = LL^T$ and low rank QTT structure for the long column-vectors in L simultaneously. This allows avoiding the prohibitive increase of the QTT matrix rank. See additional detail in [25].

Table 12.8 illustrates that for the TDA model applied to single molecules and to molecular chains, the average QTT ranks computed for the columns in the L_V factor in (12.5), and for each of $m_0 = 30$ TDA-eigenvectors (corresponding to the smallest eigenvalues), are almost equal or even smaller than the number of occupied molecular orbitals N_o in the system under consideration. Notice that these results are obtained by the QTT compression of each column from L_V or eigenvectors separately.

Figure 12.10 indicates that the behavior of the QTT ranks in the columns of the L_V-factor reproduces the system size N_{ov} in terms of N_o on the logarithmic scale.

Recall that in the case of single molecules the commonly used number of GTO basis functions satisfy the relation $N_b/N_o \geq C_{\mathrm{GTO}} \approx 10$ (see examples below), which implies the asymptotic behavior $N_{ov} \approx C_{\mathrm{GTO}} N_o^2$. Hence, the QTT rank estimate $r_{\mathrm{QTT}} \approx N_o$ obtained above leads to the following asymptotic complexity of the QTT-based tensor

Table 12.8: Average QTT ranks of the column vectors in L_V and the m_0 eigenvectors (corresponding to the smallest eigenvalues) in the TDA problem.

Molecular syst.	H_2O	H_{16}	N_2H_4	C_2H_5OH	H_{32}	$C_2H_5\,NO_2$	$C_3H_7\,NO_2$
N_o	5	8	9	13	16	20	24
QTT ranks of L_V	5.4	7	9.1	12.7	14	17.5	21
QTT ranks of eigenvect.	5.3	7.6	9.1	12.7	13.6	17.2	20.9
N_{ov}	180	448	657	1430	1792	3000	4488

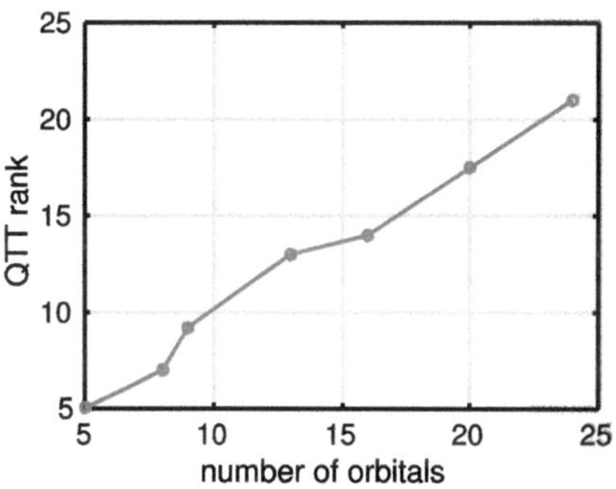

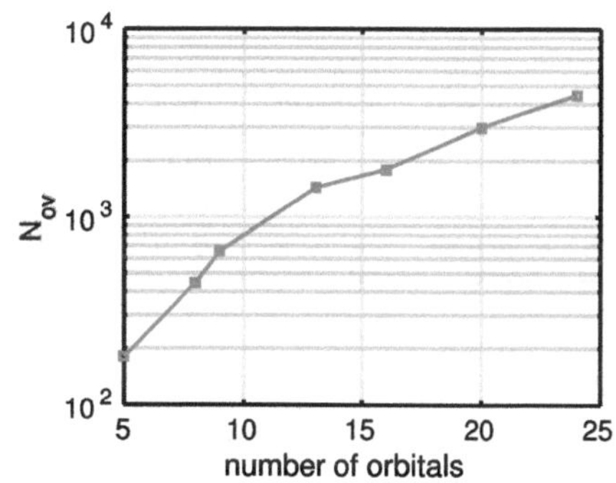

Figure 12.10: QTT ranks (left) and N_{ov} on logarithmic scale (right) vs. N_o.

solver:

$$\mathcal{W}_{\text{BSE}} = \mathcal{O}(\log(N_{ov})r_{\text{QTT}}^2) = \mathcal{O}(\log(N_o)N_o^2), \tag{12.32}$$

which is asymptotically on the same scale (but with smaller prefactor) as that for the data-structured algorithms based on full-vector arithmetics (see Sections 12.6 and 12.7).

The high-precision Hartree–Fock calculations may require much larger GTO basis sets so that the constant C_{GTO} may increase considerably. In this situation, the QTT-based tensor approach seems to outperform the algorithms in full-vector arithmetics.

An even more important consequence of (12.32) is that the rank behavior $r_{\text{QTT}} \approx N_o$ indicates that the QTT tensor-based algorithm has memory requirements and algebraic complexity of order $\mathcal{O}(\log(N_o)N_o^2)$, depending only on the fundamental physical characteristics of the molecular system; the number of occupied molecular orbitals N_o (but not on the system size N_{ov}^2). This remarkable property traces back to the similar feature observed in [157, 150]; that is, QTT ranks of the column vectors in the low-rank Cholesky factors in the TEI matrix are proportional to N_o (about $3\,N_o$).

Based on the previous discussion, we introduce the following hypothesis.

Hypothesis 1. Estimate (12.32) determines the irreducible lower bound on the asymptotic algebraic complexity of the large-scale BSE eigenvalue problems.

The CPU times for QTT calculations are comparable or smaller than the time of the best Sherman–Morrison–Woodbury inversion methods in the previous sections, as demonstrated in Table 12.9 (cf. Table 12.7). Recall that the row referred to as "absolute error" in Table 12.9 represents the quantity $\|\mu_{\mathrm{qtt}} - \mu_\star\| = (\sum_{m=1}^{m_0}(\mu_{\mathrm{qtt},m} - \mu_{\star,m})^2)^{1/2}$ characterizing the total absolute error in the first m_0 eigenvalues calculated in the Euclidean norm. The QTT format provides also a considerable reduction of memory needed to store eigenvectors.

Table 12.9: Time (s) and absolute error (eV) for QTT-DMRG eigensolvers for TDA matrix.

Molecular syst.	C_2H_5OH	H_{32}	$C_2H_5NO_2$	H_{48}	$C_3H_7NO_2$
TDA size	1430^2	1792^2	3000^2	4032^2	4488^2
time QTT eig	0.14	0.23	0.32	0.28	0.63
abs. error (eV)	0.08	0.19	0.17	0.14	0.00034

We now summarize the important result of this section: Lower bound on the asymptotic algebraic complexity $O(N_0^2)$, confirmed by extensive numerical experiments, means that solving the BSE system in the QQT tensor format leads to numerical complexity $O(N_0^2)$, which explicitly indicates the dependence on the number of electrons in a system. This seems to be the asymptotically optimal cost for solving large-scale BSE eigenvalue problems.

Notice that in recent years the analysis of eigenvalue problem solvers for large structured matrices has been widely discussed in the linear algebra community [20, 19, 22]. Tensor-structured approximation of elliptic equations with quasi-periodic coefficients has been considered in [180, 181].

13 Density of states for a class of rank-structured matrices

In this section, we discuss the new numerical approach to approximation of the density of states (DOS) of large rank-structured symmetric matrices. This approach was recently introduced in [27] in the application to estimation of the optical spectra of molecules in the framework of the BSE and TDA calculations; see the discussion in Section 12.1. In this application, the block-diagonal plus low-rank matrix structures arise in the representation of the symmetric TDA matrix. Here, we sketch the techniques for fast DOS calculation applied to the general class of rank-structured matrices.

Several methods for calculating the density of states were originally developed in condensed matter physics [74, 301, 285, 73, 296], and now this topic is also considered in numerical linear algebra community [290, 98, 283]. We refer to a recent survey on commonly used methodology for approximation of DOS for large matrices of general structure [204]. The traditional methods for approximating DOS are usually based on a polynomial or fractional-polynomial interpolation of the exact DOS function, regularized by Gaussians or Lorentzians and subsequent computing traces of certain matrix-valued functions, for example, matrix resolvents or polynomials calculated at a large set of interpolation points within the spectral interval of interest. The trace calculations are typically executed by using the heuristic stochastic sampling over a large number of random vectors [204].

The sizes of matrices arising in quantum chemistry and molecular dynamics computations are usually large, scaling polynomially in the size of a molecular system, whereas the DOS for these matrices often exhibits very complicated shapes. Hence, the traditional approaches mentioned above become prohibitively expensive. Moreover, the algorithms based on polynomial type or trigonometric interpolants have poor approximating properties when the spectrum of a matrix exhibits gaps or highly oscillating non-regular shapes, as is often the case in electronic structure calculations. Furthermore, stochastic sampling relies on Monte Carlo-type error estimates characterized by slow convergence rates and, as result, by low accuracy.

The method presented in [27] to approximate the DOS of Tumm–Dankoff (TDA) Hamiltonian applies to a class of rank-structured matrices, particularly, to the block-diagonal plus low-rank BSE/TDA matrix structures described in [23, 25]. It is based on the Lorentzian blurring [124] such that the most computationally expensive part of the calculation is reduced to the evaluation of traces of the shifted matrix inverses. Fast method is presented for calculating traces of parametric matrix resolvents at interpolation points by taking an advantage of the block-diagonal plus low-rank matrix structure. This allows us to overcome the computational difficulties of the traditional schemes and avoid the need of stochastic sampling.

https://doi.org/10.1515/9783110365832-013

Furthermore, it is shown in [27] that a regularized DOS can be accurately approximated by a low rank QTT tensor [167] that can be determined through a least squares procedure. As a result, an accurate approximation to the regularized DOS living on a large representation grid on the whole spectral interval is realized by the low-rank *QTT tensor interpolation*, calculated by the adaptive cross approximation in the TT format [230]. In what follows, we show that similar techniques can be applied for the efficient computation of DOS for the general classes of rank-structural matrices arising, in particular, in the numerical simulations of lattice-type and quasi-periodic systems.

13.1 Regularized density of states for symmetric matrices

Here we consider the class of symmetric matrices. Following [204], we use the standard definition of the DOS function for symmetric matrices

$$\phi(t) = \frac{1}{n}\sum_{j=1}^{n}\delta(t-\lambda_j), \quad t,\lambda_j \in [0,a], \tag{13.1}$$

where δ is the Dirac delta, and λ_j's are the eigenvalues of the symmetric matrix $A = A^T \in \mathbb{R}^{n\times n}$ ordered as $\lambda_1 \le \lambda_2 \le \cdots \le \lambda_n$.

Several classes of blurring approximations to $\phi(t)$ have been considered in the literature. One can replace each Dirac-δ by a Gaussian function with a small width $\eta > 0$, i. e.,

$$\delta(t) \rightsquigarrow g_\eta(t) = \frac{1}{\sqrt{2\pi}\eta}\exp\left(-\frac{t^2}{2\eta^2}\right),$$

where the choice of the regularization parameter η depends on the particular problem setting. As a result, (13.1) can be approximated by

$$\phi(t) \mapsto \phi_\eta(t) := \frac{1}{n}\sum_{j=1}^{n} g_\eta(t-\lambda_j), \tag{13.2}$$

on the whole energy interval $[0,a]$. Another option is the replacement of each Dirac-δ by a Lorentzian with a small width $\eta > 0$, i. e.,

$$\delta(t) \rightsquigarrow L_\eta(t) := \frac{1}{\pi}\frac{\eta}{t^2+\eta^2} = \frac{1}{\pi}\mathrm{Im}\left(\frac{1}{t-i\eta}\right) \tag{13.3}$$

providing an approximate DOS in the form

$$\phi(t) \mapsto \phi_\eta(t) := \frac{1}{n}\sum_{j=1}^{n} L_\eta(t-\lambda_j). \tag{13.4}$$

As $\eta \to 0_+$, both Gaussians and Lorentzians converge to the Dirac distribution:

$$\lim_{\eta\to 0_+} g_\eta(t) = \lim_{\eta\to 0_+} L_\eta(t) = \delta(t).$$

Both functions $\phi_\eta(t)$ and $L_\eta(t)$ are continuous. Hence, they can be discretized by sampling on a fine grid Ω_h over $[0, a]$, which is assumed to be the uniform cell-centered N-point grid with the mesh size $h = a/N$.

In what follows, we focus on the case of Lorentzians blurring. First, we consider the class of matrices that can be accurately approximated by a block-diagonal plus low-rank ansatz (see [23, 25]), which allows efficient explicit representation of the shifted inverse matrix.

The numerical illustrations below represent the DOS for H_2O molecule broadened by Gaussians (13.2). The data correspond to the reduced basis approach via rank-structured approximation applied to the symmetric TDA model [23, 25] described by the symmetric matrix block A of the full BSE system matrix; see Section 12. Figure 13.1 (left) represents DOS for H_2O computed by using the exact TDA spectrum (blue) and its approximation based on simplified model via low-rank approximation to A (red), whereas the right figure shows the relative error. This suggests that DOS for the initial matrix A of general structure can be accurately approximated by DOS calculated for its structural diagonal plus low rank approximation.

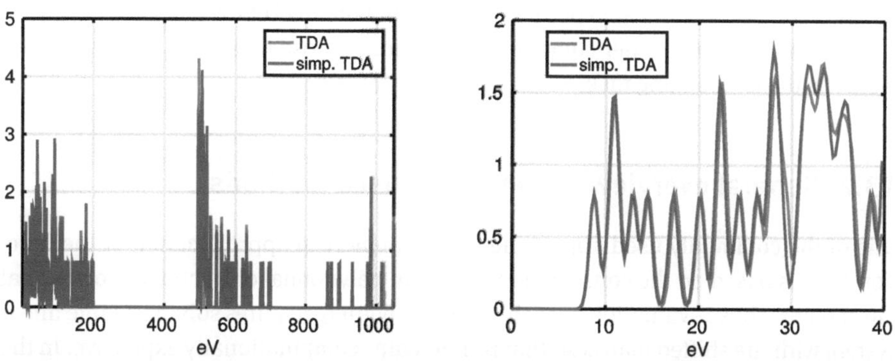

Figure 13.1: DOS for H_2O. Exact TDA vs. simplified TDA (left); zoom of the small spectral interval (right).

Let us briefly illustrate another example of DOS functions arising stochastic homogenization theory. The numerical examples below have been implemented in [158].

Spectral properties of the randomly generated elliptic operators play an important role in the analysis of average quantities in stochastic homogenization. Here, we follow [158] and present the average behavior of the density of spectrum for the family of randomly generated 2D elliptic operators $\{A_m\}$ for the large sequence of stochastic

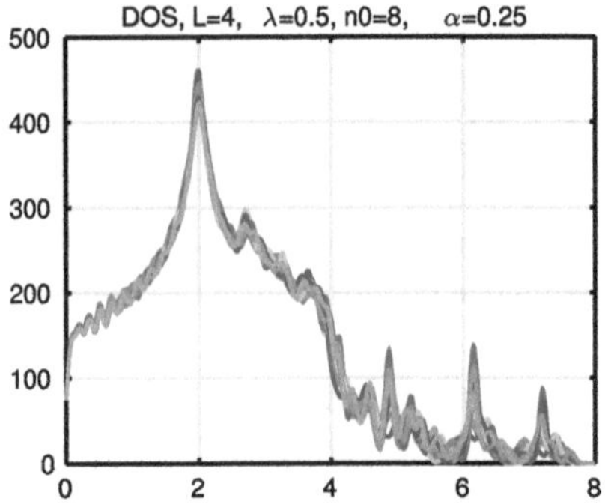

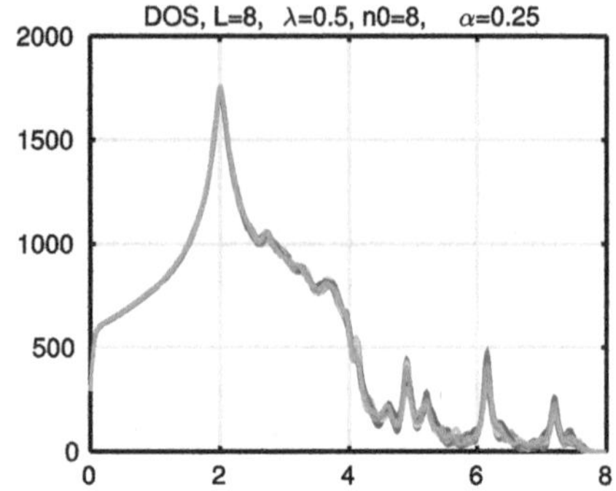

Figure 13.2: Density of states for a number of stochastic processes $M = 1, 2, \dots, 20$ with $L = 4$ (left) and $L = 8$ (right) for $\lambda = 0.5$, $n_0 = 8$, and $\alpha = 0.25$.

realizations. The DOS provides the important spectral characteristics to the differential operator that accumulates the crucial information on the static and dynamical characteristics of the complex physical or molecular system. In particular, the numerics below demonstrate the convergence of DOS to the sample average function at the limit of large number of stochastic realizations with the fixed size of the so-called representative volume element L; see [158] for more detail.

Figure 13.2 represents DOS for a sequence of $M = 1, 2, \dots, 20$ stochastic realizations on $L \times L$ lattice with $L = 4, 8$ from left to right, corresponding to the fixed model parameters. The numerical experiments show that the DOS of the stochastic operator is represented by the rather complicated functions whose numerical approximation might be the challenging task.

13.2 General overview of commonly used methods

One of the commonly used approaches to the numerical approximation of the function $L_\eta(t)$ is based on the construction of certain polynomial or fractional polynomial interpolant whose evaluation at each sampling point t_k requires solving a large linear system with the shifted matrix A, that is, it remains computationally expensive. In the case of Lorentzians broadening (13.4), the regularized DOS takes the form [204]

$$\phi(t) \mapsto \phi_\eta(t) := \frac{1}{n\pi} \sum_{j=1}^{n} \operatorname{Im}\left(\frac{1}{(t - \lambda_j) - i\eta} \right) = \frac{1}{n\pi} \operatorname{Im} \operatorname{Trace}[(tI - A - i\eta I)^{-1}]. \tag{13.5}$$

To keep real-valued arithmetics, one can use the equivalent form

$$\phi_\eta(t) := \frac{1}{n\pi} \sum_{j=1}^{n} \frac{\eta}{(t - \lambda_j)^2 + \eta^2} = \frac{1}{n\pi} \operatorname{Trace}[((tI - A)^2 + \eta^2 I)^{-1}], \tag{13.6}$$

which includes only the real-valued matrix functions.

The advantage of representations (13.5) and (13.6) is that in both cases computing the DOS in the form $\phi_\eta(t)$ allows avoiding the explicit information on the matrix spectra. Indeed, the initial task reduces to approximating the trace of the matrix resolvent

$$f_1(A) = (tI - A - i\eta I)^{-1} \quad \text{or} \quad f_2(A) = ((tI - A)^2 + \eta^2 I)^{-1}. \tag{13.7}$$

The traditional approach [204] to approximately computing the traces of the matrix-valued analytic function $f(A)$ reduces this task to the estimation of the mean of $\mathbf{v}_m^T f(A)\mathbf{v}_m$ over a sequence of random vectors $\mathbf{v}_m$, $m = 1, \ldots, m_r$ that satisfy

$$\mathbb{E}[\mathbf{v}_m] = 0, \quad \mathbb{E}[\mathbf{v}_m \mathbf{v}_m^T] = I.$$

That is, Trace$[f(A)]$ is approximated by

$$\operatorname{Trace}[f(A)] \approx \frac{1}{m_r} \sum_{m=1}^{m_r} \mathbf{v}_m^T f(A)\mathbf{v}_m. \tag{13.8}$$

The calculation of (13.8) for $f_1(A)$ and $f_2(A)$, given by (13.7), reduces to solving linear systems in the form of

$$(tI - i\eta I - A)\mathbf{x} = \mathbf{v}_m \quad \text{for } m = 1, \ldots, m_r \tag{13.9}$$

or

$$(\eta^2 I + (tI - A)^2)\mathbf{x} = \mathbf{v}_m \quad \text{for } m = 1, \ldots, m_r. \tag{13.10}$$

These linear systems have to be solved for many target points $t = t_k \in [a, b]$ in the course of a chosen interpolation scheme and the subset of spectrum of interest.

In the case of rank-structured matrices A, the solution of equations (13.9) or (13.10) can be implemented with a lower cost. However, even in this favorable situation one requires a relatively large number m_r of stochastic realizations to obtain satisfactory mean value approximation. Indeed, following the central limit theorem, the convergence rate is expected to be of order $O(1/\sqrt{m_r})$ at the limit of large number of stochastic realizations. On the other hand, with the limited number of interpolation points, the polynomial type of interpolation schemes applied to highly non-regular shapes as shown, for example, in Figure 13.1 (left), can only provide the poor resolution and it is unlikely to reveal spectral gaps and many local peaks of interest.

13.3 Computing trace of a rank-structured matrix inverse

In what follows, we discuss an approach that is based on evaluating the trace term in (13.5) directly (i. e., without stochastic sampling) provided that the target matrix allows special rank structured representation (approximation).

Definition 13.1. We consider the class of $n \times n$ rank-structured matrices in the form

$$A = E + PP^T \quad \text{with } P \in \mathbb{R}^{n\times R}, \tag{13.11}$$

where the symmetric matrix E allows an efficient computation of
- the traces of inverse matrices, $\operatorname{trace}[E^{-1}]$ and $\operatorname{trace}[E^{-2}]$ and
- the matrix–vector product with E^{-1}

both at the cost $O(n)$ up to some logarithmic factor. For numerical efficiency, the rank parameter R is supposed to be small compared with the matrix size, that is, $R \ll n$.

The rank-structured matrices (13.11) in Definition 13.1 arise in various applications.

Remark 13.2. Definition 13.1 applies, in particular, to the following classes of matrices E in (13.11):
(A) $E = \operatorname{blockdiag}\{B_0, D_0\}$, which arises when using the low-rank BSE matrix structure as in [23, 25] (see Section 12.1).
(B) E is the multilevel block circulant matrix arising in Hartree–Fock calculations for slightly perturbed periodic lattice-structured systems, [154].
(C) E represents homogenized matrix for the FEM-Galerkin approximation to elliptic operators with quasi-periodic coefficients arising, for example, in geometric/stochastic homogenization theory; see [181, 158].

In what follows, we use the notation $\mathbf{1}_m$ for a length-m vector of all ones. The following simple result that generalizes [27] Theorem 3.1 to the more general class of matrices, describes an efficient numerical scheme for calculation of traces of rank-structured matrices specified by Definition 13.1 and asserts that the corresponding cost is estimated by $O(nR^2)$.

Lemma 13.3. *For the matrix A of the form* (13.11), *the trace of the matrix inverse A^{-1} can be calculated explicitly by*

$$\operatorname{trace}[A^{-1}] = \operatorname{trace}[E^{-1}] - \mathbf{1}_n^T (U \odot V)\mathbf{1}_R$$

where $U = E^{-1}PK^{-1} \in \mathbb{R}^{n\times R}$, $V = E^{-1}P \in \mathbb{R}^{n\times R}$, and a small $R \times R$ symmetric core matrix is given by

$$K = I_R + P^T E^{-1} P.$$

Let $K \geq 0$. Then the "symmetric" representation of the trace reads

$$\operatorname{trace}[A^{-1}] = \operatorname{trace}[E^{-1}] - \mathbf{1}_n^T (U \odot U)\mathbf{1}_R,$$

where

$$U = E^{-1}PK^{-1/2} \in \mathbb{R}^{n\times R}.$$

The numerical cost is estimated by $O(nR^2)$ up to low-order term.

Proof. The proof follows the arguments similar to that in Theorem 3.1, [27]. The analysis relies on the particular favorable structure of the matrix E described in Definition 13.1. Indeed, we use the direct trace representation for both rank-R and inverse matrices E^{-1}. The argument is based on the simple observation that the trace of a rank-R matrix UV^T, where $U, V \in \mathbb{R}^{n\times R}$, $U = [\mathbf{u}_1, \ldots, \mathbf{u}_R]$, $V = [\mathbf{v}_1, \ldots, \mathbf{v}_R]$, $\mathbf{u}_k, \mathbf{v}_k \in \mathbb{R}^n$, can be calculated in terms of skeleton vectors by

$$\operatorname{trace}[UV^T] = \sum_{k=1}^{R} \langle \mathbf{u}_k, \mathbf{v}_k \rangle = \mathbf{1}_n^T (U \odot V) \mathbf{1}_R \tag{13.12}$$

at the expense $O(Rn)$. Now define the rank-R matrices by

$$U = E^{-1}PK^{-1}, \quad V = E^{-1}Q.$$

Then the Sherman–Morrison–Woodbury scheme leads to the representation

$$A^{-1} = E^{-1} - UV^T = E^{-1} - E^{-1}PK^{-1}Q^T E^{-1},$$

where $U = E^{-1}PK^{-1}$ and $V = Q^T E^{-1}$. In the symmetric version, we have $U = V = E^{-1}PK^{-1/2}$. Now we apply formula (13.12) representing the trace of a rank-R matrix to obtain the desired representation.

The complexity estimate follows by the assumptions on the matrix E. □

The above representation has to be applied many times for calculating the trace of

$$B = B(t) = tI - A - i\eta I \quad \text{or} \quad B = B(t) = (tI - A)^2 + \eta^2 I$$

at each interpolating point $t = t_m$, $m = 1, \ldots, M$.

We consider the case of real arithmetics that corresponds to the choice

$$B = B(t) = (tI - A)^2 + \eta^2 I.$$

We notice that the price to pay for the real arithmetics in equation (13.10) is the computation with squared matrices, which, however, does not deteriorate the asymptotic complexity since there is no increase of the rank parameter in the rank-structured representation of the target matrix; see Lemma 13.4, which is the respective modification of Theorem 3.2 in [27]. In what follows, we denote by $[U, V]$ the concatenation of two matrices of compatible size.

Lemma 13.4. *Given the matrix $B(t) = (tI - A)^2 + \eta^2 I$, where A is defined by* (13.11), *the trace of the real-valued matrix resolvent $B^{-1}(t)$ can be calculated explicitly by*

$$\operatorname{trace}[B^{-1}] = \operatorname{trace}[\widehat{E}^{-1}] - \mathbf{1}_n^T (\widehat{U} \odot \widehat{V}) \mathbf{1}_{2R} \tag{13.13}$$

with

$$\widehat{U} = \widehat{E}^{-1}\widehat{P}K^{-1} \in \mathbb{R}^{n\times 2R} \quad \textit{and} \quad \widehat{V} = \widehat{E}^{-1}\widehat{Q} \in \mathbb{R}^{n\times 2R},$$

where the real-valued matrix $\widehat{E}$ is given by

$$\widehat{E}(t) = (\eta^2 + t^2)I - 2tE + E^2,$$

and the rank-2R matrices $\widehat{P}, \widehat{Q}$ are represented via concatenation

$$\widehat{P} = [-2tQ + EQ + QE + Q(Q^T Q), Q] \in \mathbb{R}^{n\times 2R}, \quad \widehat{Q} = [Q, EQ] \in \mathbb{R}^{n\times 2R},$$

such that the small core matrix $K(t) \in \mathbb{R}^{2R\times 2R}$ takes the form $K(t) = I_R + \widehat{Q}^T \widehat{E}^{-1}(t)\widehat{P}$.
The numerical cost is estimated by $O(nR^2)$ up to a low-order term.

Proof. Given the block-diagonal plus low-rank matrix A in the form (13.11), we obtain

$$B = (tI - A)^2 + \eta^2 I = \widehat{E} + \widehat{P}\widehat{Q}^T, \tag{13.14}$$

where the block-diagonal matrix $\widehat{E}$ and the rank-2R matrix $\widehat{P}\widehat{Q}^T$ are defined as above. We apply the Sherman–Morrison–Woodbury scheme to the structured matrix B, then Theorem 13.3 implies the desired representation. Now we take into account that $\widehat{E}$ is the matrix polynomial in $E(t)$ of degree 2; then the assumptions on the trace properties of E prove the complexity bound. □

Based on Lemmas 13.3 and 13.4, the calculation of DOS can be implemented efficiently in real arithmetics. Notice that a similar statement to Lemma 13.4 holds in the case of complex arithmetics; see the discussion in [27], Theorem 3.1.

The following numerics demonstrate the efficiency of DOS calculations for the rank-structured TDA matrix in the form (13.13) implemented in real arithmetics (MATLAB). In this case the initial block-diagonal matrix E is given by $E = \text{blockdiag}\{B_0, D_0\}$ as described in Section 12.1.

Figure 13.3 illustrates that using only the structure-based trace representation (13.13) in Lemma 13.4, we obtain the approximation that resolves perfectly the DOS

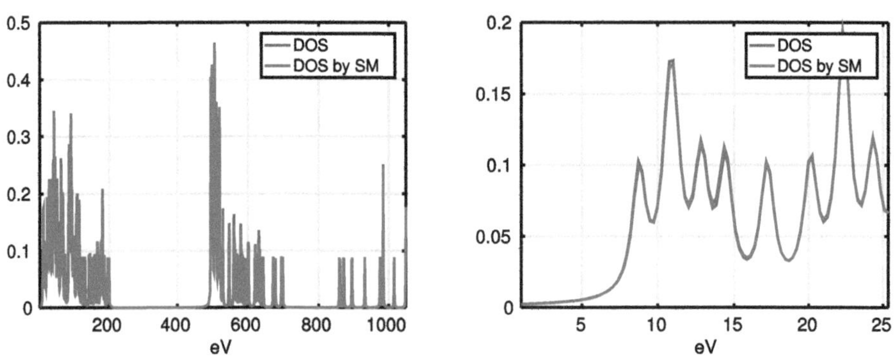

Figure 13.3: Left: DOS for H_2O vs. its recovering by using the trace of matrix resolvents; Right: zoom in the small energy interval.

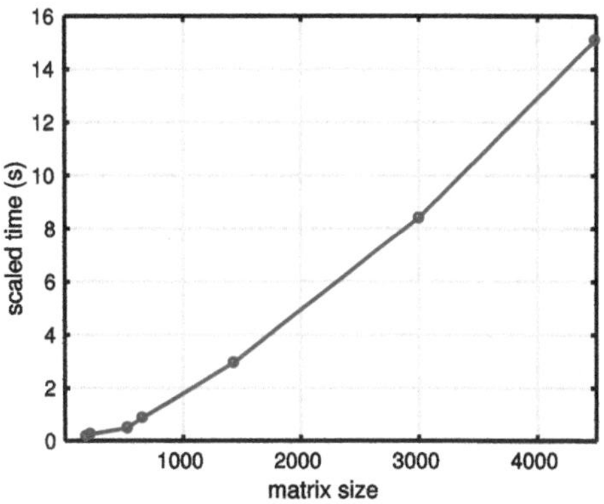

Figure 13.4: The rescaled CPU time vs. $n = N_{ov}$ for algorithm in Theorem 13.4.

function on the examples of H_2O molecule. See [27] for numerical examples for several moderate size molecules.

Figure 13.4 shows the rescaled CPU time, that is, T/R, where T denotes the total CPU time for computing DOS by the algorithm implementing (13.13). We applied the algorithm to the different system size n (i. e., the size of TDA matrices considered in Section 12.1), varying from $n = 180$ till $n = 4488$. In all cases, the N-point representation grid with fixed $N = 2^{14}$ was used. This indicates that the numerical performance of the algorithm is even better than the theoretical complexity $O(nR^2)$; see more numerics in [27].

13.4 QTT approximation of DOS via Lorentzians: rank bounds

In what follows, we outline the perspectives of QTT approximation to DOS along the line of the discussion in [27].

In the case of large grid size N, the number of representation parameters for the corresponding high-order QTT tensor can be reduced to the logarithmic scale $\mathcal{O}(\log N)$, which allows the QTT tensor interpolation of the target N-vector by using only $\mathcal{O}(\log N) \ll N$ functional calls. We demonstrate how to apply this approximation technique to long N-vectors representing the DOS sampled over the fine representation grid Ω_h.

The QTT approximant can be viewed as the rank-structured interpolant to the highly non-regular function ϕ_η regularizing the exact DOS. In this case, the application of traditional polynomial or trigonometric-type interpolation is inefficient. We apply the QTT approximation method to the DOS regularized by Lorentzians and sampled on fine representation grid of size $N = 2^d$.

In [27] it is shown that the QTT approach provides a good approximation to ϕ_η on the whole spectral interval and requires only a moderate number of representation parameters $r_{qtt}^2 \log N \ll N$, where the average QTT rank r_{qtt} is a small rank parameter depending on the truncation error $\epsilon > 0$.

In the following numerical examples, we use a sampling vector defined on a fine grid of size $N \approx 2^{14}$. We fix the QTT truncation error to $\epsilon_{QTT} = 0.04$ (if not explicitly indicated). For ease of interpretation, we set the pre-factor in (13.1) equal to 1. It is worth noting that the QTT-approximation scheme is applied to the full TDA spectrum. Our results demonstrate that the QTT-approximant renders good resolution in the whole range of energies (in eV), including large "zero gaps".

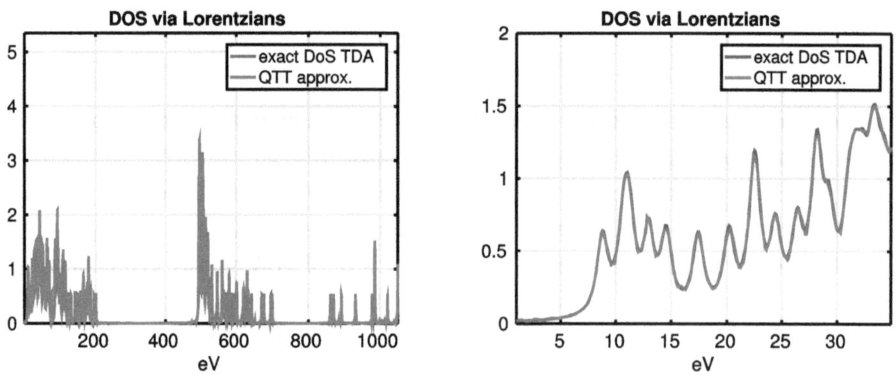

Figure 13.5: DOS for H_2O molecule via Lorentzians (blue) and its QTT approximation (red) (left). Zoom in the small energy interval (right).

Figure 13.5 (left) represents the TDA DOS (blue line) for the H_2O computed via the Lorentzian blurring with the parameter $\eta = 0.4$ and the corresponding rank-9.4 QTT tensor approximation (red line) to the discretized function $\phi_\eta(t)$. For this example, the number of eigenvalues is given by $n = N_{BSE}/2 = 180$. Figure 13.5 (right) provides a zoom of the corresponding DOS and its QTT approximant within the small energy interval $[0, 40]$ eV.

This means that for a fixed η, the QTT-rank remains rather modest, relative to the molecular size. This observation confirms the QTT ranks estimates in Section 13.6. The moderate size of QTT ranks in Figure 13.5 clearly demonstrates the potential of QTT interpolation for modeling DOS of large lattice-type clusters.

We observe several gaps in the spectral densities with complicated shapes (see Figures 13.5 and 13.6) indicating that the polynomial, rational, or trigonometric interpolation can be applied only to a small energy sub-intervals, but not in the whole interval $[0, a]$. It is remarkable that the QTT approximant resolves well the DOS function in the whole energy interval, including nearly zero values within the spectral gaps (hardly possible for polynomial/rational based interpolation).

13.5 Interpolation of the DOS function by using the QTT format

In the previous section we demonstrated that the QTT tensor approximation provides good resolution for the DOS function also observed for a number of larger molecules, see [27]. In what follows, we sketch a tensor-based heuristic QTT interpolation of DOS by using only an incomplete set of sampling points introduced in [27], that is, QTT representation by adaptive cross approximation (ACA) [230, 256]. This allows us to recover the spectral density with controllable accuracy by using $M \ll N$ interpolation points, where asymptotically M scales logarithmically in the grid size N.

This heuristic approach can be viewed as a kind of an "adaptive QTT ε-interpolation". In particular, we demonstrate by numerical experiments that the low-rank QTT adaptive cross interpolation provides a good resolution of the target DOS with the number of functional calls that asymptotically scales logarithmically $O(\log N)$ in the size of representation grid N; see Figure 13.7. In the case of large N, this beneficial feature allows us to compute the QTT approximation by requiring much less than N computationally expensive functional evaluations of $\phi_\eta(t)$.

The QTT interpolation via ACA tensor approximation serves to recover the representation parameters of the QTT tensor approximant and requires asymptotically about

$$M = C_s r_{\mathrm{qtt}}^2 \log_2 N \ll N \tag{13.15}$$

samples of the target N-vector[1] with a small pre-factor C_s, usually satisfying $C_s \le 10$, that is independent of the fine interpolation grid size $N = 2^d$; see, for example, [183]. This cost estimate seems promising in the perspective of extended or lattice-type molecular systems, requiring large spectral intervals and, as a result, a large interpolation grid of size N. Here, the QTT rank parameter r_{qtt} naturally depends on the required truncation threshold $\varepsilon > 0$, characterizing the L_2-error between the exact DOS and its QTT interpolant. The QTT tensor interpolation adaptively reduces the number of functional calls, that is, $M < N$, *if the QTT rank parameters (or threshold* $\varepsilon > 0$*) are chosen to satisfy condition* (13.15). The expression on the right-hand side of (13.15) provides a rather accurate estimate on the number of functional evaluations.

To complete this discussion, we present numerical tests on the low-rank QTT tensor interpolation applied to the long vector discretizing the Lorentzian-DOS on large representation grid.

Figure 13.6 represents the results of the QTT interpolating approximation to the discretized DOS function for NH_3 molecule. We use the QTT cross approximation algorithm based on [167, 230, 256] and implemented in the MATLAB TT-toolbox [232]. Here, we set $\varepsilon = 0.08$, $\eta = 0.1$, and $N = 2^{14}$, providing $r_{\mathrm{QTT}} = 9.8$, see [27] for more numerical examples.

---

1 In our application, this is the functional N-vector corresponding to representation of DOS via matrix resolvents in (13.6).

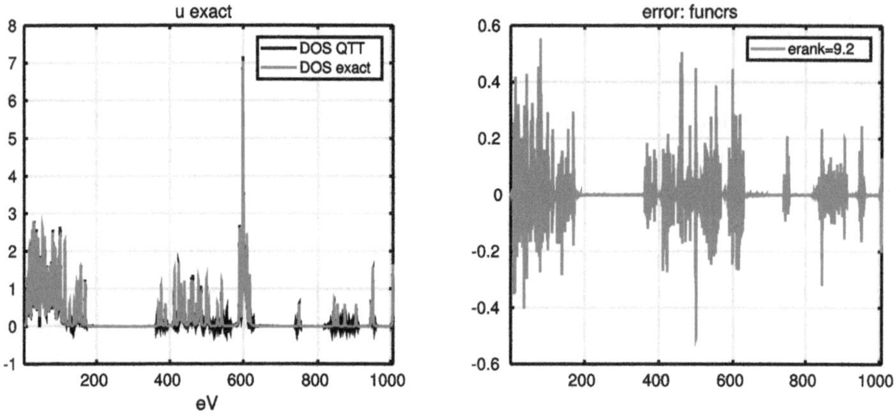

Figure 13.6: QTT ACA interpolation of DOS for NH_3 molecule (left) and its error on the whole spectrum.

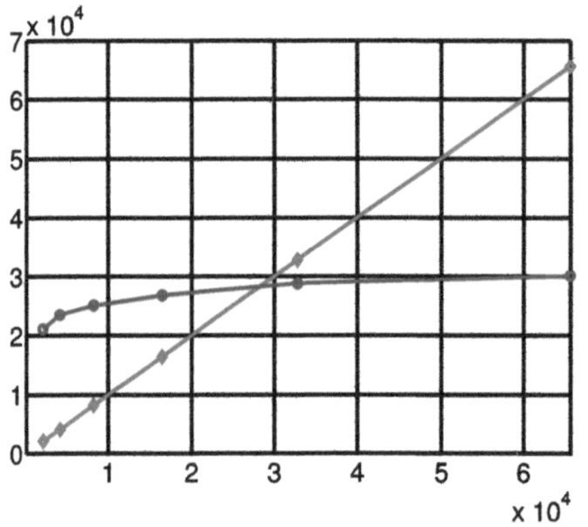

Figure 13.7: DOS for H_2O via Lorentzians: the number of functional calls for QTT cross approximation (blue) vs. the full grid size N (red).

Figure 13.7 (see [27]) illustrates the logarithmic increase in the number of samples required for the QTT interpolation of DOS (for the H_2O molecule) represented on the grid-size $N = 2^d$ with the different quantics dimensions $d = 11, 12, \ldots, 16$. The rank truncation threshold is chosen by $\epsilon = 0.05$ and the regularization parameter is $\eta = 0.2$. In this example the effective pre-factor in (13.15) is estimated by $C_s \leq 10$. This pre-factor characterizes the average number of samples required for the recovery of each of $r_{\text{qtt}}^2 \log N$ representation parameters involved in the QTT tensor ansatz.

We observe that the QTT tensor interpolant recovers the complicated shape of the exact DOS with a high precision. The logarithmic asymptotic complexity scaling $M = O(\log N)$ (i. e., the number of functional calls required for the QTT tensor interpolation) vs. the grid size N can be observed in Figure 13.7 (blue line) for large representation grids.

13.6 Upper bounds on the QTT ranks of DOS function

In this section, we sketch the analysis of the upper bounds on the QTT ranks of the discretized DOS obtained by Gaussian broadening presented in [27]. The numerical tests indicate that Lorentzian blurring leads to a similar QTT rank compared with Gaussians blurring when both are applied to the same grid, and the same truncation threshold $\varepsilon > 0$ is used in the QTT approximation. For technical reasons, we consider the case of symmetric spectral interval, that is, $t, \lambda_j \in [-a, a]$.

Assume that the function $\phi_\eta(t) = \frac{1}{n}\sum_{j=1}^{n} g_\eta(t - \lambda_j)$, $t \in [-a, a]$, in equation (13.2) is discretized by sampling over the uniform N-grid Ω_h with $N = 2^d$, where the generating Gaussian is given by $g_\eta(t) = \frac{1}{\sqrt{2\pi}\eta}\exp(-\frac{t^2}{2\eta^2})$. Denote the corresponding N-vector by $\mathbf{g} = \mathbf{g}_\eta \in \mathbb{R}^N$ and the resulting discretized density vector by

$$\phi_\eta(t) \mapsto \mathbf{p} = \mathbf{p}_\eta = \frac{1}{n}\sum_{j=1}^{n} \mathbf{g}_{\eta,j} \in \mathbb{R}^N,$$

where the shifted Gaussian is assigned by the vector $g_\eta(t - \lambda_j) \mapsto \mathbf{g}_j = \mathbf{g}_{\eta,j}$.

Without loss of generality, we suppose that all eigenvalues are situated within the set of grid points, that is, $\lambda_j \in \Omega_h$. Otherwise, we can slightly relax their positions provided that the mesh size h is small enough. This is not a severe restriction for the QTT approximation of functional vectors since storage and complexity requests depend only logarithmically on N.

Lemma 13.5 ([27]). *Assume that the effective support of the shifted Gaussians $g_\eta(t - \lambda_j)$, $j = 1, \ldots, n$, is included in the computational interval $[-a, a]$. Then the QTT ε-rank of the vector $\mathbf{p}_\eta$ is bounded by*

$$\operatorname{rank}_{\mathrm{QTT}}(\mathbf{p}_\eta) \leq Ca\log^{3/2}(|\log\varepsilon|),$$

where the constant $C = O(|\log\eta|) > 0$ depends only logarithmically on the regularization parameter η.

Proof. The main argument of the proof is similar to that in [148, 68]: the sum of discretized Gaussians, each represented in Fourier basis, can be expanded with merely the same number m_0 of Fourier harmonics as the individual Gaussian function (uniform basis).

Given exponent parameter η, we first estimate the number of essential Fourier coefficients of the Gaussian vectors $\mathbf{g}_{\eta,j}$,

$$m_0 = O(a|\log\eta|\log^{3/2}(|\log\varepsilon|))$$

taking into account their exponential decay. Notice that m_0 depends only logarithmically on η. Since each Fourier harmonic has the exact rank-2 QTT representation (see Section 4.2), we arrive at the desired bound. □

A similar QTT rank bound can be derived for the case of Lorentzian blurred DOS. Indeed, we observe that the Fourier transform of the Lorentzian in (13.3) is given by [27],

$$\mathcal{F}(L_\eta(t)) = e^{-|k|\eta}.$$

This leads to the logarithmic bound in the number m_0 of essential Fourier coefficients in the Lorentzian vectors.

Table 13.1: QTT ranks of Lorentzians-DOS for TDA matrices of some molecules with parameters $\varepsilon = 0.04$, $\eta = 0.4$, $N = 16\,384$.

Molecule	**H_2O**	**NH_3**	**H_2O_2**	**N_2H_4**	**C_2H_5OH**	**$C_2H_5NO_2$**	**$C_3H_7NO_2$**
$n = N_{ov}$	180	215	531	657	1430	3000	4488
QTT ranks	11	11	12	11	15	16	13

Table 13.1 shows that the average QTT tensor rank of Lorentzians-DOS for various TDA matrices remains almost independent of the molecular size, which confirms previous observations. The weak dependence of the rank parameter on the molecular geometry can be observed.

14 Tensor-based summation of long-range potentials on finite 3D lattices

In Chapter 9 we described the method for *direct tensor summation* of the electrostatic potentials used for calculation of the nuclear potential operator for molecules [156], which reduces the volume summation of the potentials to one-dimensional rank-structured operations. However, the rank of the resulting canonical tensor increases linearly in a number of potentials and this growth may become crucial for larger multi-particle systems. Favorably, the tensor approach often suggests new concepts to solution of classical problems.

In this chapter, we discuss the ***assembled tensor method*** for summation of the long-range potentials on finite rectangular $L \times L \times L$ lattices introduced recently by the authors in [148, 153]. This technique requires only $O(L)$ computational work for calculation of the collective electrostatic potential of large lattice systems and $O(L^2)$ for computation of their interaction energy, instead of $O(L^3 \log L)$ when using the traditional Ewald summation techniques. Surprisingly, the assembled tensor summation technique does not increase the tensor rank: the rank of the tensor for collective potential of large 3D lattice clusters equals to the rank of a single 3D reference potential. The approach was initiated by our former numerical observations in [173, 146] that the Tucker tensor rank of a sum of Slater potentials placed at nodes of a three-dimensional finite lattice remains the same as the rank of a single Slater function.

A single three-dimensional potential function (electrostatic potential $\frac{1}{\|x\|}$ or other type interactions generated by radial basis function) sampled on a large $N \times N \times N$ representation grid in a bounding box is approximated with a guaranteed precision by a low-rank Tucker/canonical reference tensor. This tensor provides the values of the discretized potential at any point of this fine auxiliary 3D grid, but needs only $O(N)$ storage. Then each 3D singular kernel function involved in the summation is represented on the same grid by a shift of the reference tensor along lattice vectors. Directional vectors of the Tucker/canonical tensor defining a full lattice sum are assembled by the 1D summation of the corresponding univariate skeleton vectors specifying the shifted tensor. The lattice nodes are not required to exactly coincide with the grid points of the global $N \times N \times N$ representation grid since the accuracy of the resulting tensor sum is well controlled due to easy availability of large grid size N (e. g., fine resolution).

The key advantage of the assembled tensor method is that the summation of potentials is implemented within the skeleton vectors of the generating canonical tensor, thus not affecting the resulting tensor rank; the number of canonical vectors representing the total tensor sum remains the same as for a single reference kernel. For a sum of electrostatic potentials over $L \times L \times L$ lattice embedded in a box, the required storage scales linearly in the one-dimensional grid-size, that is, as $O(N)$, whereas the numerical cost is estimated by $O(NL)$. The important benefit of this summation tech-

https://doi.org/10.1515/9783110365832-014

nique is that the resultant low-rank tensor representation of the total sum of potentials can be evaluated at any grid point at the cost $O(1)$.

In the case of periodic boundary conditions, the tensor approach leads to further simplifications. Indeed, the respective lattice summation is reduced to 1D operations on short canonical vectors of size $n = N/L$, which is the restriction (projection) of the global N-vectors onto the unit cell. Here, n denotes merely the number of grid points per unit cell. In this case, storage and computational costs are reduced to $O(n)$ and $O(Ln)$, respectively, whereas the traditional FFT-based approach scales at least cubically in L, $O(L^3 \log L)$, and in N. Notice that due to low cost of the tensor method at the limit of large lattice size L, the conditionally convergent sums in periodic setting can be regularized by subtraction of the constant term, which can be evaluated numerically by the Richardson extrapolation on a sequence of lattice parameters L, $2L$, $4L$, etc. (see Section 14.3). Hence, in the new framework, the analytic treatment of the conditionally convergent sums is no longer required.

We notice that the numerical treatment of long-range potentials in large lattice-type systems was always considered as a computational challenge (see [72, 235, 199] and [295, 207, 47, 208, 253]). Tracing back to Ewald summation techniques [79], the development of lattice-sum methods has led to a number of established algorithms for evaluating long-range electrostatic potentials of multiparticle systems; see for example [58, 236, 278, 139, 63, 205] and references therein. These methods usually combine the original Ewald summation approach with the fast Fourier transform (FFT). The commonly used Ewald summation algorithms [79] are based on a certain specific local-global analytical decomposition of the interaction potential. In the case of electrostatic potentials, the Newton kernel is represented by

$$\frac{1}{r} = \frac{\tau(r)}{r} + \frac{1-\tau(r)}{r},$$

where the traditional choice of the cutoff function τ is the complementary error function

$$\tau(r) = \operatorname{erfc}(r) := \frac{2}{\sqrt{\pi}} \int_r^\infty \exp(-t^2)dt.$$

The Ewald summation techniques were shown to be particularly attractive for computing the potential energies and forces of many-particle systems with long-range interaction potential in periodic boundary conditions. They are based on the spacial separation of a sum of potentials into two parts, then the short-range part is treated in the real space, and the long-range part (whose sum converges in the reciprocal space) requires grid-based FFT calculations with unreducible $O(L^3 \log L)$ computational work.

It is worth noting that the presented tensor method is applicable to the lattice sums generated by rather general class of radial basis functions, which allow an efficient local-plus-separable approximation. In particular, along with Coulombic systems, it can be applied to a wide class of commonly used interaction potentials, for example, to the Slater, Yukawa, Stokeslet, Lennard-Jones, or van der Waals interactions. In all these cases, the existence of low-rank grid-based tensor approximation can be proved and this approximation can be constructed numerically by analytic-algebraic methods as in the case of the Newton kernel; see the detailed discussion in [153, 171].

The tensor approach is advantageous in other functional operations with the lattice potential sums represented on a 3D grid such as integration, differentiation, or force and energy calculations using tensor arithmetics of 1D complexity [174, 146, 152, 240, 24]. Notice that the summation cost in the Tucker/canonical formats $O(L\,N)$ can be reduced to the logarithmic scale in the grid size $O(L\log N)$ by using the low-rank quantized tensor approximation (QTT) [167] of long canonical/Tucker vectors as it was suggested and analyzed in [148].

14.1 Assembled tensor summation of potentials on finite lattices

In this section, following [148], we present the efficient scheme for fast assembled tensor summation of electrostatic potentials for a finite 3D lattice system in a box. Given the unit reference cell $\Omega = [-b/2, b/2]^d$, $d = 3$, of size $b \times b \times b$, we consider an interaction potential in a bounded box

$$\Omega_L = B_1 \times B_2 \times B_3$$

consisting of a union of $L_1 \times L_2 \times L_3$ unit cells $\Omega_{\mathbf{k}}$, obtained by a shift of Ω that is a multiple of b in each variable, and specified by the lattice vector $b\mathbf{k}$, $\mathbf{k} = (k_1, k_2, k_3) \in \mathbb{Z}^d$, $0 \le k_\ell \le L_\ell - 1$ for $L_\ell \in \mathbb{N}$ ($\ell = 1, 2, 3$). Here, $B_\ell = [-b/2, b/2 + (L_\ell - 1)b]$, so that the case $L_\ell = 1$ corresponds to one-layer systems in the variable x_ℓ. Recall that $b = nh$ by construction, where $h > 0$ is the mesh-size (same for all spacial variables).

In the case of an extended system in a box, the summation problem for the total potential $v_{c_L}(x)$ is formulated in the rectangular volume $\Omega_L = \bigcup_{k_1,k_2,k_3=1}^{L} \Omega_{\mathbf{k}}$, where for ease of exposition, we consider a lattice of equal sizes $L_1 = L_2 = L_3 = L$. For implementational reasons, the computational volume box is chosen slightly larger than Ω_L by a size of several Ω (see (14.3) and Figure 14.1). On each $\Omega_{\mathbf{k}} \subset \Omega_L$ the potential sum of interest $v_{\mathbf{k}}(x) = (v_{c_L})_{|\Omega_{\mathbf{k}}}$ is obtained by summation over all unit cells in Ω_L,

$$v_{\mathbf{k}}(x) = \sum_{k_1,k_2,k_3=0}^{L-1} \sum_{\nu=1}^{M_0} \frac{Z_\nu}{\|x - a_\nu(k_1, k_2, k_3)\|}, \quad x \in \Omega_{\mathbf{k}}, \tag{14.1}$$

where $a_\nu(k_1, k_2, k_3) = a_\nu + b\mathbf{k}$. This calculation is performed at each of L^3 elementary cells $\Omega_{\mathbf{k}} \subset \Omega_L$, which presupposes substantial numerical costs for large L. In the presented approach, these costs are essentially reduced as described further.

Figure 14.1 shows an example of a computational box with a 3D lattice-type molecular structure of $6 \times 4 \times 6$ atoms.

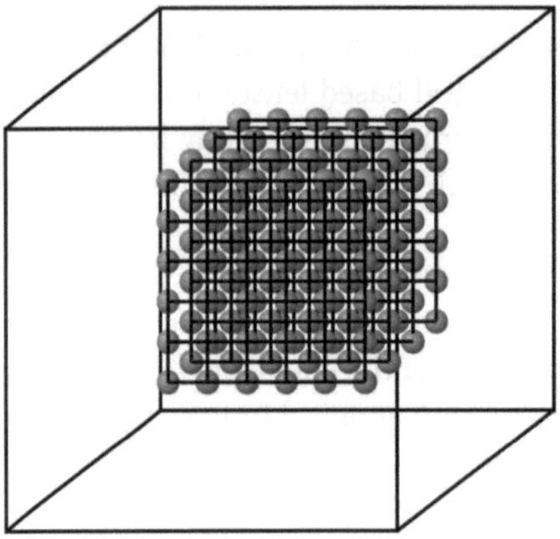

Figure 14.1: Rectangular 6 × 4 × 6 lattice in a box.

Let Ω_{N_L} be the $N_L \times N_L \times N_L$ uniform grid on Ω_L with the same mesh-size h as above, and introduce the corresponding space of piecewise constant basis functions of the dimension N_L^3. In this construction, we have

$$N_L = n + n(L-1) = Ln. \tag{14.2}$$

In practice, the computational box Ω_L and the grid size N_L can be taken larger than (14.2) by some "dummy" distance with the grid size N_0, so that

$$N_L = Ln + 2N_0. \tag{14.3}$$

Similarly to (11.9), we employ the rank-R reference tensor defined on the auxiliary box $\widetilde{\Omega}_L$ by scaling Ω_L with factor 2,

$$\widetilde{\mathbf{P}}_{L,R} = \sum_{q=1}^{R} \mathbf{p}_q^{(1)} \otimes \mathbf{p}_q^{(2)} \otimes \mathbf{p}_q^{(3)} \in \mathbb{R}^{2N_L \times 2N_L \times 2N_L}, \tag{14.4}$$

and let $\mathcal{W}_{\nu(k_i)}$, $i = 1, 2, 3$, be the directional windowing operators associated with the lattice vector $\mathbf{k}$. The next theorem proves the storage and numerical costs for the lattice sum of single potentials, each represented by a canonical rank-R tensor, which corresponds to the choice of $M_0 = 1$ and $a_1 = 0$ in (14.1). The Ω_L-windowing operator $\mathcal{W} = \mathcal{W}_{(\mathbf{k})}$ (tracing onto $N_L \times N_L \times N_L$ window) is rank-1 separable:

$$\mathcal{W}_{(\mathbf{k})} = \mathcal{W}_{(k_1)}^{(1)} \otimes \mathcal{W}_{(k_2)}^{(2)} \otimes \mathcal{W}_{(k_3)}^{(3)}$$

specifying the shift by the lattice vector $b\mathbf{k}$.

Theorem 14.1 ([148]). *Given a canonical rank-R tensor representation of a single long-range potential* (14.4), *the projected tensor of the interaction potential* $v_{c_L}(x)$, $x \in \Omega_L$, *representing the collective potential sum over* L^3 *charges of a rectangular lattice, can be* *presented by the canonical tensor* $\mathbf{P}_{c_L}$ *with the same rank R:*

$$\mathbf{P}_{c_L} = \sum_{q=1}^{R}\left(\sum_{k_1=0}^{L-1}\mathcal{W}_{(k_1)}\mathbf{p}_q^{(1)}\right)\otimes\left(\sum_{k_2=0}^{L-1}\mathcal{W}_{(k_2)}\mathbf{p}_q^{(2)}\right)\otimes\left(\sum_{k_3=0}^{L-1}\mathcal{W}_{(k_3)}\mathbf{p}_q^{(3)}\right). \tag{14.5}$$

The numerical cost and storage size are estimated by $O(RLN_L)$ *and* $O(RN_L)$*, respectively, where* N_L *is the univariate grid size as in* (14.2).

Proof. We fix the index $\nu = 1$ in (14.1) and consider only the second sum defined on the complete domain Ω_L,

$$v_{c_L}(x) = \sum_{k_1,k_2,k_3=0}^{L-1}\frac{Z}{\|x-b\mathbf{k}\|}, \quad x\in\Omega_L. \tag{14.6}$$

Then the projected tensor representation of $v_{c_L}(x)$ takes the form (setting $Z = 1$)

$$\mathbf{P}_{c_L} = \sum_{k_1,k_2,k_3=0}^{L-1}\mathcal{W}_{\nu(\mathbf{k})}\mathbf{P}_{L,R} = \sum_{k_1,k_2,k_3=0}^{L-1}\sum_{q=1}^{R}\mathcal{W}_{(\mathbf{k})}(\mathbf{p}_q^{(1)}\otimes\mathbf{p}_q^{(2)}\otimes\mathbf{p}_q^{(3)})\in\mathbb{R}^{N_L\times N_L\times N_L},$$

where $\mathbf{p}_q^{(\ell)}$, $\ell = 1,2,3$, are vectors of the reference tensor (14.4) and the 3D shift vector is defined by $\mathbf{k}\in\mathbb{Z}^{L\times L\times L}$. Now, the above summation can be represented by

$$\mathbf{P}_{c_L} = \sum_{q=1}^{R}\sum_{k_1,k_2,k_3=0}^{L-1}\mathcal{W}_{(k_1)}\mathbf{p}_q^{(1)}\otimes\mathcal{W}_{(k_2)}\mathbf{p}_q^{(2)}\otimes\mathcal{W}_{(k_3)}\mathbf{p}_q^{(3)}. \tag{14.7}$$

To simplify the large sum over the full 3D lattice, we use the following property of a sum of canonical tensors with equal ranks R and with two coinciding factor matrices: the concatenation in the third mode ℓ can be reduced to point-wise summation ("assembling") of the respective canonical vectors

$$C^{(\ell)} = [\mathbf{a}_1^{(\ell)}+\mathbf{b}_1^{(\ell)},\dots,\mathbf{a}_{R_a}^{(\ell)}+\mathbf{b}_{R_b}^{(\ell)}], \tag{14.8}$$

thus preserving the same rank parameter R for the resulting sum. Notice that, for each fixed q, the inner sum in (14.7) satisfies the above property. By repeatedly applying this property to all canonical tensors for $q = 1,\dots,R$, the 3D sum (14.7) can be simplified to a rank-R tensor obtained by 1D summations only:

$$\begin{aligned}\mathbf{P}_{c_L} &= \sum_{q=1}^{R}\left(\sum_{k_1=0}^{L-1}\mathcal{W}_{(k_1)}\mathbf{p}_q^{(1)}\right)\otimes\left(\sum_{k_2,k_3=0}^{L-1}\mathcal{W}_{(k_2)}\mathbf{p}_q^{(2)}\otimes\mathcal{W}_{(k_3)}\mathbf{p}_q^{(3)}\right)\\ &= \sum_{q=1}^{R}\left(\sum_{k_1=0}^{L-1}\mathcal{W}_{(k_1)}\mathbf{p}_q^{(1)}\right)\otimes\left(\sum_{k_2=0}^{L-1}\mathcal{W}_{(k_2)}\mathbf{p}_q^{(2)}\right)\otimes\left(\sum_{k_3=0}^{L-1}\mathcal{W}_{(k_3)}\mathbf{p}_q^{(3)}\right).\end{aligned}$$

The cost can be estimated by following the standard properties of canonical tensors. □

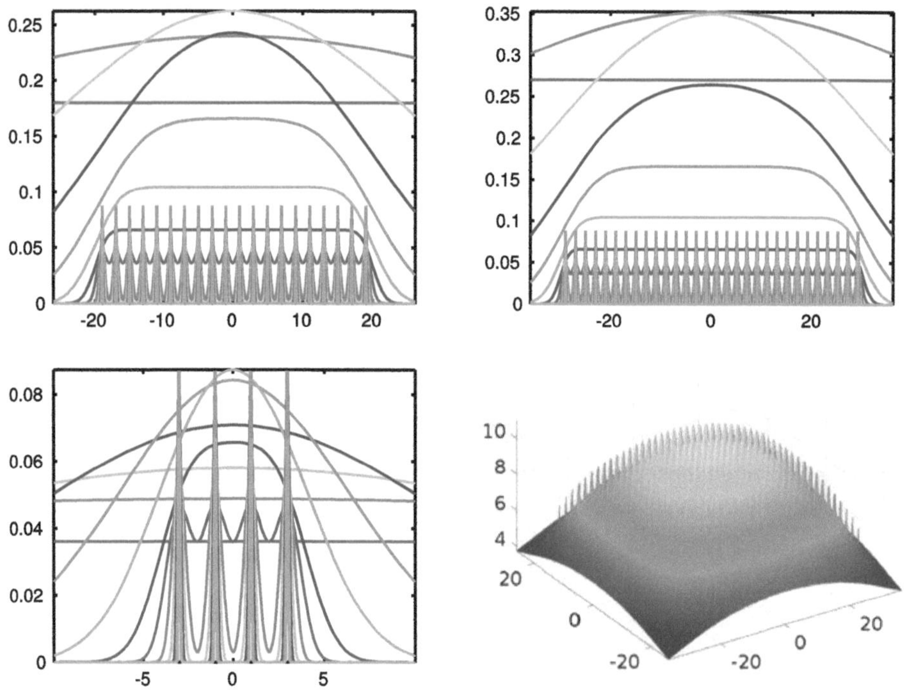

Figure 14.2: Assembled canonical vectors for a sum of electrostatic potentials for a cluster of $20 \times 30 \times 4$ Hydrogen atoms in a rectangular box of size $\sim 55.4 \times 33.6 \times 22.4$ au^3. Top left-right: vectors in x- and y-axes, respectively; bottom left: vectors along z-axis. Bottom right: the resulting sum of 2400 nuclei potentials at the middle cross-section with $z = 11.2$ au.

Remark 14.2. For the general case $M_0 > 1$, the weighted summation over M_0 charges leads to the low-rank tensor representation, that is, $\operatorname{rank}(\mathbf{P}_{c_L}) \leq M_0 R$, and

$$\mathbf{P}_{c_L} = \sum_{\nu=1}^{M_0} Z_\nu \sum_{q=1}^{R} \left(\sum_{k_1=0}^{L-1} \mathcal{W}_{\nu(k_1)} \mathbf{p}_q^{(1)} \right) \otimes \left(\sum_{k_2=0}^{L-1} \mathcal{W}_{\nu(k_2)} \mathbf{p}_q^{(2)} \right) \otimes \left(\sum_{k_3=0}^{L-1} \mathcal{W}_{\nu(k_3)} \mathbf{p}_q^{(3)} \right). \quad (14.9)$$

The previous construction applies to the uniformly spaced positions of charges. However, our tensor summation method remains valid for a non-equidistant $L \times L \times L$ tensor lattice.

Here we sketch some numerical examples presented in [148]. Figure 14.2 illustrates the shape of assembled canonical vectors for the $32 \times 16 \times 8$ lattice sum in a box (summation of 4096 potentials). Here, the canonical rank is $R = 25$, and $\varepsilon = 10^{-6}$. It demonstrates how the assembled vectors composing the tensor lattice sum incorporate simultaneously the canonical vectors of shifted Newton kernels. It can be seen that canonical vectors capture the local, intermediate, and long-range contributions to the total sum. Figure 14.3 represents agglomerated canonical vectors in x-, y-, and

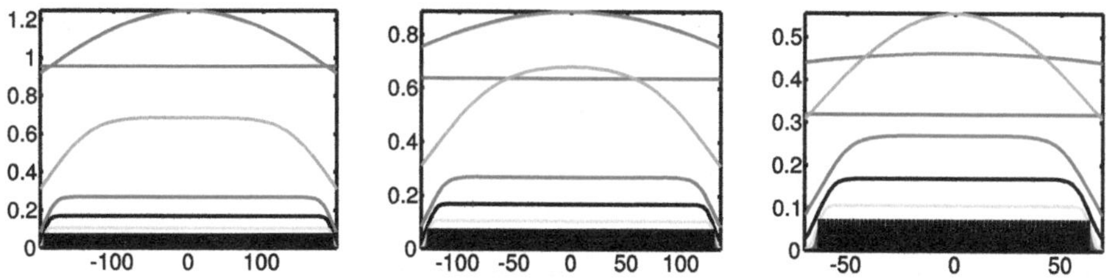

Figure 14.3: Assembled canonical vectors in x-, y-, and z-axes for a sum of 1 572 864 nuclei potentials.

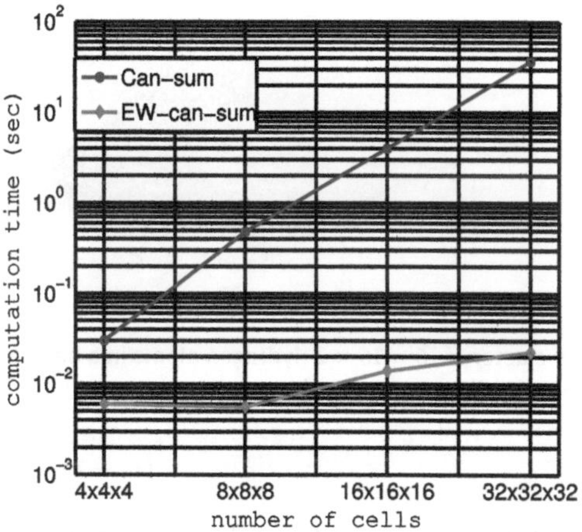

Figure 14.4: CPU times (log scaling) for calculating the sum of Coulomb potentials over 3D $L \times L \times L$ lattice by using direct canonical tensor summation (blue line) and assembled lattice summation (red line).

z-axes for a sum of 1 572 864 nuclei potentials for a cluster of $192 \times 128 \times 64$ Hydrogen atoms in a box of size $\approx 19.8 \times 13.4 \times 7\,\mathrm{nm}^3$.

The canonical tensor representation (14.5) reduces dramatically the numerical costs and storage consumptions. Figure 14.4 compares the direct and assembled tensor summation methods (grid-size of a unit cell, $n = 256$). Contrary to the direct canonical summation of the nuclear potentials on a 3D lattice, which scales at least linearly in the size of the cubic lattice as $N_L^3 L^3$ (blue line), the CPU time for directionally agglomerated canonical summation in a box via (14.5) scales as $N_L L$ (red line).

Table 14.1 presents the times for assembled computation of the sum of potentials positioned in nodes of $L \times L \times L$ lattice clusters. Approximate sizes of finite clusters are given in nanometers. This table shows that computation time for the tensor approach scales logarithmically in the cluster size. We refer to [148] for the more detailed presentation of numerical experiments.

Figure 14.5 compares the tensor sum obtained by the assembled canonical vectors with the results of direct tensor sum for the same configuration. The absolute difference of the corresponding sums for a cluster of $16 \times 16 \times 2$ cells (here a cluster of 512 Hydrogen atoms) is close to machine accuracy $\sim 10^{-14}$.

Table 14.1: CPU times (sec) vs. the lattice size for the assembled calculation of their sum $\mathbf{P}_{c_L}$ over the $L \times L \times L$ clusters. Approximate sizes of finite clusters are given in nanometers.

L	32	64	128	256
Total L^3	32 768	262 144	2 097 152	16 777 216
Cluster size	3.8^3	7^3	13.4^3	26.2^3
Summation time (sec)	0.2	0.27	0.83	3.87

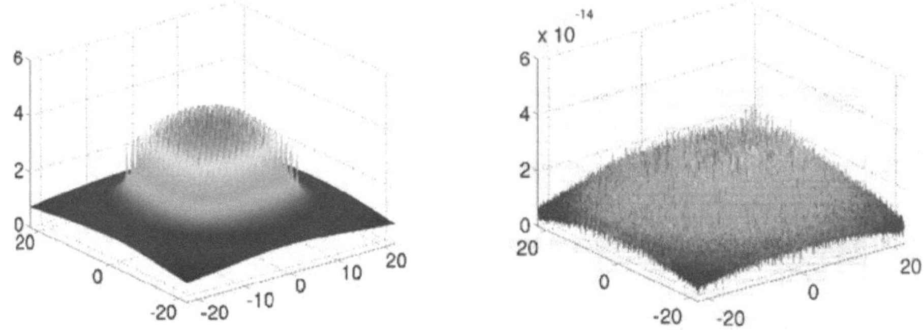

Figure 14.5: Left: The electrostatic potential of the cluster of 16 × 16 × 2 Hydrogen atoms in a box (512 atoms). Right: the absolute error of the assembled tensor sum on this cluster by (14.5) with respect to the direct tensor summation (11.10).

14.2 Assembled summation of lattice potentials in Tucker tensor format

Similar to (14.4), we introduce the rank-$\mathbf{r}$ "reference" Tucker tensor $\widetilde{\mathbf{T}}_{L,\mathbf{r}} \in \mathbb{R}^{2N_L \times 2N_L \times 2N_L}$ defined on the auxiliary domain $\widetilde{\Omega}_L$.

The following theorem provides theoretical background for the fast tensor methods of grid-based computation of the large sum of long-range potentials on a 3D lattice. It generalizes [148, Theorem 3.1], which was applied to the Newton kernel, to a rather general class of functions $p(\|x\|)$ in (14.1) and to the case of Tucker tensor decompositions. It justifies the low storage and numerical costs for the total potential sum in terms of lattice size.

Theorem 14.3 ([153]). *Given the rank-$\mathbf{r}$ "reference" Tucker tensor $\widetilde{\mathbf{T}}_{L,\mathbf{r}} \in \mathbb{R}^{2N_L \times 2N_L \times 2N_L}$ approximating the potential function $p(\|x\|)$, the rank-$\mathbf{r}$ Tucker approximation of a lattice-sum v_{c_L} can be computed in the form*

$$\mathbf{T}_{c_L} = \sum_{\mathbf{m}=\mathbf{1}}^{\mathbf{r}} b_{\mathbf{m}} \Big(\sum_{k_1 \in \mathcal{K}} \mathcal{W}_{(k_1)} \widetilde{\mathbf{t}}_{m_1}^{(1)} \Big) \otimes \Big(\sum_{k_2 \in \mathcal{K}} \mathcal{W}_{(k_2)} \widetilde{\mathbf{t}}_{m_2}^{(2)} \Big) \otimes \Big(\sum_{k_3 \in \mathcal{K}} \mathcal{W}_{(k_3)} \widetilde{\mathbf{t}}_{m_3}^{(3)} \Big). \tag{14.10}$$

The numerical cost and storage are estimated by $O(3rLN_L)$ and $O(3rN_L)$, respectively.

Proof. We apply a similar argument as in Theorem 14.1 to obtain

$$\mathbf{T}_{c_L} = \sum_{k_1,k_2,k_3\in\mathcal{K}} \mathcal{W}_{(\mathbf{k})}\widetilde{\mathbf{T}}_{L,\mathbf{r}}$$
$$= \sum_{\mathbf{m}=\mathbf{1}}^{\mathbf{r}} b_{\mathbf{m}}\Big(\sum_{k_1\in\mathcal{K}} \mathcal{W}_{(k_1)}\widetilde{\mathbf{t}}^{(1)}_{m_1}\Big)\otimes\Big(\sum_{k_2\in\mathcal{K}} \mathcal{W}_{(k_2)}\widetilde{\mathbf{t}}^{(2)}_{m_2}\Big)\otimes\Big(\sum_{k_3\in\mathcal{K}} \mathcal{W}_{(k_3)}\widetilde{\mathbf{t}}^{(3)}_{m_3}\Big).$$

Simple complexity estimates complete the proof. □

Figure 14.6 illustrates the shape of several assembled Tucker vectors obtained by assembling vectors $\widetilde{\mathbf{t}}^{(1)}_{m_1}$ along x_1-axis. It can be seen that assembled Tucker vectors simultaneously accumulate the contributions of all single potentials involved in the total sum. Note that the assembled Tucker vectors do not preserve the initial orthogonality of directional vectors $\{\widetilde{\mathbf{t}}^{(\ell)}_{m_\ell}\}$. In this case the simple Gram–Schmidt orthogonalization can be applied. Next remark generalizes Theorems 14.1 and 14.3.

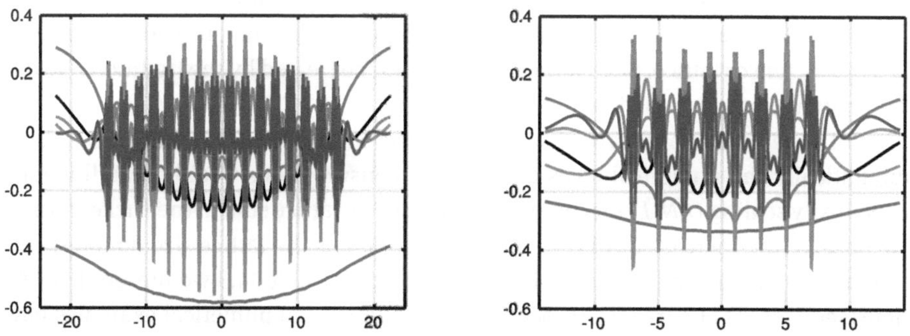

Figure 14.6: Assembled Tucker vectors by using $\widetilde{\mathbf{t}}^{(1)}_{m_1}$ and $\widetilde{\mathbf{t}}^{(2)}_{m_1}$ along the x- and y-axes, respectively, for a sum over $16\times 8\times 1$ lattice.

Remark 14.4. In the general case $M_0 > 1$, the weighted summation over M_0 charges leads to the rank-R_c canonical tensor representation on the "reference" domain $\widetilde{\Omega}_L$, which can be used to obtain the rank-R_c representation of a sum in the whole $L\times L\times L$ lattice (cf. Remark 14.2 and Theorem 14.1):

$$\mathbf{P}_{c_L} = \sum_{q=1}^{R_c}\Big(\sum_{k_1\in\mathcal{K}} \mathcal{W}_{(k_1)}\widetilde{\mathbf{p}}^{(1)}_q\Big)\otimes\Big(\sum_{k_2\in\mathcal{K}} \mathcal{W}_{(k_2)}\widetilde{\mathbf{p}}^{(2)}_q\Big)\otimes\Big(\sum_{k_3\in\mathcal{K}} \mathcal{W}_{(k_3)}\widetilde{\mathbf{p}}^{(3)}_q\Big). \tag{14.11}$$

Likewise, the rank-$\mathbf{r}_c$ Tucker approximation of a lattice potential sum v_{c_L} can be computed in the form [153]

$$\mathbf{T}_{c_L} = \sum_{\mathbf{m}=\mathbf{1}}^{\mathbf{r}_0} b_{\mathbf{m}}\Big(\sum_{k_1\in\mathcal{K}} \mathcal{W}_{(k_1)}\widetilde{\mathbf{t}}^{(1)}_{m_1}\Big)\otimes\Big(\sum_{k_2\in\mathcal{K}} \mathcal{W}_{(k_2)}\widetilde{\mathbf{t}}^{(2)}_{m_2}\Big)\otimes\Big(\sum_{k_3\in\mathcal{K}} \mathcal{W}_{(k_3)}\widetilde{\mathbf{t}}^{(3)}_{m_3}\Big). \tag{14.12}$$

Table 14.2: MATLAB calculations: time (sec.) vs. the total number of potentials L^3 for the assembled Tucker representation of the lattice sum $\mathbf{T}_{c_L}$ on the fine $N_L \times N_L \times N_L$ grid with the mesh size $h = 0.0034$ Å.

L^3	16^3	32^3	64^3	128^3	256^3	512^3
Time	0.33	1.25	5.22	19.56	85.17	439.9
N_L^3	5632^3	9728^3	$17\,920^3$	$34\,304^3$	$67\,072^3$	$132\,608^3$

Table 14.3: Times in MATLAB for computation of the 3D FFT for a sequence of n^3 grids. Times for grids $n \geq 2048$ are estimated by extrapolation.

n^3	512^3	1024^3	2048^3	4096^3	8192^3	$16\,384^3$
FFT_3	5.4	51.6	~500	~1 hour	~10 hours	~100 hours

The previous construction applies to the uniformly spaced positions of charges. However, the agglomerated tensor summation method in both canonical and Tucker formats applies, with slight modification of the windowing operator, to a non-equidistant $L_1 \times L_2 \times L_3$ tensor lattice. Such lattice sums cannot be treated by the traditional Ewald summation methods based on the FFT transform.

Both the Tucker and canonical tensor representations (14.10) and (14.5) reduce dramatically the numerical costs and storage consumptions.[1] Table 14.2 illustrates complexity scaling $O(N_L L)$ for computation of $L \times L \times L$ lattice sum in the Tucker format. We observe the increase of CPU time in a factor of 4 as the lattice size doubles, confirming our theoretical estimates. For comparison, in Table 14.3 we present the CPU time (sec.) for 3D FFT transform, see [153] where the initial numerical examples have been presented.

Figure 14.7 shows the sum of Newton kernels on a lattice $8 \times 4 \times 1$ and the respective Tucker summation error achieved with the rank $\mathbf{r} = (16, 16, 16)$ Tucker tensor defined on the large 3D representation grid with the mesh size about 0.002 atomic units (0.001 Å). Figure 14.8 represents the Tucker vectors obtained from the canonical-to-Tucker (C2T) approximation of the assembled canonical tensor sum of potentials on an $8 \times 4 \times 1$ lattice. In this case, the Tucker vectors are orthogonal.

14.3 Assembled tensor sums in a periodic setting

Here we sketch the results in [148, 153]. In the periodic case, we introduce the periodic cell $\mathcal{R} = b\mathbb{Z}^d$, $d = 1, 2, 3$, and consider a 3D T-periodic supercell of size $T \times T \times T$ with

1 Note that the total number of potentials on a lattice 256^3 is more than 16 millions. The cluster size in every space dimension is $2(256 + 6) = 516$ au, or ~26 nanometers. (Here 2 au is the inter-atomic distance, and 6 is the gap between the lattice and the boundary of a box.)

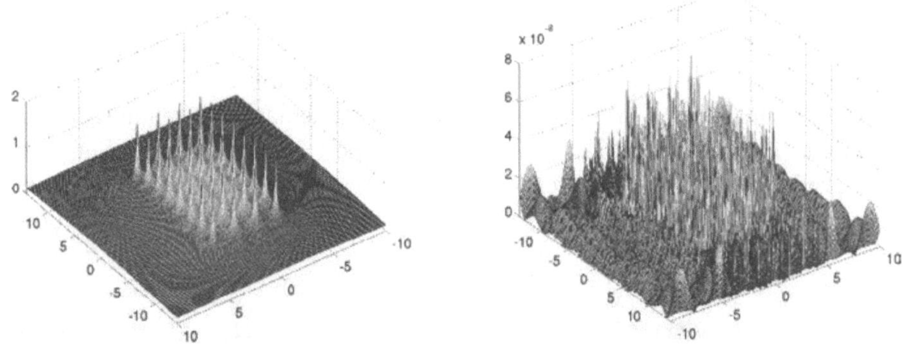

Figure 14.7: Left: Sum of Newton potentials on an $8 \times 4 \times 1$ lattice generated in a volume with the 3D grid of size $14\,336 \times 10\,240 \times 7168$. Right: the absolute approximation error (about $8 \cdot 10^{-8}$) of the rank-r Tucker representation.

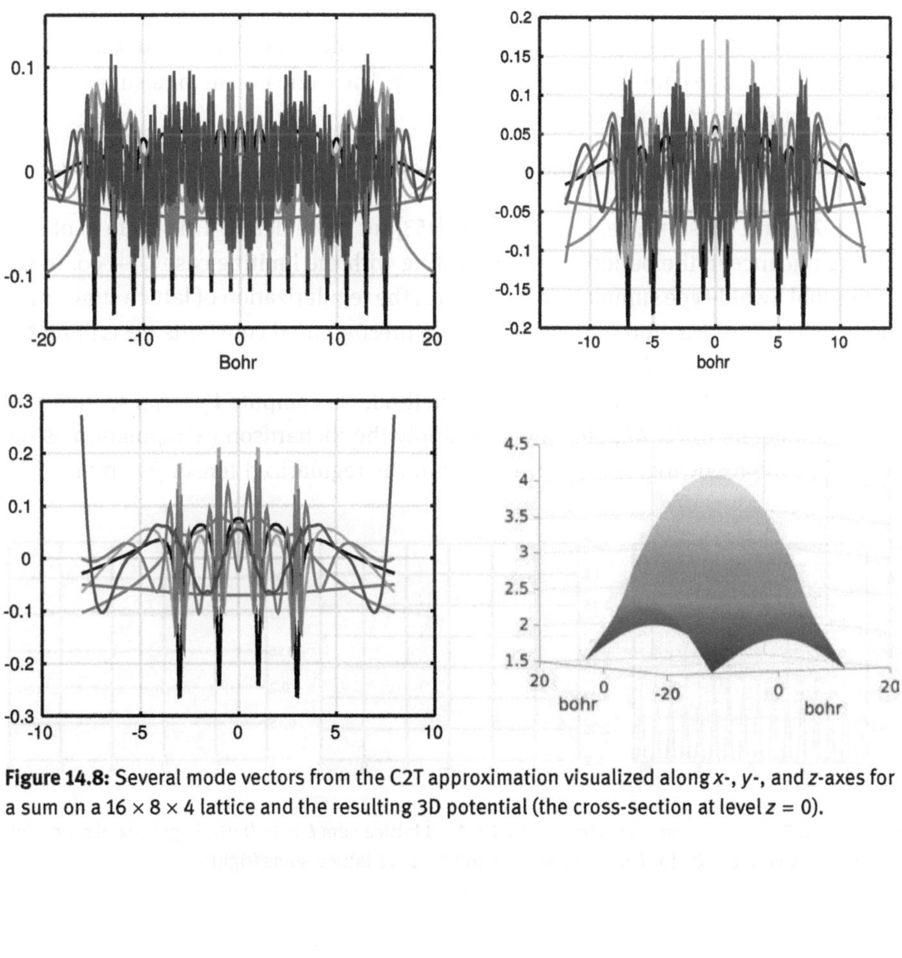

Figure 14.8: Several mode vectors from the C2T approximation visualized along x-, y-, and z-axes for a sum on a $16 \times 8 \times 4$ lattice and the resulting 3D potential (the cross-section at level $z = 0$).

$T = bL$. The total electrostatic potential in Ω_L is obtained by the respective summation over the supercell Ω_L for possibly large L. Then the electrostatic potential in any of T-periods is obtained by replication of the respective data from Ω_L.

The potential sum $v_{c_L}(x)$ is designated at each elementary unit-cell in Ω_L by the same value (**k**-translation invariant). Consider the case $d = 3$. Supposing for simplicity that L is odd, $L = 2p + 1$, the reference value of $v_0(x) = v_{c_L}(x)$ will be computed at the central cell $x \in \Omega_0$, indexed by $(p+1, p+1, p+1)$, by summation over all the contributions from L^3 elementary sub-cells in Ω_L:

$$v_0(x) = \sum_{\nu=1}^{M_0} \sum_{k_1,k_2,k_3=1}^{L} \frac{Z_\nu}{\|x - a_\nu(k_1,k_2,k_3)\|}, \qquad x \in \Omega_0. \tag{14.13}$$

Now the discretized potential can be computed as a tensor sum in (14.11).

Lemma 14.5 ([153]). *The discretized potential v_{c_L} for the full sum over M_0 charges can be presented by rank-(M_0R) canonical tensor. The computational cost is estimated by $O(M_0RnL)$, whereas the storage size is bounded by $O(M_0Rn)$.*

Figure 14.9 (left) shows the assembled canonical vectors for a lattice structure in a periodic setting. Recall that in the limit of large L the lattice sum $\mathbf{P}_{c_L}$ of the Newton kernels is known to converge only conditionally. The same is true for a sum in a box. The maximum norm increases as

$$C_1 \log L, \quad C_2 L, \quad \text{and} \quad C_3 L^2 \tag{14.14}$$

for 1D, 2D, and 3D sums, respectively; see [153] for more detail. This issue is of special significance in the periodic setting dealing with the limiting case $L \to \infty$. In the traditional Ewald-type summation techniques the regularization of lattice sums is implemented by subtraction of the analytically precomputed constants describing the asymptotic behavior in L.

To approach the limiting case, in our method, we compute $\mathbf{P}_{c_L}$ on a sequence of large parameters L, $2L$, $4L$, etc. and then apply the Richardson extrapolation as described in the following. As result, we obtain the regularized tensor $\widehat{p}_L$ obtained by

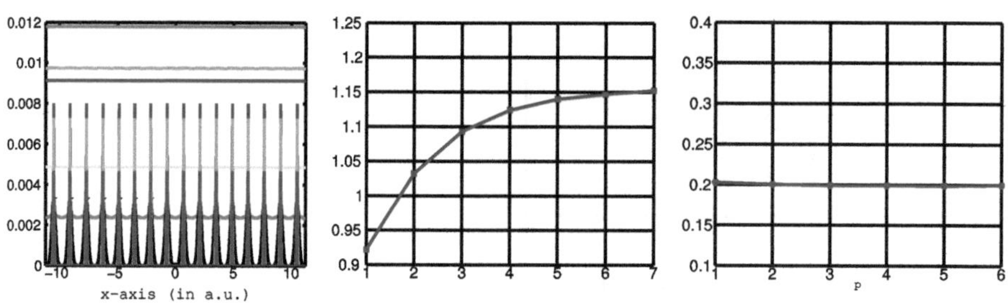

Figure 14.9: Periodic canonical vectors in the $L \times 1 \times 1$ lattice sum $L = 16$ (left). Regularized potential sum $\widehat{p}_L$ vs. m with $L = 2^m$ for $L \times L \times 1$ (middle) and $L \times L \times L$ lattice sums (right).

subtraction of the leading terms in (14.14) and restricted to the reference unit cell Ω_0. Denoting the target value of the potential by p_L, the extrapolation formulas for the linear ($d = 2$) and quadratic ($d = 3$) behavior take form

$$\widehat{p}_L := 2p_L - p_{2L} \quad \text{and} \quad \widehat{p}_L := (4p_L - p_{2L})/3,$$

respectively.

The effect of Richardson extrapolation is illustrated in Figure 14.9. This figure indicates that the potential sum computed at the same point as for the previous example (in the case of $L \times L \times 1$ and $L \times L \times L$ lattices) converges to the limiting values of $\widehat{p}_L$ after applying the Richardson extrapolation (regularized sum).

14.4 QTT ranks of the assembled canonical vectors in the lattice sum

Assembled canonical vectors in the rank-R tensor representation (14.5) are defined over large uniform grid of size N_L. Hence, the numerical cost for evaluating each of these $3R$ vectors scales as $O(N_L L)$, which might become too expensive for large L (recall that $N_L = nL$ scales linearly in L). Using the QTT approximation [167], this cost can be reduced to the logarithmic scale in N_L, whereas the storage need will become $O(\log N_L)$ only. The QTT-rank estimates are based on three main ingredients:

- the global canonical tensor representation of $1/\|x\|$, $x \in \mathbb{R}^3$, on a supercell [111, 30];
- QTT approximation to the Gaussian function (Proposition 14.6); and
- the rank estimate for the block QTT decomposition (Lemma 14.7).

The next statement presents the QTT-rank estimate for Gaussian vector obtained by uniform sampling of $e^{-\frac{x^2}{2p^2}}$ on the finite interval [68]; see also Section 4.2.

Proposition 14.6. *For the given the uniform grid $-a = x_0 < x_1 < \cdots < x_N = a$, $x_i = -a + hi$, $N = 2^L$ on an interval $[-a, a]$, and the vector $\mathbf{g} = [g_i] \in \mathbb{R}^N$ defined by its elements $g_i = e^{-\frac{x_i^2}{2p^2}}$, $i = 0, \ldots, N-1$, and fixed $\varepsilon > 0$, assume that $e^{-\frac{a^2}{2p^2}} \le \varepsilon$. Then there exists the QTT approximation $\mathbf{g}_r$ of the accuracy $\|\mathbf{g} - \mathbf{g}_r\|_\infty \le c\varepsilon$, with the ranks bounded by*

$$\operatorname{rank}_{QTT}(\mathbf{g}_r) \le c \log\left(\frac{p}{\varepsilon}\right),$$

where c does not depend on a, p, ε, or N.

The next lemma proves the important result that the QTT-rank of a weighted sum of regularly shifted bumps (see for example Figure 14.9, left) does not exceed the product of the QTT-rank of an individual sample and the weighting factor.

Lemma 14.7 ([153]). *Let $N = 2^L$ with $L = L_1 + L_2$, where $L_1, L_2 \geq 1$, and assume that the index set $I := \{1, 2, \ldots, N\}$ is split into $n_2 = 2^{L_2}$ equal non-overlapping subintervals $I = \bigcup_{k=1}^{n_2} I_k$, each of length $n_1 = 2^{L_1}$. Given n_1-vector $\mathbf{x}_0$ that obeys the rank-r_0 QTT representation, define N-vectors $\mathbf{x}_k$, $k = 1, \ldots, L_2$, as*

$$\mathbf{x}_k(i) = \begin{cases} \mathbf{x}_0(:) & \textit{for } i \in I_k, \\ 0 & \textit{for } i \in I \setminus I_k, \end{cases} \tag{14.15}$$

and denote $\mathbf{x} = \mathbf{x}_1 + \cdots + \mathbf{x}_{L_2}$. Then for any choice of N-vector $\mathbf{f}$, we have

$$\operatorname{rank}_{\mathrm{QTT}}(\mathbf{f} \odot \mathbf{x}) \leq \operatorname{rank}_{\mathrm{QTT}}(\mathbf{f}) r_0.$$

Notice that Lemma 14.7 provides a constructive algorithm and rigorous proof of the low-rank QTT decomposition for certain class of Bloch functions [37] and Wannier-type functions.

Figure 14.10 (left) illustrates shapes of the assembled canonical vectors modulated by a sin-harmonics.

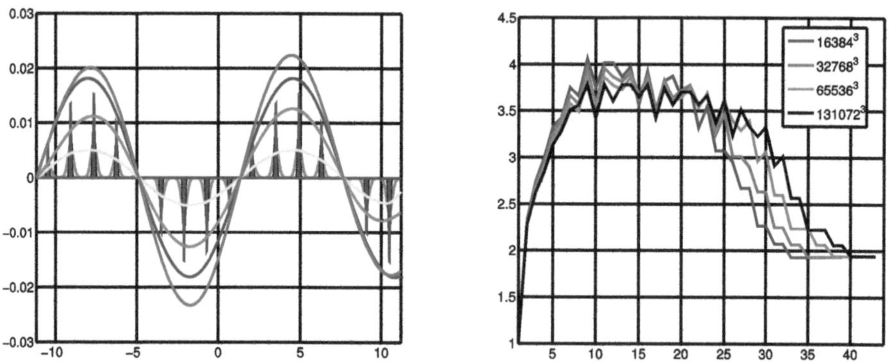

Figure 14.10: Canonical vectors of the lattice sum modulated by a sin-function (left). Right: QTT-ranks of the canonical vectors of a 3D Newton kernel discretized on a cubic grids of size $n^3 = 16\,384^3, 32\,768^3, 65\,536^3$, and $131\,072^3$.

The following Lemma estimates the bounds for the average QTT ranks of the assembled vectors in $\mathbf{P}_{C_L}$ in a periodic setting.

Lemma 14.8 ([153]). *For given tolerance $\varepsilon > 0$, suppose that the set of Gaussian functions $S := \{g_k = e^{-t_k^2\|x\|^2}\}$, $k = 0, 1, \ldots, M$, representing canonical vectors in tensor decomposition $\mathbf{P}_R$, is specified by parameters in (6.3) and set $e^{-t_k^2\|x\|^2} = e^{-\|x\|^2/2p_k^2}$. Let us split the set S into two subsets $S = S_{\mathrm{loc}} \cup S_{\mathrm{glob}}$ such that*

$$S_{\mathrm{loc}} := \{g_k : a_\varepsilon(g_k) \leq b\} \quad \textit{and} \quad S_{\mathrm{glob}} = S \setminus S_{\mathrm{loc}},$$

where $a_\varepsilon(g_k) = \sqrt{2}p_k \log^{1/2}(1/\varepsilon)$. Then the QTT-rank of each canonical vector v_q, $q = 1,\dots,R$ in (14.5), where $R = M + 1$, corresponding to S_{loc}, obeys the uniform in L rank bound

$$r_{\mathrm{QTT}} \le C\log(1/\varepsilon).$$

For vectors in S_{glob}, we have the rank estimate

$$r_{\mathrm{QTT}} \le C\log(L/\varepsilon).$$

Proof. In our notation, we have $1/(\sqrt{2}p_k) = t_k = (k\log M)/M$, $k = 1,\dots,M$ ($k = 0$ is the trivial case). We omit the constant factor $\sqrt{2}$ to obtain $p_k = M/(k\log M)$.

For functions $g_k \in S_{\mathrm{loc}}$, the condition $e^{-\frac{a^2}{2p^2}} \le \varepsilon$ implies

$$O(1) = b \ge a_\varepsilon(g_k) = \sqrt{2}p_k \log^{1/2}(1/\varepsilon),$$

justifying the uniform bound $p_k \le C$, and then the rank estimate $r_{\mathrm{QTT}} \le C\log(1/\varepsilon)$ in view of Proposition 14.6. Now we apply Lemma 14.7 to obtain the uniform in L rank bound.

For globally supported functions in S_{glob}, we have $bL \ge a_\varepsilon \simeq p_k \log^{1/2}(1/\varepsilon) \ge b$. Hence, we consider all these functions on the maximal support of the size of super-cell bL and set $a = bL$. Using the trigonometric representation as in the proof of Lemma 2 in [68], we conclude that for each fixed k, the shifted Gaussians $g_{k,\ell}(x) = e^{-t_k^2\|x-\ell b\|^2}$ ($\ell = 1,\dots,L$) can be approximated by shifted trigonometric series

$$G_r(x-b\ell) = \sum_{m=0}^{M} C_m p e^{-\frac{\pi^2 m^2 p^2}{2a^2}} \cos\left(\frac{\pi m(x-b\ell)}{a}\right), \quad a = bL,$$

which all have the common trigonometric basis containing about

$$\operatorname{rank}_{\mathrm{QTT}}(G_r) = O\left(\log\left(\frac{p_k}{\varepsilon}\right)\right) = O\left(\log\left(\frac{bL}{\varepsilon}\right)\right)$$

terms. Hence, the sum of shifted Gaussian vectors over L unit cells will be approximated with the same QTT-rank bound as each individual term in this sum, which proves the assertion. □

Based on the previous statements, we arrive at the following result.

Theorem 14.9 ([153]). *The tensor representation of v_{c_L} for the full lattice sum generated by a single charge can be presented by the rank-R QTT-canonical tensor*

$$\mathbf{P}_{c_L} = \sum_{q=1}^{R}\left(\mathcal{Q}\sum_{k_1=1}^{L} \mathcal{W}_{\nu(k_1)}\mathbf{p}_q^{(1)}\right) \otimes \left(\mathcal{Q}\sum_{k_2=1}^{L} \mathcal{W}_{\nu(k_2)}\mathbf{p}_q^{(2)}\right) \otimes \left(\mathcal{Q}\sum_{k_3=1}^{L} \mathcal{W}_{\nu(k_3)}\mathbf{p}_q^{(3)}\right), \tag{14.16}$$

where $\mathcal{Q}\mathbf{p}_q^{(\ell)}$ denotes the QTT tensor approximations of the canonical vector $\mathbf{p}_q^{(\ell)}$. Here the QTT-rank of each canonical vector is bounded by $r_{\mathrm{QTT}} \le C\log(L/\varepsilon)$. The computational cost and storage are estimated by $O(RLr_{\mathrm{QTT}}^3)$ and $O(R\log^2(L/\varepsilon))$, respectively.

Figure 14.10, right, represents QTT-ranks of the canonical vectors of a single 3D Newton kernel discretized on a large cubic grids.

Figures 14.10 and 14.11 demonstrate that the average QTT-ranks of the assembled canonical vectors for $k = 1, \dots, R$ scale logarithmically both in L and in the total grid-size $n = N_L$.

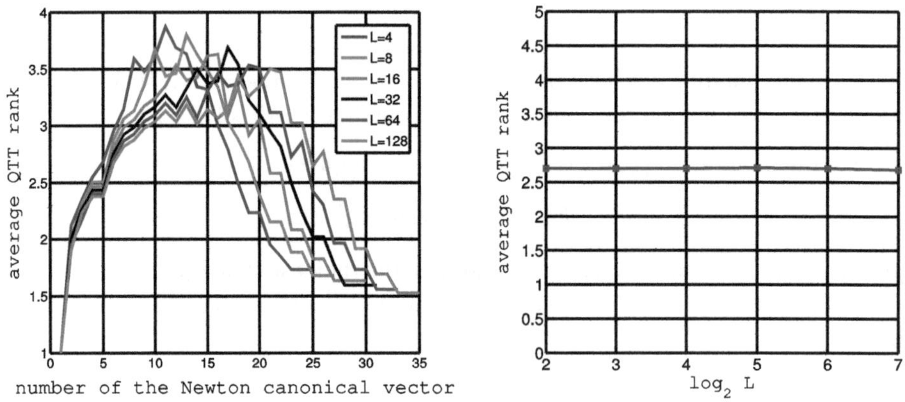

Figure 14.11: Left: QTT-ranks of the assembled canonical vectors vs. L for fixed grid size $N^3 = 16\,384^3$. Right: Average QTT-ranks over R canonical vectors vs. $\log L$ for 3D evaluation of the $L \times 1 \times 1$ chain of Hydrogen atoms on $N \times N \times N$ grids, $N = 2048, 4096, 8192, 16\,384$.

14.5 Summation of long-range potentials on 3D lattices with defects

In this section, we describe a tensor method introduced in [149, 153] for fast summation of long-range potentials on 3D lattices with multiple defects, such as vacancies, impurities, and in the case of hexagonal symmetries. The resulting lattice sum is calculated as a Tucker or canonical representation whose directional vectors are assembled by the 1D summation of the generating vectors for the shifted reference tensor, once precomputed on large $N \times N \times N$ representation grid in a 3D bounding box.

For lattices with defects, the overall potential is obtained as an algebraic sum of several tensors, each representing the contribution of a certain cluster of individual defects. This leads to increase in the tensor rank of the resultant potential sum. For rank reduction in the canonical format, the canonical-to-Tucker decomposition is applied based on the RHOSVD approximation [174]; see Section 3.3. For the RHOSVD approximation to a sum of canonical/Tucker tensors, the stable error bounds in the relative norm in terms of discarded singular values of the side matrices are proven in [153].

14.5.1 Sums of potentials on defected lattices in canonical format

We consider the sum of canonical tensors on a lattice with defects located at S sources. The canonical rank of the resultant tensor may increase at a factor of S. The effective rank of the perturbed sum may be reduced by using the RHOSVD approximation via Can $\mapsto$ Tuck $\mapsto$ Can algorithm (see [174]). This approach basically provides the compressed tensor with the canonical rank quadratically proportional to those of the respective Tucker approximation to the sum with defects.

Here, for the readers convenience, we recall shortly the basics of the RHOSVD and C2T decomposition described in detail in Section 3.3.2. In what follows, we focus on the stability conditions for RHOSVD approximation and their applicability in the summation of spherically symmetric interaction potentials. The canonical rank-R tensor representation (2.13) can be written as the rank-(R,R,R) Tucker tensor by introducing the diagonal Tucker core tensor $\boldsymbol{\xi} := \operatorname{diag}\{\xi_1,\dots,\xi_R\} \in \mathbb{R}^{R\times R\times R}$ such that $\xi_{\nu_1,\nu_2,\nu_3} = 0$ except when $\nu_1 = \cdots = \nu_3$ with $\xi_{\nu,\dots,\nu} = \xi_\nu$, $\nu = 1,\dots,R$ (see Figure 3.12)

$$\mathbf{A} = \boldsymbol{\xi} \times_1 A^{(1)} \times_2 A^{(2)} \times_d A^{(3)}. \tag{14.17}$$

Given the rank parameter $\mathbf{r} = (r_1, r_2, r_3)$, to define the reduced rank-$\mathbf{r}$ HOSVD-type Tucker approximation to the tensor in (2.13), we set $n_\ell = n$ and suppose for definiteness that $n \le R$, so that SVD of the side-matrix $A^{(\ell)}$ is given by

$$A^{(\ell)} = Z^{(\ell)} D_\ell V^{(\ell)^T} = \sum_{k=1}^{n} \sigma_{\ell,k} \mathbf{z}_k^{(\ell)} \mathbf{v}_k^{(\ell)^T}, \quad \mathbf{z}_k^{(\ell)} \in \mathbb{R}^n,\ \mathbf{v}_k^{(\ell)} \in \mathbb{R}^R,$$

with the orthogonal matrices $Z^{(\ell)} = [\mathbf{z}_1^{(\ell)},\dots,\mathbf{z}_n^{(\ell)}]$ and $V^{(\ell)} = [\mathbf{v}_1^{(\ell)},\dots,\mathbf{v}_n^{(\ell)}]$, $\ell = 1,2,3$. Given rank parameters $r_1,\dots,r_\ell < n$, introduce the truncated SVD of the side-matrix $A^{(\ell)}, Z_0^{(\ell)} D_{\ell,0} V_0^{(\ell)^T}$ $(\ell = 1,2,3)$, where $D_{\ell,0} = \operatorname{diag}\{\sigma_{\ell,1},\sigma_{\ell,2},\dots,\sigma_{\ell,r_\ell}\}$, and $Z_0^{(\ell)} \in \mathbb{R}^{n\times r_\ell}$ and ${V_0}^{(\ell)} \in \mathbb{R}^{R\times r_\ell}$ represent the orthogonal factors being the respective sub-matrices in the SVD factors of $A^{(\ell)}$. Here, we recall the definition of RHOSVD tensor approximation (see Section 3.3): the RHOSVD approximation of $\mathbf{A}$, further denoted as $\mathbf{A}^0_{(\mathbf{r})}$, is defined as the rank-$\mathbf{r}$ Tucker tensor obtained by the projection of $\mathbf{A}$ in the form (14.17) onto the orthogonal matrices of the dominating singular vectors in $Z_0^{(\ell)}$ $(\ell = 1,2,3)$.

The stability of RHOSVD approximation is formulated in the following assertion.

Lemma 14.10 ([174]). *Let the canonical decomposition* (2.13) *satisfy the stability condition*

$$\sum_{\nu=1}^{R} \xi_\nu^2 \le C\|\mathbf{A}\|^2. \tag{14.18}$$

Then the quasi-optimal RHOSVD approximation is robust in the relative norm

$$\|\mathbf{A} - \mathbf{A}^0_{(\mathbf{r})}\| \le C\|\mathbf{A}\| \sum_{\ell=1}^{3} \left(\sum_{k=r_\ell+1}^{\min(n,R)} \sigma_{\ell,k}^2 \right)^{1/2},$$

where $\sigma_{\ell,k}$ $(k = r_\ell + 1,\dots,n)$ denote the truncated singular values.

Notice that the stability condition (14.18) is fulfilled, in particular, if

(a) All canonical vectors in (2.13) are non-negative, which is the case for *sinc*-quadrature based approximations to Green's kernels based on integral transforms (6.8)–(6.11), since $a_k > 0$.

(b) The partial orthogonality of the canonical vectors holds, that is, rank-1 tensors $\mathbf{a}_\nu^{(1)} \otimes \cdots \otimes \mathbf{a}_\nu^{(d)}$ ($\nu = 1, \dots, R$) are mutually orthogonal. We refer to [192] for various definitions of orthogonality for canonical tensors.

14.5.2 Tucker tensor format in summation on defected lattices

In this section, following [153], we analyze the assembled summation of Tucker/canonical tensors on the defected lattices in the algebraic framework. Denote the perturbed Tucker tensor by $\widehat{\mathbf{U}}$. Let us introduce a set of $\mathbf{k}$-indices on the lattice, $\mathcal{S} =: \{\mathbf{k}_1, \dots, \mathbf{k}_S\}$, where the unperturbed Tucker tensor $\mathbf{U}_0 := \mathbf{T}_{c_L}$ initially given by summation over the full rectangular lattice (14.10) is defected at positions associated with $\mathbf{k} \in \mathcal{S}$ by the Tucker tensor $\mathbf{U}_\mathbf{k} = \mathbf{U}_{\mathbf{k}_s} = \mathbf{U}_s$ ($s = 1, \dots, S$) given by

$$\mathbf{U}_s = \sum_{\mathbf{m}=\mathbf{1}}^{\mathbf{r}_s} b_{s,\mathbf{m}} \mathbf{u}_{s,m_1}^{(1)} \otimes \mathbf{u}_{s,m_2}^{(2)} \otimes \mathbf{u}_{s,m_3}^{(3)}, \quad s = 1, \dots, S. \tag{14.19}$$

Without loss of generality, all Tucker tensors $\mathbf{U}_s$ ($s = 0, 1, \dots, S$) can be assumed orthogonal.

Now the perturbed Tucker tensor $\widehat{\mathbf{U}}$ is obtained from the non-perturbed one $\mathbf{U}_0$ by adding a sum of all defects $\mathbf{U}_\mathbf{k}$, $\mathbf{k} \in \mathcal{S}$:

$$\mathbf{U}_0 \mapsto \widehat{\mathbf{U}} = \mathbf{U}_0 + \sum_{s=1}^{S} \mathbf{U}_s, \tag{14.20}$$

which implies the simple upper rank estimates for best Tucker approximation of $\widehat{\mathbf{U}}$,

$$\widehat{r}_\ell \le r_{0,\ell} + \sum_{s=1}^{S} r_{s,\ell} \quad \text{for } \ell = 1, 2, 3.$$

If the number of perturbed cells S is large enough, then the numerical computations with the Tucker tensor of rank $\widehat{r}_\ell$ become prohibitive, and the rank reduction procedure is required.

In the case of Tucker sum (14.20), we define the assembled side matrices $\widehat{U}^{(\ell)}$ by concatenation of the directional side-matrices of individual tensors $\mathbf{U}_s$, $s = 0, 1, \dots, S$:

$$\widehat{U}^{(\ell)} = [\mathbf{u}_1^{(\ell)} \cdots \mathbf{u}_{r_{0,\ell}}^{(\ell)}, \mathbf{u}_1^{(\ell)} \cdots \mathbf{u}_{r_{1,\ell}}^{(\ell)}, \dots, \mathbf{u}_1^{(\ell)} \cdots \mathbf{u}_{r_{S,\ell}}^{(\ell)}] \in \mathbb{R}^{n \times (r_{0,\ell} + \sum_{s=1,\dots,S} r_{s,\ell})}, \quad \ell = 1, 2, 3. \tag{14.21}$$

Given the rank parameter $\mathbf{r} = (r_1, r_2, r_3)$, introduce the truncated SVD of $\widehat{U}^{(\ell)}$,

$$\widehat{U}^{(\ell)} \approx Z_0^{(\ell)} D_{\ell,0} V_0^{(\ell)^T}, \quad Z_0^{(\ell)} \in \mathbb{R}^{n\times r_\ell}, \quad V_0{}^{(\ell)} \in \mathbb{R}^{(r_{0,\ell}+\sum_{s=1,\dots,S} r_{s,\ell})\times r_\ell},$$

where $D_{\ell,0} = \operatorname{diag}\{\sigma_{\ell,1}, \sigma_{\ell,2}, \dots, \sigma_{\ell,r_\ell}\}$. Here, instead of fixed rank parameter, the truncation threshold $\varepsilon > 0$ can be chosen.

The stability criteria for RHOSVD approximation, as in Lemma 14.10, allows natural extension to the case of generalized RHOSVD approximation applied to a sum of Tucker tensors in (14.20).

The following theorem proven in [153] provides an error estimate for the generalized RHOSVD approximation, converting a sum of Tucker tensors to a single Tucker tensor with fixed rank bounds or subject to the given tolerance $\varepsilon > 0$.

Theorem 14.11 (Tucker-sum-to-Tucker). *Given a sum of Tucker tensors* (14.20) *and the rank truncation parameter* $\mathbf{r} = (r_1, \dots, r_d)$:

(a) *Let* $\sigma_{\ell,1} \geq \sigma_{\ell,2} \geq \cdots \geq \sigma_{\ell,\min(n,R)}$ *be the singular values of the* ℓ*-mode side-matrix* $\widehat{U}^{(\ell)} \in \mathbb{R}^{n\times R}$ $(\ell = 1,2,3)$ *defined in* (14.21). *Then the generalized RHOSVD approximation* $\mathbf{U}^0_{(\mathbf{r})}$ *obtained by the projection of* $\widehat{\mathbf{U}}$ *onto the dominating singular vectors* $Z_0^{(\ell)}$ *of the Tucker side-matrices* $\widehat{U}^{(\ell)} \approx Z_0^{(\ell)} D_{\ell,0} V_0^{(\ell)^T}$ *exhibits the error estimate*

$$\|\widehat{\mathbf{U}} - \mathbf{U}^0_{(\mathbf{r})}\| \leq |\widehat{\mathbf{U}}| \sum_{\ell=1}^{d} \left(\sum_{k=r_\ell+1}^{\min(n,\widehat{r}_\ell)} \sigma_{\ell,k}^2 \right)^{1/2}, \quad \text{where } |\widehat{\mathbf{U}}|^2 = \sum_{s=0}^{S} \|\mathbf{U}_s\|^2. \tag{14.22}$$

(b) *Assume the stability condition* $\sum_{s=0}^{S} \|\mathbf{U}_s\|^2 \leq C\|\widehat{\mathbf{U}}\|^2$ *for the sum* (14.20). *Then the generalized RHOSVD approximation provides the quasi-optimal error bound*

$$\|\widehat{\mathbf{U}} - \mathbf{U}^0_{(\mathbf{r})}\| \leq C\|\widehat{\mathbf{U}}\| \sum_{\ell=1}^{d} \left(\sum_{k=r_\ell+1}^{\min(n,\widehat{r}_\ell)} \sigma_{\ell,k}^2 \right)^{1/2}.$$

The resultant Tucker tensor $\mathbf{U}^0_{(\mathbf{r})}$ can be considered as the initial guess for the ALS iteration to compute best Tucker ε-approximation of a sum of Tucker tensors.

Figure 14.12 (left) visualizes the result of assembled Tucker summation of the three-dimensional grid-based Newton potentials on a $16\times16\times1$ lattice, with a vacancy and impurity, each of $2\times2\times1$ lattice size. Figure 14.12 (right) shows the corresponding Tucker vectors along x-axis, which distinctly display the local shapes of vacancies and impurities.

14.5.3 Numerical examples for non-rectangular and composite lattices

Though the rectangular structures with lattice-type vacancies and impurities are the most representative structures in crystalline-type systems, in many practically interesting cases, the physical lattice may have a non-rectangular geometry that does not

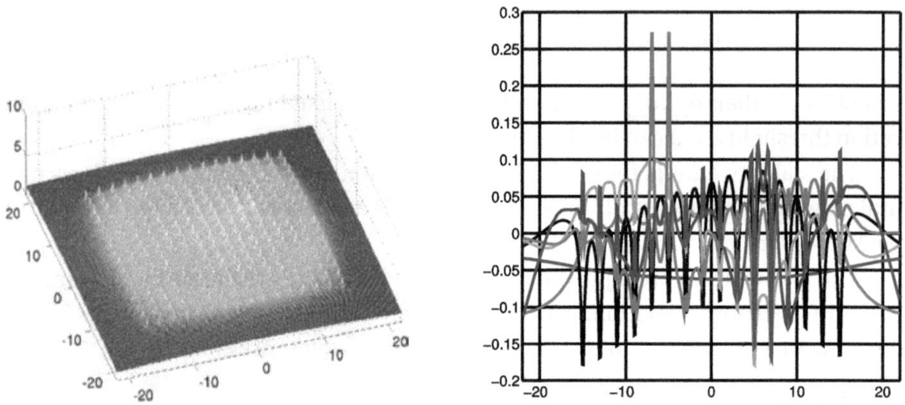

Figure 14.12: Left: assembled grid-based Tucker sum of 3D Newton potentials on a lattice $16 \times 16 \times 1$ with an impurity and a vacancy, both of size $2 \times 2 \times 1$. Right: the Tucker vectors along x-axis.

fit exactly the tensor-product structure of the canonical/Tucker data arrays. For example, the hexagonal or parallelepiped-type lattices can be considered. Here, following [153], we discuss how to apply tensor summation methods to certain classes of non-rectangular geometries and show a few numerical examples demonstrating the required (minor) modifications of the basic assembled summation schemes.

It is worth noting that most of interesting lattice structures (say, arising in crystalline modeling) inherit a number of spacial symmetries, which allow us, first, to classify and then simplify the computational schemes for each particular case of symmetry. In this regard, we mention the following class of lattice topologies, which can be efficiently treated by our tensor summation techniques:

- The target lattice $\mathcal{L}$ can be split into the union of several (few) sub-lattices $\mathcal{L} = \bigcup \mathcal{L}_q$ such that each sub-lattice $\mathcal{L}_q$ allows a 3D rectangular grid-structure.
- Defects in the target composite lattice may be distributed over rectangular subdomains (clusters) represented on a coarser scale. Numerically, it reduces to summation of tensors corresponding to each of $\mathcal{L}_q$.
- Defects in the target lattice are distributed over rectangular subdomains (clusters) represented on a coarser scale.

For such lattice topologies, the assembled tensor summation algorithm applies independently to each rectangular sub-lattice $\mathcal{L}_q$, and then the target tensor is obtained as a direct sum of tensors associated with $\mathcal{L}_q$ accomplished with the subsequent rank-reduction procedure. The example of such a geometry is given by hexagonal lattice presented in Figure 14.13 (rectangular in the third axis), which can be split into a union of two rectangular sub-lattices $\mathcal{L}_1$ (red) and $\mathcal{L}_2$ (blue).

Numerically it is implemented by summation of two tensors via concatenation of the canonical vectors corresponding to “blue” and “green” lattices, both living on the same fine 3D Cartesian grid.

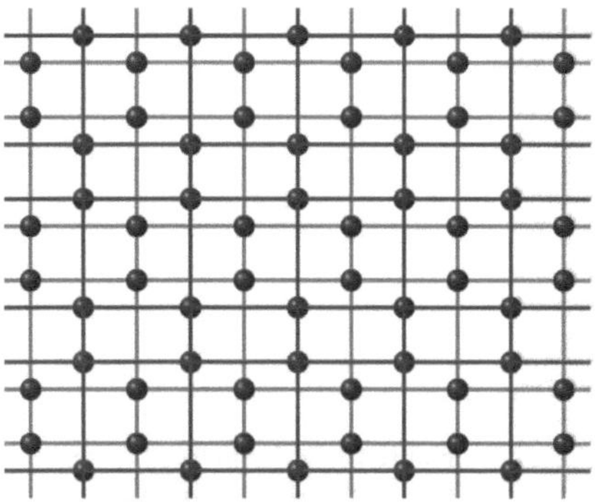

Figure 14.13: Hexagonal lattice is a union of two rectangular lattices, "green" and "blue".

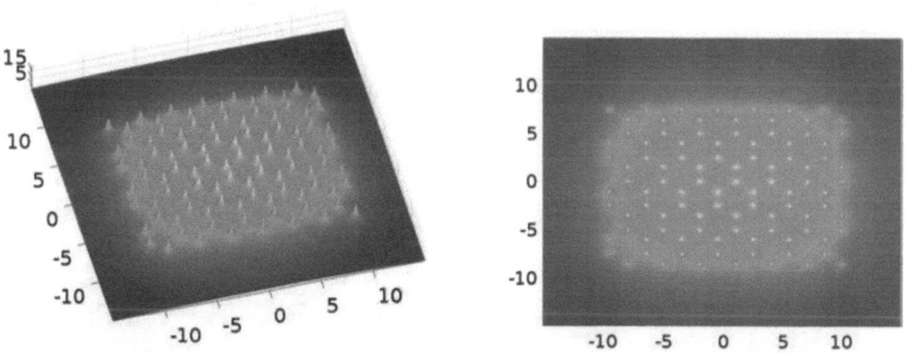

Figure 14.14: Left: Sum of potentials over the hexagonal lattice of the type shown in Figure 14.13. Right: rotated view.

The following numerical results basically reproduce those in [153]. Figures 14.14 (left and right) show the resulting potential sum for the hexagonal lattice structure composed of a sum of $7\times7\times1$ "blue" and $7\times7\times1$ "green" potentials. The rank of the tensor representing the sum is two times larger than the rank of the single reference Newton kernel.

In the case of regularly positioned vacancies, as in Figure 14.15, showing the result of assembled canonical summation of the grid-based Newton potentials on a lattice $24\times24\times1$ with $6\times6\times1$ vacancies (two-level lattice), the resulting tensor rank is only two times larger than the rank of a single Newton potential.

Figure 14.16 illustrates the situation when defects are located in a compact subdomain. It represents the result of assembled canonical sum of the Newton potentials on L-shaped (left) and O-shaped (right) lattices. The resulting potentials sum for L-shape lattice is a difference of a full $24\times18\times1$ lattice and a sublattice of size $12\times9\times1$. For O-shape, the resultant tensor is obtained as the difference between the full lattice sum over $12\times12\times1$ and central $6\times6\times1$ clusters. In both cases, the total canonical tensor rank is two times larger than the rank of the single reference potential.

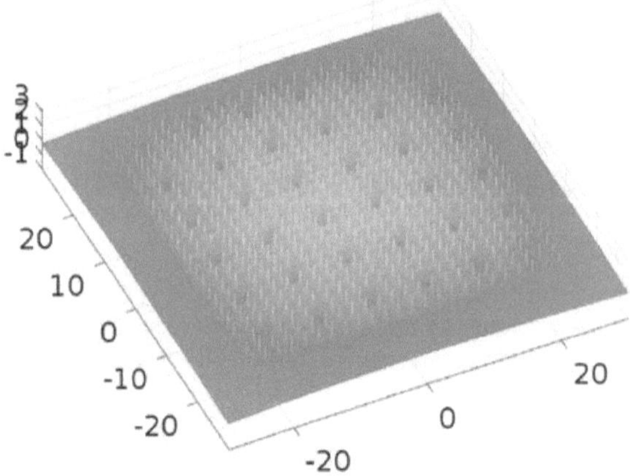

Figure 14.15: Assembled summation of 3D grid-based Newton potentials in canonical format on a $24 \times 24 \times 1$ lattice with regular $6 \times 6 \times 1$ vacancies.

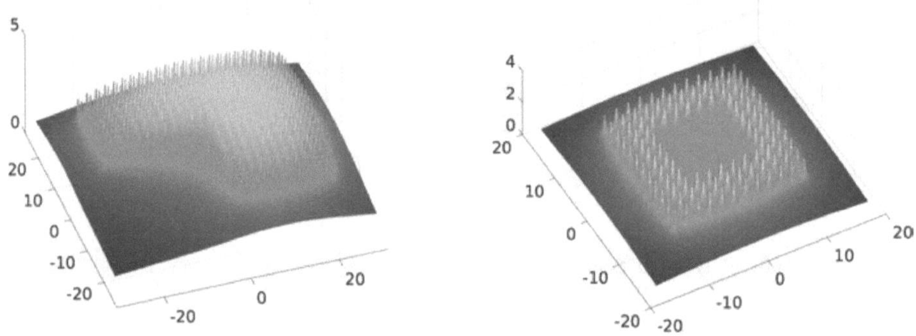

Figure 14.16: Assembled summation of 3D grid-based Newton potentials in canonical format on a $24 \times 18 \times 1$ lattice with L-shaped geometry (left) and O-shaped geometry $12 \times 12 \times 1$ lattice with subtracted vacancy sub-lattice $6 \times 6 \times 1$.

For composite shapes of lattice geometries, one can use the canonical-to-Tucker transform to reduce the canonical rank. In the case of complicated geometries, the Tucker reference tensor for the Newton kernel may be preferable. For example, in the case of O-shaped domain, the maximal Tucker rank of the resultant tensor is 25, whereas the respective ranks for rectangular compounds are 17 and 15.

Since the lattice is not necessarily aligned with the 3D representation grid, it is easy to assemble potentials centered independently on the lattice nodes, for example, for modeling lattices with insertions having other inter-atomic displacements compared with the main lattice. Figure 14.17 represents the result of assembled canonical summation of 3D grid-based Newton potentials on a lattice $12 \times 12 \times 1$ with an impurity of size $2 \times 2 \times 1$ with the interatomic distances different from the main lattice. Since the impurity potentials are determined on the same fine $N_L \times N_L \times N_L$ representation grid,

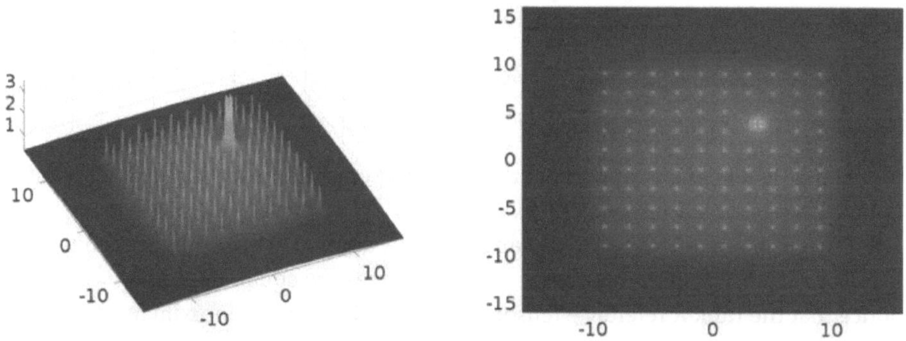

Figure 14.17: Left: assembled canonical summation of 3D grid-based Newton potentials on a lattice $10 \times 10 \times 1$ with an impurity of size $2 \times 2 \times 1$. Right: the vertical projection.

variations in the inter-potential distances do not influence the numerical treatment of defects.

We conclude that in all cases discussed above, the tensor summation approach [153] can be gainfully applied. The overall numerical cost may depend on the geometric structure and symmetries of the system under consideration, since violation of the tensor-product rectangular structure of the lattice may lead to the increase in the Tucker/canonical rank. This is clearly observed in the case of moderate number of defects distributed randomly. In all such cases, the RHOSVD approximation, combined with the ALS iteration, serves for the robust rank reduction in the Tucker format. We also note that many of the crystalline-type structures belong to types of face-cubic or hexagonal symmetries. Taking into account the facilities for easy numerical treatments of multiple defects, we could expect that the tensor summation method of long-range potentials can be used in a wide range of applications.

Finally we notice that our approach has the natural extension to the higher-dimensional lattices in $\mathbb{R}^d$, $d > 3$, so that along with the canonical and Tucker tensors, the TT tensor format [226] can be adapted.

14.6 Interaction energy of the long-range potentials on finite lattices

Fast and accurate computation of the interaction energy of the long-range potentials on finite lattices is one of the challenging tasks in computer modeling of macromolecular structures such as quantum dots, nanostructures, and biological systems. In this section, we recall the efficient scheme for tensor-based calculation of the interaction energy of the long-range potentials on finite lattices proposed in [152]. For the nuclear charges $\{Z_{\mathbf{k}}\}$ centered at points $x_{\mathbf{k}}$, $\mathbf{k} \in \mathcal{K}^3$, which are located on the $L \times L \times L$ lattice $\mathcal{L}_L = \{x_{\mathbf{k}}\}$ with the step-size b, the total interaction energy of these charges is defined as

$$E_L = \frac{1}{2} \sum_{\mathbf{k},\mathbf{j}\in\mathcal{K},\mathbf{k}\neq\mathbf{j}} \frac{Z_\mathbf{k} Z_\mathbf{j}}{\|x_\mathbf{j} - x_\mathbf{k}\|}, \quad \text{i. e., for } \|x_\mathbf{j} - x_\mathbf{k}\| \geq b. \tag{14.23}$$

Notice that local density approximation for long-range and short-range energy functionals have been addressed in [279].

The tensor summation scheme can be directly applied to this computational problem. For this discussion, we assume that all charges are equal, that is, $Z_\mathbf{k} = Z$. First, notice that the rank-R reference tensor $h^{-3}\widetilde{\mathbf{P}}$ defined in (14.4) approximates with high accuracy $O(h^2)$ the Coulomb potential $\frac{1}{\|x\|}$ in $\widetilde{\Omega}_L$ (for $\|x\| \geq b$ that is required for the energy expression) on the fine $2n\times 2n\times 2n$ representation grid with mesh size h. Likewise, the tensor $h^{-3}\mathbf{P}_{c_L}$ approximates the potential sum $v_{c_L}(x)$ on the same fine representation grid including the lattice points $x_\mathbf{k}$.

We evaluate the energy expression (14.23) by using tensor sums as in (14.5), but now applied to a small sub-tensor of the rank-R canonical reference tensor $\widetilde{\mathbf{P}}$, that is, $\widetilde{\mathbf{P}}_L := [\widetilde{\mathbf{P}}_{|x_\mathbf{k}}] \in \mathbb{R}^{2L\times 2L\times 2L}$ obtained by tracing of $\widetilde{\mathbf{P}}$ at the accompanying lattice of the double size $2L \times 2L \times 2L$, that is, $\widetilde{\mathcal{L}}_L = \{x_\mathbf{k}\} \cup \{x_{\mathbf{k}'}\} \in \widetilde{\Omega}_L$. Here, $\widetilde{\mathbf{P}}_{|x_\mathbf{k}}$ denotes the tensor entry corresponding to the $\mathbf{k}$th lattice point designating the atomic center $x_\mathbf{k}$.

We are interested in the computation of the rank-R tensor $\widehat{\mathbf{P}}_{c_L} = [\mathbf{P}_{c_L|x_\mathbf{k}}]_{\mathbf{k}\in\mathcal{K}} \in \mathbb{R}^{L\times L\times L}$, where $\mathbf{P}_{c_L|x_\mathbf{k}}$ denotes the tensor entry corresponding to the $\mathbf{k}$th lattice point on $\mathcal{L}_L$. The tensor $\widehat{\mathbf{P}}_{c_L}$ can be computed at the expense $O(L^2)$ by

$$\widehat{\mathbf{P}}_{c_L} = \sum_{q=1}^{R} \left(\sum_{k_1\in\mathcal{K}} \mathcal{W}_{(k_1)} \widetilde{\mathbf{p}}^{(1)}_{L,q} \otimes \sum_{k_2\in\mathcal{K}} \mathcal{W}_{(k_2)} \widetilde{\mathbf{p}}^{(2)}_{L,q} \otimes \sum_{k_3\in\mathcal{K}} \mathcal{W}_{(k_3)} \widetilde{\mathbf{p}}^{(3)}_{L,q} \right).$$

This leads to the representation of the energy sum (14.23) (with accuracy $O(h^2)$) in a form

$$E_{L,T} = \frac{Z^2 h^{-3}}{2} \left(\langle \widehat{\mathbf{P}}_{c_L}, \mathbf{1} \rangle - \sum_{\mathbf{k}\in\mathcal{K}} \widetilde{\mathbf{P}}_{|x_\mathbf{k}=0} \right),$$

where the first term in brackets represents the full canonical tensor lattice sum restricted to the $\mathbf{k}$-grid composing the lattice $\mathcal{L}_L$, whereas the second term introduces the correction at singular points $x_\mathbf{j} - x_\mathbf{k} = 0$. Here, $\mathbf{1} \in \mathbb{R}^{L\times L\times L}$ is the all-ones tensor. By using the rank-1 tensor $\mathbf{P}_{0L} = \widetilde{\mathbf{P}}_{|x_\mathbf{k}=0}\mathbf{1}$, the correction term can be represented by a simple tensor operation

$$\sum_{\mathbf{k}\in\mathcal{K}} \widetilde{\mathbf{P}}_{|x_\mathbf{k}=0} = \langle \mathbf{P}_{0L}, \mathbf{1} \rangle.$$

Finally, the interaction energy E_L allows the approximate representation

$$E_L \approx E_{L,T} = \frac{Z^2 h^{-3}}{2} (\langle \widehat{\mathbf{P}}_{c_L}, \mathbf{1} \rangle - \langle \mathbf{P}_{0L}, \mathbf{1} \rangle), \tag{14.24}$$

Table 14.4: Comparison of times for the full (T_{full}), ($O(L^6)$), and tensor-based ($T_{tens.}$) calculation of the interaction energy sum for the lattice electrostatic potentials.

L^3	T_{full}	$T_{tens.}$	$E_{L,T}$	abs. err.
24^3	37	1.2	$3.7 \cdot 10^6$	$2 \cdot 10^{-8}$
32^3	250	1.5	$1.5 \cdot 10^7$	$1.5 \cdot 10^{-9}$
48^3	3374	2.8	$1.12 \cdot 10^8$	0
64^3	–	5.7	$5.0 \cdot 10^8$	–
128^3	–	13.5	$1.6 \cdot 10^{10}$	–
256^3	–	68.2	$5.2 \cdot 10^{11}$	–

which can be implemented in $O(L^2) \ll L^3 \log L$ complexity by tensor operations with the rank-R canonical tensors in $\mathbb{R}^{L\times L\times L}$.

Table 14.4 illustrates the performance of the algorithm described above. We compare the exact value computed by (14.23), of complexity $O(L^6)$, with the calculation time obtained by using our scheme with the approximate tensor representation (14.24) on the fine representation grid with $n = n_0 L$, $n_0 = 128$. The tested lattice systems are composed of Hydrogen atoms with interatomic distance 2.0 bohr. The geometric size of the largest 3D lattice with 256^3 potentials is about 524^3 bohr^3 or 26^3 cubic nanometers, see [152] for further details.

15 Range-separated tensor format for many-particle systems

Numerical modelling of long-range electrostatic potentials in many-particle systems leads to challenging computational problems, as it was already mentioned in Section 14. Well recognized traditional approaches based on the Ewald summation [79], fast Fourier transform or fast multipole expansion [103] methods usually apply to calculating the interaction energy of a system (including the evaluation of the potential only at N points s_ν) scale as $O(N \log N)$ for N-particle systems. These approaches need large computer facilities for meshing up the result of Ewald sums. Computation of long-range interaction potentials of large multiparticle systems is discussed for example in [58, 199, 278, 139], and using grid-based approaches in [16, 312]. Ewald-type splitting of the Coulomb interaction into long- and short-range components was applied in density functional theory calculations [280].

A novel range-separated (RS) canonical/Tucker tensor format was recently introduced in [24] for modeling of the multidimensional long-range interaction potentials in multi-particle systems of general type. The main idea of the RS tensor format is the independent grid-based low-rank representation of the localized and global parts in the target tensor, which allows the efficient numerical approximation of N-particle interaction potentials. The single reference potential such as $1/\|x\|$ is split into a sum of localized and long-range low-rank canonical tensors represented on a fine 3D $n \times n \times n$ Cartesian grid. The smooth long-range contribution to the total potential sum is represented in the form of low-rank canonical/Tucker tensor in $O(n)$ storage. It is proven that the resultant rank parameters depend only logarithmically on the number of particles N and the grid-size n. Agglomeration of the short-range part in the sum is reduced to an independent treatment of N localized terms with almost disjoint effective supports, calculated in $O(N)$ operations. Last but not the least, the RS-tensor format allows to represents the collective potential of a multiparticle system at any point of fine $n \times n \times n$ grid at $O(1)$ cost.

The RS canonical/Tucker tensor representations reduce the cost of multi-linear algebraic operations on the 3D potential sums arising in multi-dimensional data modeling by radial basis functions, for example, in computation of the electrostatic potential of a protein, in 3D integration and convolution transform, computation of gradients, forces, and the interaction energy of a many-particle systems, and in the approximation of d-dimensional scatters data by reducing all of them to 1D calculations.

The presentation here mainly follows [24] (see also the recent publication [28]). For a given non-local generating kernel $p(\|x\|)$, $x \in \mathbb{R}^3$, the calculation of a weighted sum of interaction potentials in the large N-particle system with the particle locations at $s_\nu \in \mathbb{R}^3$, $\nu = 1, \ldots, N$,

$$P(x) = \sum_{\nu=1}^{N} Z_\nu \, p(\|x - s_\nu\|), \quad Z_\nu \in \mathbb{R}, \quad s_\nu, x \in \Omega = [-b, b]^3, \tag{15.1}$$

https://doi.org/10.1515/9783110365832-015

leads to computationally intensive numerical task. Indeed, the generating radial basis function $p(\|x\|)$ is allowed to have a slow polynomial decay in $1/\|x\|$ so that each individual term in (15.1) contributes essentially to the total potential at each point in Ω, thus predicting the $O(N)$ complexity for the straightforward summation at every fixed target $x \in \mathbb{R}^3$. Moreover, in general, the function $p(\|x\|)$ has a singularity or a cusp at the origin $x = 0$. Typical examples of the radial basis function $p(\|x\|)$ are given by the Newton $1/\|x\|$, Slater $e^{-\lambda\|x\|}$, Yukawa $e^{-\lambda\|x\|}/\|x\|$, and other Green's kernels (see examples in Section 15.3.1).

The important ingredient of the RS approach is the splitting of a single reference potential, say $p(\|x\|) = 1/\|x\|$, into a sum of localized and long-range low-rank canonical tensors represented on the grid Ω_n. In this regard, it can be shown that the explicit sinc-based canonical tensor decomposition of the generating reference kernel $p(\|x\|)$ by a sum of Gaussians implies the distinct separation of its long- and short-range parts.

Such range separation techniques can be gainfully applied to summation of a large number of generally distributed potentials in (15.1). Indeed, a sum of the long-range contributions can be represented by a single tensor leaving on the $\Omega_n \subset \Omega$ grid by using the canonical-to-Tucker transform [174], which returns this part in the form of a low-rank Tucker tensor. Hence, the smooth long-range contribution to the overall sum is represented on the fine $n \times n \times n$ grid Ω_n in $O(n)$ storage via the global canonical or Tucker tensor with the separation rank that only weakly (logarithmically) depends on the number of particles N. This important feature is confirmed by numerical tests for the large clusters of generally distributed potentials in 3D; see Section 15.2.

In turn, the short-range contribution to the total sum is constructed by using a *single* low-rank reference tensor with a small local support selected from the "short-range" canonical vectors in the tensor decomposition of $p(\|x\|)$. To that end, the whole set of N short-range clusters is represented by replication and rescaling of the small-size localized reference tensor, thus reducing the storage to the $O(1)$-parametrization of the reference canonical tensor, and the list of coordinates and charges of particles. Representation of the short-range part over $n \times n \times n$ grid needs $O(N\,n)$ computational work for N-particle system. Such cumulated sum of the short-range components allows the "local operations" in the RS-canonical format, making it particularly efficient for tensor multilinear algebra.

The RS tensor formats provide a tool for the efficient numerical treatment of interaction potentials in many-particle systems, which, in some aspects, can be considered as an alternative to the well-established multipole expansion method [103]. The particular benefit is the low-parametric representation of the collective interaction potential on large 3D Cartesian grid in the whole computational domain in the linear cost $O(n)$, thus outperforming the grid-based summation techniques based on the full-grid $O(n^3)$-representation in the volume. Both global and local summation schemes are quite easy in program implementation. The prototype algorithms in MATLAB applied on a laptop allow computing the electrostatic potential of large many-particle systems on fine grids of size up to $n^3 = 10^{12}$.

15.1 Tensor splitting of the kernel into long- and short-range parts

From the definition of the sinc-quadrature (6.6), (6.3), we can easily observe that the full set of approximating Gaussians includes two classes of functions: those with small "effective support" and the long-range functions. Clearly, functions from different classes may require different tensor-based schemes for their efficient numerical treatment. Hence, the idea of the new approach is the constructive implementation of a range separation scheme that allows the independent efficient treatment of both the long- and short-range parts in the approximating radial basis functions.

Without loss of generality, we further consider the case of the Newton kernel, so that the sum in (6.6) reduces to $k = 0, 1, \ldots, M$ (due to symmetry argument). From (6.3), we observe that the sequence of quadrature points $\{t_k\}$ can be split into two subsequences

$$\mathcal{T} := \{t_k \mid k = 0, 1, \ldots, M\} = \mathcal{T}_l \cup \mathcal{T}_s$$

with

$$\mathcal{T}_l := \{t_k \mid k = 0, 1, \ldots, R_l\} \quad \text{and} \quad \mathcal{T}_s := \{t_k \mid k = R_l + 1, \ldots, M\}. \tag{15.2}$$

The set $\mathcal{T}_l$ includes quadrature points t_k condensed "near" zero, hence generating the long-range Gaussians (low-pass filters), and $\mathcal{T}_s$ accumulates the increasing in $M \to \infty$ sequence of "large" sampling points t_k with the upper bound $C_0^2 \log^2(M)$, corresponding to the short-range Gaussians (high-pass filters). The quasi-optimal choice of the constant $C_0 \approx 3$ was determined numerically in [30]. We further denote

$$\mathcal{K}_l := \{k \mid k = 0, 1, \ldots, R_l\} \quad \text{and} \quad \mathcal{K}_s := \{k \mid k = l + 1, \ldots, M\}.$$

Splitting (15.2) generates the additive decomposition of the canonical tensor $\mathbf{P}_R$ onto the short- and long-range parts:

$$\mathbf{P}_R = \mathbf{P}_{R_s} + \mathbf{P}_{R_l},$$

where

$$\mathbf{P}_{R_s} = \sum_{t_k \in \mathcal{T}_s} \mathbf{p}_k^{(1)} \otimes \mathbf{p}_k^{(2)} \otimes \mathbf{p}_k^{(3)}, \quad \mathbf{P}_{R_l} = \sum_{t_k \in \mathcal{T}_l} \mathbf{p}_k^{(1)} \otimes \mathbf{p}_k^{(2)} \otimes \mathbf{p}_k^{(3)}. \tag{15.3}$$

The choice of the critical number $R_l = \#\mathcal{T}_l - 1$ (or equivalently, $R_s = \#\mathcal{T}_s = M - R_l$) that specifies the splitting $\mathcal{T} = \mathcal{T}_l \cup \mathcal{T}_s$, is determined by the *active support* of the short-range components such that one can cut off the vectors $\mathbf{p}_k^{(\ell)}$, $t_k \in \mathcal{T}_s$, outside of the sphere B_σ of radius $\sigma > 0$, subject to a certain threshold $\delta > 0$. For fixed $\delta > 0$, the choice of R_s is uniquely defined by the (small) parameter σ and vise versa. Given σ, the two basic criteria corresponding to (A) the max-norm and (B) L^1-norm estimates can be applied:

$$\text{(A)} \quad \mathcal{T}_s = \{t_k : a_k e^{-t_k^2 \sigma^2} \le \delta\} \quad \Leftrightarrow \quad R_l = \min k : a_k e^{-t_k^2 \sigma^2} \le \delta \tag{15.4}$$

or

$$\text{(B)}\quad \mathcal{T}_s := \left\{t_k : a_k \int_{B_\sigma} e^{-t_k^2 x^2} dx \leq \delta\right\} \quad \Leftrightarrow \quad R_l = \min k : a_k \int_{B_\sigma} e^{-t_k^2 x^2} dx \leq \delta. \tag{15.5}$$

Clearly, the sphere B_σ can be subsituted by the small box of the corresponding size. The quantitative estimates on the value of R_l can be easily calculated by using the explicit equation (6.3) for the quadrature parameters. For example, in case $C_0 = 3$ and $a(t) = 1$, criteria (A) implies that R_l solves the equation

$$\left(\frac{3R_l \log M}{M}\right)^2 \sigma^2 = \log\left(\frac{\mathfrak{h}_M}{\delta}\right).$$

Criteria (15.4) and (15.5) can be slightly modified, depending on the particular applications to many-particles systems. For example, in electronic structure calculations, the parameter σ can be associated with the typical inter-atomic distance in the molecular system of interest (Van der Waals distance).

Figures 15.1 and 15.2 illustrate the splitting (15.2) for the tensor $\mathbf{P}_R$ computed on the $n \times n \times n$ grid with the parameters $R = 20$, $R_l = 12$ and $R_s = 8$, respectively. Figure 15.1 shows the long-range canonical vectors from $\mathbf{P}_{R_l}$ in (15.3), whereas Figure 15.2 displays the short-range part described by $\mathbf{P}_{R_s}$. Following criteria (A) with $\delta \approx 10^{-4}$, the effective support for this splitting is determined by $\sigma = 0.9$. The complete Newton kernel simultaneously resolves both the short- and long-range behavior, whereas the function values of the tensor $\mathbf{P}_{R_s}$ vanish exponentially fast apart from the effective support, as can be seen in Figure 15.2.

Inspection of the quadrature point distribution in (6.3) shows that the short- and long-range subsequences are distributed nearly equally balanced, so that one can expect approximately

$$R_s \approx R_l = M/2. \tag{15.6}$$

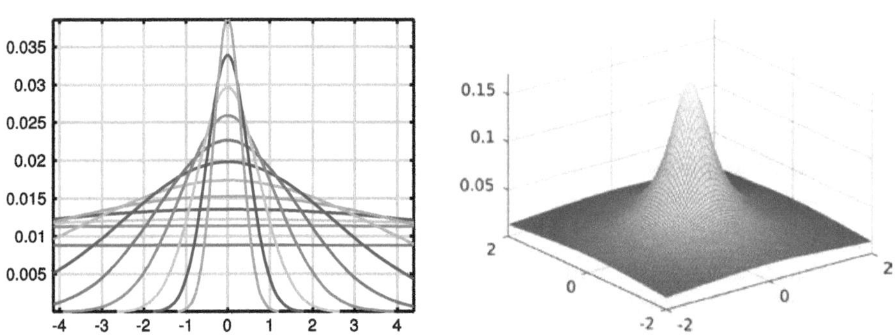

Figure 15.1: Long-range canonical vectors for $n = 1024$, $R = 20$, $R_l = 12$, and the corresponding potential.

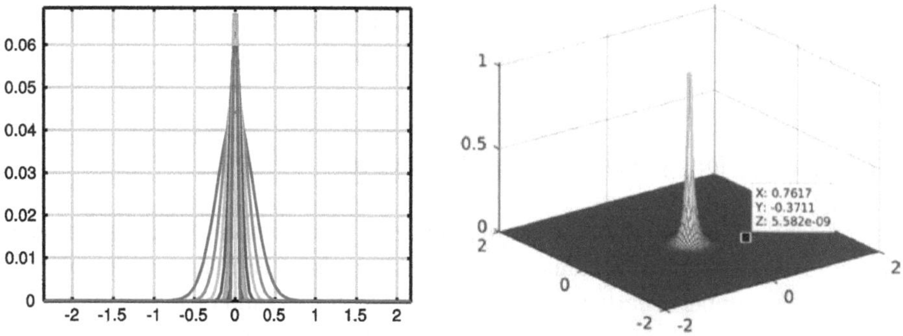

Figure 15.2: Short-range canonical vectors for $n = 1024$, $R = 20$, $R_s = 8$, and the corresponding potential.

The optimal choice may depend on the particular application specified by the separation parameter $\sigma > 0$ and the required accuracy.

The main advantage of the range separation in the splitting to the canonical tensor $\mathbf{P}_R$ in (15.3) is the opportunity for independent tensor representations of both subtensors $\mathbf{P}_{R_s}$ and $\mathbf{P}_{R_l}$, which leads to simultaneous reduction of their complexity and storage demands. Indeed, the effective *local* support characterized by $\sigma > 0$ includes a much smaller number of grid points $n_s \ll n$ compared with the global grid size. Hence, the storage cost $\mathrm{Stor}(\mathbf{P}_{R_s})$ for the canonical tensor representation of the short-range part is estimated by

$$\mathrm{Stor}(\mathbf{P}_{R_s}) \le R_s n_s \ll Rn.$$

Furthermore, the long-range part $\mathbf{P}_l$ approximates a *global* smooth function, which can be represented in Ω on a coarser grid with the number of grid points $n_l \ll n$. Hence, we gain from the reduced complexity estimate

$$\mathrm{Stor}(\mathbf{P}_{R_l}) \le R_l n_l \ll Rn.$$

It is worth noting that the advantage of separate treatment of smooth-nonlocal and non-smooth but locally supported tensor components, allows not only the dramatic reduction of the storage costs $R_s n_s + R_l n_l = O(Rn)$, but also efficient bilinear tensor operations preserving the individual storage complexities as will be shown in the next sections.

15.2 Tensor summation of range-separated potentials

In this section, following [24], we describe how the range-separated tensor representation of the generating potential function can be applied to the fast and accurate grid-based computation of a large sum of non-local potentials centered at arbitrary locations in the 3D volume. This task leads to the bottleneck computational problem in the modeling of large stationary and dynamical N-particle systems.

15.2.1 Quasi-uniformly separable point distributions

One of the main limitations for the use of direct grid-based canonical/Tucker approximations to the large potential sums is due to the strong increase in tensor rank proportionally to the number of particles N_0 in a system. Figures 15.3 and 15.5 show the Tucker ranks for electrostatic potential in the protein-type system consisting of $N_0 = 783$ atoms.

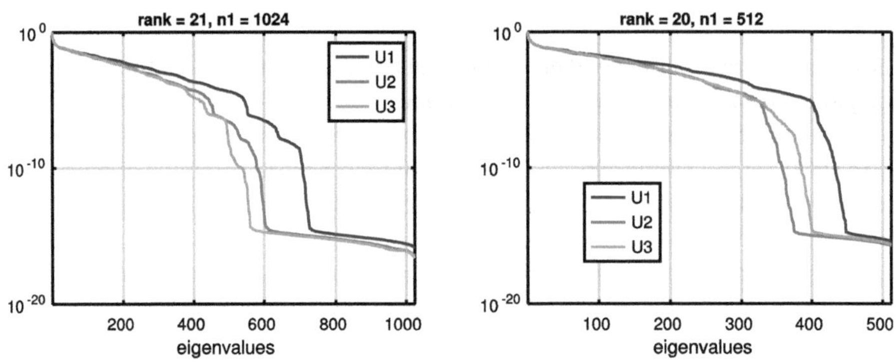

Figure 15.3: The directional Tucker ranks computed by RHOSVD for a protein-type system with $n = 1024$ (left) and $n = 512$ (right).

Given the generating kernel $p(\|x\|)$, we consider the problem of efficiently calculating the weighted sum of a large number of single potentials located in a set $\mathcal{S}$ of separable distributed points (sources) $s_\nu \in \mathbb{R}^3$, $\nu = 1, \dots, N_0$, embedded into the fixed bounding box $\Omega = [-b, b]^3$,

$$P_0(x) = \sum_{\nu=1}^{N_0} z_\nu p(\|x - s_\nu\|), \quad z_\nu \in \mathbb{R}. \tag{15.7}$$

The function $p(\|x\|)$ is allowed to have slow polynomial decay in $1/\|x\|$ so that each individual source contributes essentially to the total potential at each point in Ω.

Definition 15.1 (Well-separable point distribution). Given a constant $\sigma_* > 0$, a set $\mathcal{S} = \{s_\nu\}$ of points in $\mathbb{R}^d$ is called σ_*-separable if

$$d(s_\nu, s_{\nu'}) := \|s_\nu - s_{\nu'}\| \geq \sigma_* \quad \text{for all } \nu \neq \nu'. \tag{15.8}$$

A family of point sets $\{\mathcal{S}_1, \dots, \mathcal{S}_m\}$ is called uniformly σ_*-separable if (15.8) holds for every set $\mathcal{S}_{m'}$, $m' = 1, 2, \dots, m$, independently of the number of particles in a set $\mathcal{S}_{m'}$.

Condition (15.8) can be reformulated in terms of the so-called separation distance $q_{\mathcal{S}}$ of the point set $\mathcal{S}$:

$$q_{\mathcal{S}} := \min_{s \in \mathcal{S}} \min_{s_\nu \in \mathcal{S} \setminus s} d(s_\nu, s) \geq \sigma_*. \tag{15.9}$$

Definition 15.1 on separability of point distributions is fulfilled, particularly, in the case of large molecular systems (proteins, crystals, polymers, nano-clusters), where all atomic centers are strictly separated from each other by a certain fixed *inter-atomic distance* (e. g., Van der Waals distance). The same happens for lattice-type structures, where each atomic cluster within the unit cell is separated from the neighbors by a distance proportional to the lattice step-size.

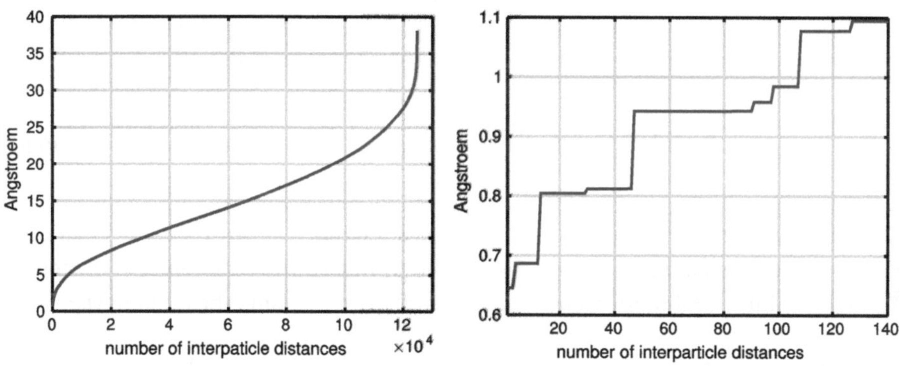

Figure 15.4: Inter-particle distances in an ascendant order for protein-type structure with 500 particles (left); zoom for the first 100 smallest inter-particle distances (right).

Figure 15.4 (left) shows inter-particle distances in ascending order for a protein-type structure including 500 particles. The total number of distances equals to $N(N-1)/2$, where N is the number of particles. Figure 15.4 (right) indicates that the number of particles with small inter-particle distances is very moderate. In particular, for this example, the number of pairs with interparticle distances less than 1 Å is about 0.04 % (≈110) of the total number of $2.495 \cdot 10^5$ distances.

For ease of presentation, we further confine ourselves to the case of electrostatic potentials described by the Newton kernel $p(\|x\|) = \frac{1}{\|x\|}$.

15.2.2 Low-rank representation to the sum of long-range terms

First, we describe the tensor summation method for calculating the collective interaction potential of a multi-particle system that includes only the long-range contribution from the generating kernel. We introduce the $n \times n \times n$ rectangular grid Ω_n in $\Omega = [-b, b]^3$ and the auxiliary $2n \times 2n \times 2n$ grid on the accompanying domain $\widetilde{\Omega} = 2\Omega$

of double size. Conventionally, the canonical rank-R tensor representing the Newton kernel (by projecting onto the $n \times n \times n$ grid) is denoted by $\mathbf{P}_R \in \mathbb{R}^{n\times n\times n}$; see (6.6).

Consider the splitting (15.3) applied to the reference canonical tensor $\mathbf{P}_R$ and to its extended version $\widetilde{\mathbf{P}}_R = [\widetilde{p}_R(i_1, i_2, i_3)]$, $i_\ell \in I_\ell$, $\ell = 1, 2, 3$ such that

$$\widetilde{\mathbf{P}}_R = \widetilde{\mathbf{P}}_{R_s} + \widetilde{\mathbf{P}}_{R_l} \in \mathbb{R}^{2n\times 2n\times 2n}.$$

For technical reasons, we further assume that the tensor grid Ω_n is fine enough, such that all charge centers $\mathcal{S} = \{s_\nu\}$ specifying the total electrostatic potential in (15.7) belong to the set of grid points, that is, $s_\nu = (s_{\nu,1}, s_{\nu,2}, s_{\nu,3})^T = h(j_1^{(\nu)}, j_2^{(\nu)}, j_3^{(\nu)})^T \in \Omega_h$ with some indices $1 \le j_1^{(i)}, j_2^{(i)}, j_3^{(i)} \le n$.

The total electrostatic potential $P_0(x)$ in (15.7) is represented by a projected tensor $\mathbf{P}_0 \in \mathbb{R}^{n\times n\times n}$, which can be constructed by a direct sum of shift-and-windowing transforms of the reference tensor $\widetilde{\mathbf{P}}_R$ (see Chapter 14 for more detail):

$$\mathbf{P}_0 = \sum_{\nu=1}^{N_0} z_\nu \, \mathcal{W}_\nu(\widetilde{\mathbf{P}}_R) = \sum_{\nu=1}^{N_0} z_\nu \, \mathcal{W}_\nu(\widetilde{\mathbf{P}}_{R_s} + \widetilde{\mathbf{P}}_{R_l}) =: \mathbf{P}_s + \mathbf{P}_l. \tag{15.10}$$

The shift-and-windowing transform $\mathcal{W}_\nu$ maps a reference tensor $\widetilde{\mathbf{P}}_R \in \mathbb{R}^{2n\times 2n\times 2n}$ onto its sub-tensor of smaller size $n \times n \times n$, obtained by first shifting the center of the tensor $\widetilde{\mathbf{P}}_R$ to the point s_ν and then tracing (windowing) the result onto the domain Ω_n:

$$\mathcal{W}_\nu : \widetilde{\mathbf{P}}_R \mapsto \mathbf{P}^{(\nu)} = [p^{(\nu)}_{i_1,i_2,i_3}], \quad p^{(\nu)}_{i_1,i_2,i_3} := \widetilde{p}_R(i_1 + j_1^{(\nu)}, i_2 + j_2^{(\nu)}, i_3 + j_3^{(\nu)}), \quad i_\ell \in I_\ell.$$

Notice that the Tucker rank of the full tensor sum $\mathbf{P}_0$ increases almost proportionally to the number N_0 of particles in the system (see Figure 15.5) representing singular values of the side matrix in the canonical tensor $\mathbf{P}_0$. On the other hand, the canonical rank of the tensor $\mathbf{P}_0$ shows up the pessimistic bound $\le R\,N_0$.

To overcome this difficulty, in what follows, we consider the global tensor decomposition of only the *long-range part* in the tensor $\mathbf{P}_0$, defined by

$$\mathbf{P}_l = \sum_{\nu=1}^{N_0} z_\nu \, \mathcal{W}_\nu(\widetilde{\mathbf{P}}_{R_l}) = \sum_{\nu=1}^{N_0} z_\nu \, \mathcal{W}_\nu\Big(\sum_{k\in\mathcal{K}_l} \widetilde{\mathbf{p}}_k^{(1)} \otimes \widetilde{\mathbf{p}}_k^{(2)} \otimes \widetilde{\mathbf{p}}_k^{(3)} \Big). \tag{15.11}$$

The initial canonical rank of the tensor $\mathbf{P}_l$ equals to $R_l\,N_0$, and, again, it may increase dramatically for a large number of particles N_0. Since by construction the tensor $\mathbf{P}_l$ approximates rather smooth function on the domain Ω, one may expect that the large initial rank can be reduced considerably to some value R_*, which remains almost independent of N_0. The same beneficial property can be expected for the Tucker rank of $\mathbf{P}_l$. The principal ingredient of our tensor approach is the rank reduction in the initial canonical sum $\mathbf{P}_l$ by application of RHOSVD and the multigrid accelerated canonical-to-Tucker transform [174].

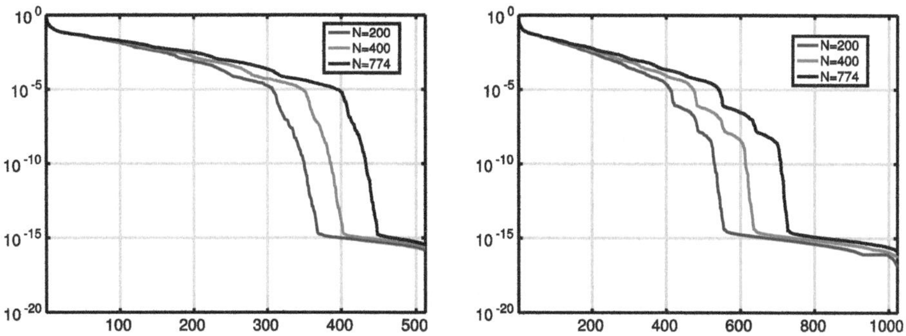

Figure 15.5: Mode-1 singular values of the side matrix in the full potential sum vs. the number of particles $N_0 = 200, 400, 774$ and grid-size n: $n = 512$ (left), $n = 1024$ (right).

To simplify the exposition, we suppose that the tensor entries in $\mathbf{P}_l$ are computed by collocation of Gaussian sums at the centers of the grid-cells. This provides the representation that is very close to that obtained by (6.6).

We consider the Gaussian in normalized form $G_p(x) = e^{-\frac{x^2}{2p^2}}$ so that the relation $e^{-t_k^2 x^2} = e^{-\frac{x^2}{2p^2}}$ holds, that is, we set $t_k = \frac{1}{\sqrt{2}p_k}$ with $t_k = k\mathfrak{h}_M$, $k = 0, 1, \ldots, M$, where $\mathfrak{h}_M = C_0 \log M/M$. Now criterion (B) on the bound of the L^1-norm (see (15.5)) reads

$$a_k \int_a^\infty e^{-\frac{x^2}{2p_k^2}} \leq \frac{\varepsilon}{2} < 1, \quad a_k = \mathfrak{h}_M.$$

The following theorem proves the important result justifying the efficiency of range-separated formats applied to a class of radial basis functions $p(r)$: the Tucker ε-rank of the long-range part in accumulated sum of potentials computed in the bounding box $\Omega = [-b, b]^3$ remains almost uniformly bounded in the number of particles N_0 (but depends on the size b of the domain).

Theorem 15.2 ([24]). *Let the long-range part $\mathbf{P}_l$ in the total interaction potential (see* (15.11)*) correspond to the choice of splitting parameter in* (15.6) *with $M = O(\log^2 \varepsilon)$. Then the total ε-rank $\mathbf{r}_0$ of the Tucker approximation to the canonical tensor sum $\mathbf{P}_l$ is bounded by*

$$|\mathbf{r}_0| := \operatorname{rank}_{\mathrm{Tuck}}(\mathbf{P}_l) \leq Cb \log^{3/2}(|\log(\varepsilon/N_0)|),$$

where the constant C does not depend on the number of particles N_0.

Proof. The proof can be sketched by the following steps: First, we represent all shifted Gaussian functions contributing to the total sum in the fixed set of basis functions by using truncated Fourier series. Second, we prove that on the "long-range" index set $k \in \mathcal{T}_l$ the parameter p_k remains uniformly bounded in N_0 from below, implying the uniform bound on the number of terms in the ε-truncated Fourier series. Finally we take into account that the summation of elements presented in the fixed Fourier basis set does not enlarge the Tucker rank, but only affects the Tucker core. The dependence on b appears in the explicit form.

Specifically, let us consider the rank-1 term in the splitting (15.3) with maximal index $k \in \mathcal{T}_l$. Taking into account the asymptotic choice $M = \log^2 \varepsilon$ (see (6.5)), where $\varepsilon > 0$ is the accuracy of the sinc-quadrature, relation (15.6) implies

$$\max_{k\in\mathcal{T}_l} t_k = R_l \mathfrak{h}_M = \frac{M}{2} C_0 \log(M)/M \approx \log(M) = 2\log(|\log(\varepsilon)|). \tag{15.12}$$

Now we consider the Fourier transform of the univariate Gaussian on $[-b, b]$,

$$G_p(x) = e^{-\frac{x^2}{2p^2}} = \sum_{m=0}^{M} \alpha_m \cos\left(\frac{\pi m x}{b}\right) + \eta, \quad \text{with } |\eta| = \left|\sum_{m=M+1}^{\infty} \alpha_m \cos\left(\frac{\pi m x}{b}\right)\right| < \varepsilon,$$

where

$$\alpha_m = \frac{\int_{-b}^{b} e^{-\frac{x^2}{2p^2}} \cos(\frac{\pi m x}{b})dx}{|C_m|^2} \quad \text{with } |C_m|^2 = \int_{-b}^{b} \cos^2\left(\frac{\pi m x}{b}\right)dx = \begin{cases} 2b & \text{if } m = 0, \\ b & \text{otherwise.} \end{cases}$$

Following arguments in [68], one obtains

$$\alpha_m = (pe^{-\frac{\pi^2 m^2 p^2}{2a^2}} - \xi_m)/|C_m|^2, \quad \text{where } 0 < \xi_m < \varepsilon.$$

The truncation coefficients α_m at $m = m_0$ such that $\alpha_{m_0} \leq \varepsilon$ lead to the bound

$$m_0 \geq \frac{\sqrt{2}\, b}{\pi\; p} \log^{0.5}\left(\frac{p}{(1 + |C_M|^2)\varepsilon}\right) = \frac{\sqrt{2}\, b}{\pi\; p} \log^{0.5}\left(\frac{p}{1+b}\frac{1}{\varepsilon}\right).$$

On the other hand, (15.12) implies

$$1/p_k \leq c\log(|\log \varepsilon|), \quad k \in \mathcal{T}_l, \quad \text{i. e.,} \quad 1/p_{R_l} \approx \log(|\log \varepsilon|),$$

which ensures the following estimate of m_0:

$$m_0 = O(b \log^{3/2}(|\log \varepsilon|)). \tag{15.13}$$

Following [148], we represent the Fourier transform of the shifted Gaussians by

$$G_p(x - x_\nu) = \sum_{m=0}^{M} \alpha_m \cos\left(\frac{\pi m(x - x_\nu)}{b}\right) + \eta_\nu, \quad |\eta_\nu| < \varepsilon,$$

which requires only the double number of terms compared with the single Gaussian analyzed above. To compensate the possible increase in $|\sum_\nu \eta_\nu|$, we refine $\varepsilon \mapsto \varepsilon/N_0$. These estimates also apply to all Gaussian functions presented in the long-range sum since they have larger values of p_k than p_{R_l}. Indeed, in view of (15.6), the number of summands in the long-range part is of order $R_l = M/2 = O(\log^2 \varepsilon)$. Combining these arguments with (15.13) proves the resulting estimate. □

Figure 15.6 illustrates the very fast decay of the Fourier coefficients for the "long-range" discrete Gaussians sampled on n-point grid (left) and the slow decay of Fourier coefficients for the "short-range" Gaussians (right). In the latter case, almost all the coefficients remain essential, resulting in the full rank decomposition. The grid size is chosen as $n = 1024$.

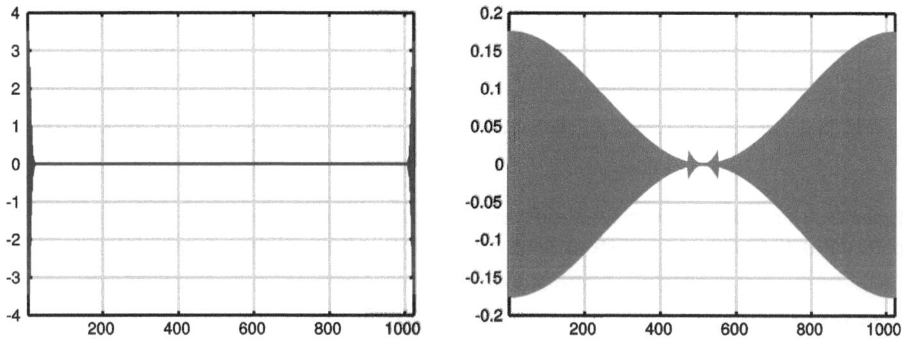

Figure 15.6: Fourier coefficients of the long- (left) and short-range (right) discrete Gaussians.

Remark 15.3. Notice that for fixed $\sigma > 0$ the σ-separability of the point distributions (see Definition 15.1) implies that the volume size of the computational box $[-b, b]^3$ should increase proportionally to the number of particles N_0, i. e., $b = O(N_0^{1/3})$. Hence, Theorem 15.2 indicates that the number of entries in the Tucker core of size $r_1 \times r_2 \times r_3$ can be estimated by CN_0. This asymptotic cost remains of the same order in N_0 as that for the short-range part in the potential sum.

Figure 15.7 (left) illustrates that the singular values of side matrices for the long-range part (by choosing $R_l = 12$) exhibit fast exponential decay with a rate independent of the number of particles $N_0 = 214, 405, 754$. Figure 15.7 (right) zooms into the first 50 singular values, which are almost identical for different values of N_0. The fast decay in these singular values guarantees the low-rank RHOSVD-based Tucker decomposition of the long-range part in the potential sum.

Table 15.1 shows the Tucker ranks of sums of long-range ingredients in the electrostatic potentials for the N-particle clusters. The Newton kernel is generated on the grid with $n^3 = 1024^3$ in the computational box of volume size $b^3 = 40^3$ Å, with accuracy $\varepsilon = 10^{-4}$ and canonical rank 21. Particle clusters with 200, 400, and 782 atoms are taken as a part of protein-like multiparticle system. The clusters of size 1728 and 4096 correspond to the lattice structures of sizes $12 \times 12 \times 12$ and $16 \times 16 \times 16$, with randomly generated charges. The line "RS-canonical rank" shows the resulting rank after the canonical-to-Tucker and Tucker-to-canonical transform with $\varepsilon_{C2T} = 4 \cdot 10^{-5}$ and $\varepsilon_{T2C} = 4 \cdot 10^{-6}$. Figures 15.8 show the accuracy of the RS-canonical tensor approximation for a multiparticle cluster of 400 particles at the middle section of the

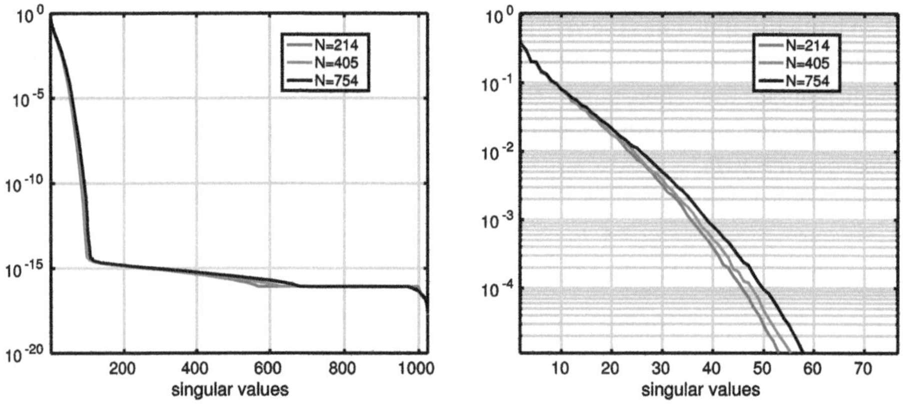

Figure 15.7: Mode-1 singular values of side matrices for the long range part ($R_l = 12$) in the total potential vs. the number of particles N (left), and zoom of the first singular values (right).

Table 15.1: Tucker ranks and the RS canonical rank of the multiparticle potential sum vs. the number of particles N for varying parameters R_ℓ and R_s. Grid size $n^3 = 1024^3$.

	N	200	400	782	1728	4096
R_ℓ/R_s	Ranks full can.	4200	8400	16 422	32 288	86 016
9/12	Ranks long range	1800	3600	7038	15 552	36 864
	RS-Tucker ranks	21,16,18	22,19,23	24,22,24	23,24,24	24,24,24
	RS-canonical rank	254	292	362	207	243

computational box $[-20, 20]^3$ Å by using an $n \times n \times n$ 3D Cartesian grid with $n = 1024$ and step size $h = 0.04$ Å. The top-left figure shows the surface of the potential at the level $z = 0$, whereas the top-right figure shows the absolute error of the RS approximation with the ranks $R_l = 15$, $R_s = 11$, and the separation distance $\sigma^* = 1.5$. The bottom figures visualize the long-range (left) and short-range (right) parts of the RS-tensor, respectively.

Figure 15.9 demonstrates the decay in singular values of the side matrices in the canonical tensor representing potential sums of long-range parts for $R_l = 10$, 11, and 12.

The proof of Theorem 15.2 indicates that the Tucker directional vectors living on large $n^{\otimes d}$ spatial grids are represented in the uniform Fourier basis with a small number of terms. Hence, following the arguments in [68] and [148], we are able to apply the low-rank QTT tensor approximation [167] to these long vectors (see [225] for the case of matrices). The QTT tensor compression makes it possible to reduce the representation complexity of the long-range part in an RS tensor to the logarithmic scale in the univariate grid size $O(\log n)$.

The most time- consuming part in our scheme is the canonical-to-Tucker algorithm for computing the long-range part of the RS format tensors. Table 15.2 indicates

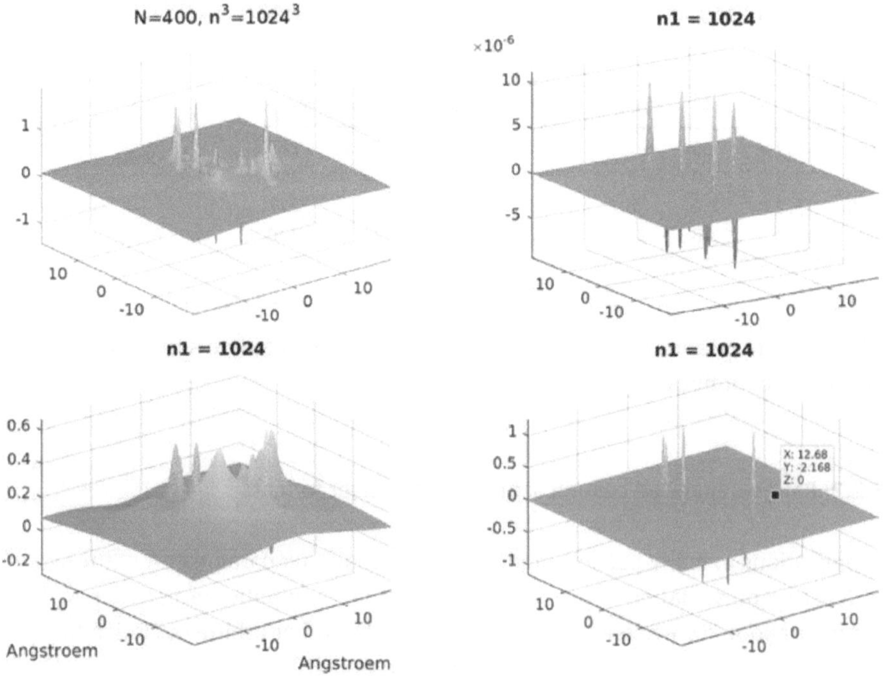

Figure 15.8: Top: the potential sum at the middle plane of a cluster with 400 atoms (left) and the error of the RS-canonical approximation (right). Bottom: long-range part of a sum (left); short range part of a sum (right).

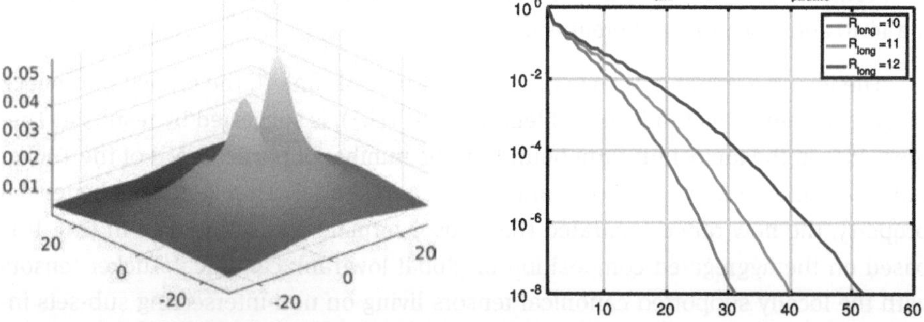

Figure 15.9: Example of potential surface at level $z = 0$ (left) for a sum of $N_0 = 200$ particles computed using only their long-range parts with $R_l = 12$. Decay in singular values of the side matrices for the canonical tensor representing sums of long-range parts for $R_l = 10, 11$, and 12.

almost linear scaling of CPU time in the number of particles and in the univariate grid-size n of the $n \times n \times n$ representation grid. The last column shows the resulting ranks of side matrices $U^{(\ell)}$ in the canonical tensor $\mathbf{U}$, (see (15.16)). The asymptotically opti-

Table 15.2: Times (sec) for canonical-to-Tucker rank reduction vs. number of particles N and grid size n^3.

N / n^3	512^3	1024^3	2048^3	4096^3	8192^3	$16\,384^3$	$R_{RS,C}$
100	0.9	1.5	2.3	4.1	6.0	12.2	183
200	2.3	3.0	4.7	7.9	14.4	23.4	214
400	5.2	7.0	8.7	16.1	32.9	71.7	227
770	12.3	13.8	18.3	32.7	67.5	147.3	290

mal complexity scaling of the RS decomposition and the required storage is the main motivation for applications of the RS tensor format.

15.2.3 Range-separated canonical and Tucker tensor formats

In applications to many-particle modeling, the initial rank parameter R of the canonical tensor representation is proportional to the (large) number of particles N_0 with pre-factor about 30, whereas the weights z_k can be rather arbitrary. Notice that the sub-class of the so-called *orthogonal canonical tensors* [192] allows stable multi-linear algebra, but suffers from the poor approximation capacity. Another important class is specified by the case of "monotone" or all positive canonical vectors (see [174, 153] for definition), which is also the case in decomposition of the elliptic Green's kernels.

Remark 15.4. The second class of all positive vectors ensures the stability of RHOSVD for problems such as (15.7) in the case of all positive (negative) weights (see Lemma 14.10 and discussions thereafter).

The idea regarding how to get rid of the "curse of ranks", the critical bottleneck in applying tensor methods to problems such as (15.7), is suggested by results in Theorem 15.2 on the almost uniform bound (in the number of particles N_0) of the Tucker rank for the long-range part of a multi-particle potential. Thanks to this beneficial property, the new range-separated (RS) tensor formats was introduced in [24]. It is based on the aggregated composition of global low-rank canonical/Tucker tensors with the locally supported canonical tensors living on non-intersecting sub-sets indices embedded into the large corporate multi-index set $\mathcal{I} = I_1 \times \cdots \times I_d$, $I_\ell = \{1, \ldots, n\}$. Such a parametrization attempts to represent the large multidimensional arrays with a storage cost linearly proportional to the number of cumulated inclusions (sub-tensors).

The structure of the *range-separated canonical/Tucker tensor formats* is specified by a composition of the local-global low parametric representations, which provide good approximation features in application to the problems of grid-based representation to many-particle interaction potentials with multiple singularities.

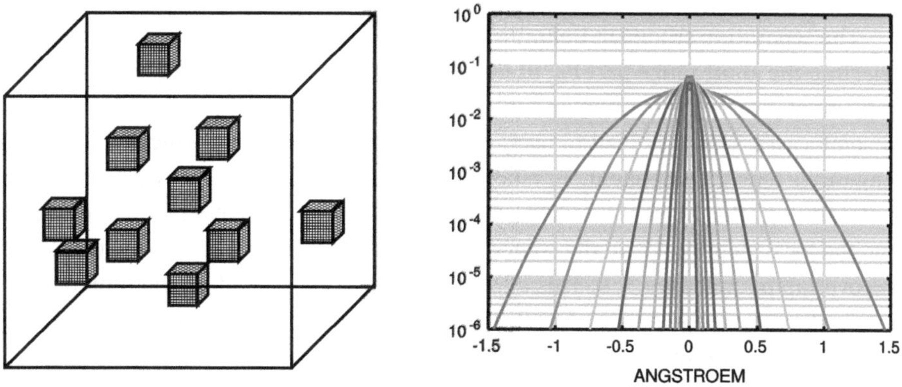

Figure 15.10: Schematic illustration of effective supports of the cumulated canonical tensor (left); short-range canonical vectors for $k = 1, \ldots, 11$, presented in logarithmic scale (right).

Definition 15.5 (Cumulated canonical tensors, [24]). Given the index set $\mathcal{I}$, a set of multi-indices (sources) $\mathcal{J} = \{\mathbf{j}^{(\nu)} := (j_1^{(\nu)}, j_2^{(\nu)}, \ldots, j_d^{(\nu)})\}$, $\nu = 1, \ldots, N_0$, $j_\ell^{(\nu)} \in I_\ell$, and the width index parameter $\gamma \in \mathbb{N}$ such that the γ-vicinity of each point $\mathbf{j}^{(\nu)} \in \mathcal{J}$, that is, $\mathcal{J}_\gamma^{(\nu)} := \{\mathbf{j} : |\mathbf{j} - \mathbf{j}^{(\nu)}| \le \gamma\}$, does not intersect all others:

$$\mathcal{J}_\gamma^{(\nu)} \cap \mathcal{J}_\gamma^{(\nu')} = \varnothing, \quad \nu \neq \nu'.$$

A rank-R_0 cumulated canonical tensor $\widehat{\mathbf{U}}$ associated with $\mathcal{J}$ and width parameter γ is defined as a set of tensors that can be represented in form

$$\widehat{\mathbf{U}} = \sum_{\nu=1}^{N_0} c_\nu \mathbf{U}_\nu \quad \text{with } \operatorname{rank}(\mathbf{U}_\nu) \le R_0, \tag{15.14}$$

where the rank-R_0 canonical tensors $\mathbf{U}_\nu = [u_\mathbf{j}]$ are vanishing beyond the γ-vicinity of $\mathbf{j}^{(\nu)}$:

$$u_\mathbf{j} = 0 \quad \text{for } \mathbf{j} \subset \mathcal{I} \setminus \mathcal{J}_\gamma^{(\nu)}, \; \nu = 1, \ldots, N_0. \tag{15.15}$$

Definition 15.5 provides the description of a sum of short-range potentials having the local (up to some threshold) non-intersecting supports. Given the particular point distribution, the effective support of the localized sub-tensors should be of the size close to the parameter σ_* (see Definition 15.1) that introduces the σ_*-separable point distributions characterized by the separation parameter $\sigma_* > 0$. In this case, we use the relation $\sigma_* \approx \gamma h$, where $h = 2b/n$ is the mesh size of the computational $(n \times \cdots \times n)$-grid.

Figure 15.10 (left) illustrates the effective supports of a cumulated canonical tensor in the non-overlapping case, whereas Figure 15.10 (right) presents the supports for the first 11 short-range canonical vectors (selected from rank-24 reference canonical tensor $\mathbf{P}_R$), which allows choosing the parameter γ in separation criteria.

The separation criteria in Definition 15.5 leads to a rather “aggressive” strategy for selecting the short-range part $\mathbf{P}_{R_s}$ in the reference canonical tensor $\mathbf{P}_R$ allowing easy implementation of the cumulated canonical tensor (non-overlapping case). However, in some cases, this may lead to overestimation of the Tucker/canonical rank in the long-range tensor component. To relax the criteria in Definition 15.5, we propose the “soft” strategy that allows including a few (i. e., $O(1)$ for large N_0) neighboring particles into the local vicinity $\mathcal{J}_\gamma^{(\nu)}$ of the source point s_ν, which can be achieved by increasing the overlap parameter $\gamma > 0$. This allows controlling the bound on the rank parameter in the long-range tensor almost uniformly in the system size N_0. The following example illustrates this issue.

Example 15.6. Assume that the separation distance is equal to $\sigma_* = 0.8$ Å, corresponding to the example in Figure 15.4 (right), and the given computational threshold is $\varepsilon = 10^{-4}$. Then we find from Figure 15.10 (right) that the “aggressive” criteria in Definition 15.5 lead to choosing $R_s = 10$, since the value of the canonical vector with $k = 11$ at point $x = \sigma_*$ is about 10^{-3}. Hence, in order to control the required rank parameter R_l, we have to extend the overlap area to larger parameter σ_* and, hence, to larger γ. This will lead to a small $O(1)$-overlap between supports of the short-range tensor components, but without asymptotic increase in the total complexity.

Table 15.3 represents the Tucker ranks $\mathbf{r} = (r_1, r_2, r_3)$ for the long-range parts of N_0-particle potentials. The reference Newton kernel is approximated on a 3D grid of size 2048^3 with the rank $R = 29$ and accuracy $\varepsilon_{\mathcal{N}} = 10^{-5}$. Here, the Tucker tensor is computed with the stopping criteria $\varepsilon_{T2C} = 10^{-5}$ in the ALS iteration. It can be seen that for fixed R_l, the Tucker ranks increase only moderately in the system size N_0.

Table 15.3: Tucker ranks $\mathbf{r} = (r_1, r_2, r_3)$ for the long-range parts of N_0-particle potentials.

N_0 / R_l	8	9	10	11	12	13
200	10,10,11	13,12,12	18,15,16	23,19,21	32,24,27	42,30,34
400	11,10,11	14,13,14	19,16,20	26,21,26	35,27,36	47,34,47
782	11,11,12	15,14,15	20,18,20	28,26,27	39,35,37	52,46,50

Below, we distinguish a special subclass of *uniform CCT tensors*.

Definition 15.7 (Uniform CCT tensors, [24]). A CCT tensor in (15.14) is called uniform if all components $\mathbf{U}_\nu$ are generated by a single rank-R_0 tensor $\mathbf{U}_0 = \sum_{m=1}^{R_0} \mu_m \hat{\mathbf{u}}_m^{(1)} \otimes \cdots \otimes \hat{\mathbf{u}}_m^{(d)}$ such that $\mathbf{U}_\nu|_{\mathcal{J}_\delta^{(\nu)}} = \mathbf{U}_0$.

Now, we are in a position to define the range separated canonical and Tucker tensor formats in $\mathbb{R}^{n_1 \times \cdots \times n_d}$. The RS canonical format is defined as follows.

Definition 15.8 (RS-canonical tensors, [24]). The RS-canonical tensor format specifies the class of d-tensors $\mathbf{A} \in \mathbb{R}^{n_1\times\cdots\times n_d}$, which can be represented as a sum of a rank-R canonical tensor $\mathbf{U} \in \mathbb{R}^{n_1\times\cdots\times n_d}$ and a (uniform) cumulated canonical tensor generated by $\mathbf{U}_0$ with $\operatorname{rank}(\mathbf{U}_0) \leq R_0$ as in Definition 15.7 (or more generally in Definition 15.5):

$$\mathbf{A} = \sum_{k=1}^{R} \xi_k \mathbf{u}_k^{(1)} \otimes \cdots \otimes \mathbf{u}_k^{(d)} + \sum_{\nu=1}^{N_0} c_\nu \mathbf{U}_\nu, \tag{15.16}$$

where $\operatorname{diam}(\operatorname{supp}\mathbf{U}_\nu) \leq 2\gamma$ in the index size.

For a given grid-point $\mathbf{i} \in \mathcal{I} = I_1 \times \cdots \times I_d$, we define the set of indices

$$\mathcal{L}(\mathbf{i}) := \{\nu \in \{1,\dots,N_0\} : \mathbf{i} \in \operatorname{supp}\mathbf{U}_\nu\},$$

which label all short-range tensors $\mathbf{U}_\nu$, including the grid-point $\mathbf{i}$ within its effective support.

Lemma 15.9 ([24]). *The storage cost of RS-canonical tensor is estimated by*

$$\operatorname{stor}(\mathbf{A}) \leq dRn + (d+1)N_0 + dR_0\gamma.$$

Given $\mathbf{i} \in \mathcal{I}$, denote by $\overline{\mathbf{u}}_{i_\ell}^{(\ell)}$ the row-vector with index i_ℓ in the side matrix $U^{(\ell)} \in \mathbb{R}^{n_\ell\times R}$, and let $\xi = (\xi_1,\dots,\xi_d)$. Then the $\mathbf{i}$th entry of the RS-canonical tensor $\mathbf{A} = [a_\mathbf{i}]$ can be calculated as a sum of long- and short-range contributions by

$$a_\mathbf{i} = (\odot_{\ell=1}^{d} \overline{\mathbf{u}}_{i_\ell}^{(\ell)})\xi^T + \sum_{\nu\in\mathcal{L}(\mathbf{i})} c_\nu \mathbf{U}_\nu(\mathbf{i})$$

at the expense $O(dR + 2d\gamma R_0)$.

Proof. Definition 15.8 implies that each RS-canonical tensor is uniquely defined by the following parametrization: rank-R canonical tensor $\mathbf{U}$, the rank-R_0 local reference canonical tensor $\mathbf{U}_0$ with mode-size bounded by 2γ, and list $\mathcal{J}$ of the coordinates and weights of N_0 particles. Hence, the storage cost directly follows. To justify the representation complexity, we notice that by well-separability assumption (see Definition 15.1), we have $\#\mathcal{L}(\mathbf{i}) = O(1)$ for all $\mathbf{i} \in \mathcal{I}$. This proves the complexity bounds. □

Now we define the class of RS-Tucker tensors.

Definition 15.10 (RS-Tucker tensors, [24]). The RS-Tucker tensor format specifies the class of d-tensors $\mathbf{A} \in \mathbb{R}^{n_1\times\cdots\times n_d}$, which can be represented as a sum of a rank-$\mathbf{r}$ Tucker tensor $\mathbf{V}$ and a (uniform) cumulated canonical tensor generated by $\mathbf{U}_0$ with $\operatorname{rank}(\mathbf{U}_0) \leq R_0$ as in Definition 15.7 (or more generally in Definition 15.5):

$$\mathbf{A} = \boldsymbol{\beta} \times_1 V^{(1)} \times_2 V^{(2)} \cdots \times_d V^{(d)} + \sum_{\nu=1}^{N_0} c_\nu \mathbf{U}_\nu, \tag{15.17}$$

where the tensor $\mathbf{U}_\nu$, $\nu = 1,\dots,N_0$, has local support, that is, $\operatorname{diam}(\operatorname{supp}\mathbf{U}_\nu) \leq 2\gamma$.

Similar to Lemma 15.9, the corresponding statement for the RS-Tucker tensors can be proven.

Lemma 15.11 ([24]). *The storage size for RS-Tucker tensor does not exceed*

$$\mathrm{stor}(\mathbf{A}) \leq r^d + drn + (d+1)N_0 + dR_0\gamma.$$

Let the r_ℓ-vector $\mathbf{v}_{i_\ell}^{(\ell)}$ be the i_ℓth row of the matrix $V^{(\ell)}$. Then the $\mathbf{i}$th element of the RS-Tucker tensor $\mathbf{A} = [a_\mathbf{i}]$ can be calculated by

$$a_\mathbf{i} = \boldsymbol{\beta} \times_1 \mathbf{v}_{i_1 l}^{(1)} \times_2 \mathbf{v}_{i_2}^{(2)} \cdots \times_d \mathbf{v}_{i_d}^{(d)} + \sum_{\nu \in \mathcal{L}(\mathbf{i})} c_\nu \mathbf{U}_\nu(\mathbf{i})$$

at the expanse $O(r^d + 2d\gamma R_0)$.

Proof. In view of Definition 15.10, each RS-Tucker tensor is uniquely defined by the following parametrization: the rank-$\mathbf{r} = (r_1, \ldots, r_d)$ Tucker tensor $\mathbf{V} \in \mathbb{R}^{n_1 \times \cdots \times n_d}$, the rank-$R_0$ local reference canonical tensor $\mathbf{U}_0$ with $\mathrm{diam}(\mathrm{supp}\mathbf{U}_0) \leq 2\gamma$, list $\mathcal{J}$ of the coordinates of N_0 centers of particles, $\{s_\nu\}$, and N_0 weights $\{c_\nu\}$. This proves the complexity bounds. □

The main computational benefits of the new range-separated canonical/Tucker tensor formats are explained by the important uniform bounds on the Tucker-rank of the long-range part in the large sum of interaction potentials (see Theorem 15.2 and numerics in Section 15.2.2). Moreover, we have the low storage cost for RS-canonical/Tucker tensors, cheap representation of each entry in an RS-tensor, possibility for simple implementation of multilinear algebra on these tensors (see the discussion at the end of this section).

The total rank of the sum of canonical tensors in $\widehat{\mathbf{U}}$ (see (15.14)) may become large for larger N_0 since the pessimistic bound $\mathrm{rank}(\widehat{\mathbf{U}}) \leq N_0 R_0$. However, cumulated canonical tensors (CCT) have two beneficial features, which are particularly useful in the low-rank tensor representation of large potential sums.

Proposition 15.12 (Properties of CCT tensors).

(A) *The rank of a CCT tensor $\widehat{\mathbf{U}}$ is bounded by R_0: $\mathrm{rank}_{\mathrm{loc}}(\widehat{\mathbf{U}}) := \max_\nu \mathrm{rank}(\mathbf{U}_\nu) \leq R_0$.*

(B) *Local components in the CCT tensor* (15.14) *are "block orthogonal" in the sense that*

$$\langle \mathbf{U}_\nu, \mathbf{U}_{\nu'} \rangle = 0, \quad \forall \nu \neq \nu'. \tag{15.18}$$

(C) $\|\mathbf{U}\| = \sum_{\nu=1}^{N_0} c_\nu \|\mathbf{U}_\nu\|$.

If $R_0 = 1$, that is, $\widehat{\mathbf{U}}$ is the conventional rank-N_0 canonical tensor, then property (B) in Proposition 15.12 leads to the definition of orthogonal canonical tensors in [192]. Hence, in case $R_0 > 1$, we arrive at the generalization further called the *block orthogonal canonical tensors*.

The rank bound $R' = \operatorname{rank}(\widehat{\mathbf{U}}) \leq N_0 R_0$ indicates that the direct summation in (15.14) in the canonical/Tucker formats may lead to practically non-tractable representations. However, the block orthogonality property in Proposition 15.12 (B) allows applying the stable RHOSVD approximation for the rank optimization (see Section 3.3). The stability of RHOSVD in the case of orthogonal canonical tensors was analyzed in [174, 153]. Concerning stability of RHOSVD for RS tensor format see Remark 15.4.

In what follows, we prove the stability of such tensor approximation applied to CCT representations.

Lemma 15.13 ([24]). *Let the local canonical tensors be stable, that is,* $\sum_{m=1}^{R_0} \mu_m^2 \leq C\|\mathbf{U}_\nu\|^2$ *(see Definition 15.7). Then the rank-$\mathbf{r}$ RHOSVD-Tucker approximation $\mathbf{U}^0_{(\mathbf{r})}$ to the CCT $\widehat{\mathbf{U}}$ provides the stable error bound*

$$\|\widehat{\mathbf{U}} - \mathbf{U}^0_{(\mathbf{r})}\| \leq C \sum_{\ell=1}^{3} \left(\sum_{k=r_\ell+1}^{\min(n,R')} \sigma_{\ell,k}^2 \right)^{1/2} \|\widehat{\mathbf{U}}\|,$$

where $\sigma_{\ell,k}$ denote the singular values of the side matrices $U^{(\ell)}$; see (3.34).

Proof. We apply the general error estimate for RHOSVD approximation [174] to obtain

$$\|\widehat{\mathbf{U}} - \mathbf{U}^0_{(\mathbf{r})}\| \leq C \sum_{\ell=1}^{3} \left(\sum_{k=r_\ell+1}^{\min(n,R')} \sigma_{\ell,k}^2 \right)^{1/2} \left(\sum_{\nu=1}^{N_0} \sum_{m=1}^{R_0} c_\nu^2 \mu_m^2 \right)^{1/2}$$

and then take into account the property (C), Proposition 15.12 to estimate

$$\sum_{\nu=1}^{N_0} \sum_{m=1}^{R_0} c_\nu^2 \mu_m^2 = \sum_{\nu=1}^{N_0} c_\nu^2 \sum_{m=1}^{R_0} \mu_m^2 \leq C \sum_{\nu=1}^{N_0} c_\nu^2 \|\mathbf{U}_\nu\|^2 = C\|\widehat{\mathbf{U}}\|^2,$$

which completes the proof. □

The stability assumption in Lemma 15.13 is satisfied in the case of the constructive canonical tensor approximation to the Newton and other types of Green's kernels obtained by sinc-quadrature based representations, where all canonical skeleton vectors are non-negative and monotone.

Remark 15.14. In the case of higher dimensions $d > 3$, the local canonical tensors can be combined with the global tensor train (TT) format [226] such that the simple canonical-to-TT transform can be applied. In this case, the RS-TT format can be introduced as a set of tensor represented as a sum of CCT term and the global TT-tensor. The complexity and structural analysis is completely similar to those in the case of the RS-Canonical and RS-Tucker formats.

We sketch the algebraic operations on the RS tensors. Multilinear algebraic operations in the format of RS-canonical/Tucker tensor parametrization can be implemented by using 1D vector operations applied to both localized and global tensor components. In particular, the following operations on RS canonical/Tucker tensors can be realized efficiently: (a) storage of a tensor; (b) real space representation on a fine rectangular grid; (c) summation of many-particle interaction potentials represented on the fine tensor grid; (d) computation of scalar products; and (e) computation of gradients and forces.

Estimates on the storage complexity for the RS-canonical and RS-Tucker formats were presented in Lemmas 15.9 and 15.11. Items (b) and (c) were addressed earlier. Calculation of the scalar product of two RS-canonical tensors in the form (15.16), defined on the same set $\mathcal{S}$ of particle centers, can be reduced to the standard calculation of the cross scalar products between all elementary canonical tensors presented in (15.16). Hence, the numerical cost can be estimated by $O(\frac{1}{2}R(R-1)dn + 2\gamma RR_0N_0)$.

15.3 Outline of possible applications

The RS tensor formats can be gainfully applied in computational problems, including functions with multiple local singularities or cusps, Green kernels with essentially non-local behavior, and in various approximation problems by means of radial basis functions. In this section, we follow [24] and sketch how the RS tensor representations can be applied to some computationally extensive problems, such as grid representation of multidimensional scattered data, interaction energy of charged many-particle system, computation of gradients and forces for nano-particle potentials, and construction of approximate boundary/interface conditions in the Poisson–Boltzmann equation describing the electrostatic potential of proteins.

15.3.1 Multidimensional data modeling

Here we and briefly describe the model reduction approach to the problem of multidimensional data fitting based on the RS tensor approximation. The problems of multidimensional scattered data modeling and data mining are known to lead to computationally intensive simulations. We refer to [42, 141, 34, 84, 129] for the discussion of most commonly used computational approaches in this field of numerical analysis.

The mathematical problems in scattered data modeling are concerned with the approximation of multi-variate function $f : \mathbb{R}^d \to \mathbb{R}$ ($d \geq 2$) by using samples given at certain finite set $\mathcal{X} = \{x_1, \ldots, x_N\} \subset \mathbb{R}^d$ of pairwise distinct points; see, e. g., [42]. The function f may describe the surface of a solid body, the solution of a PDE, many-body potential field, multiparametric characteristics of physical systems, or some other multidimensional data.

In a particular problem setting, one may be interested in recovering f from a given sampling vector $f_{|\mathcal{X}} = (f(x_1), \ldots, f(x_N)) \in \mathbb{R}^N$. One of the traditional ways to tackle this problem is based on constructing a suitable functional interpolant $P_N : \mathbb{R}^d \to \mathbb{R}$, satisfying $P_{N|\mathcal{X}} = f_{|\mathcal{X}} =: \mathbf{f}$, that is,

$$P_N(x_j) = f(x_j), \quad \forall\, 1 \le j \le N \tag{15.19}$$

or approximating the sampling vector $f_{|\mathcal{X}}$ on the set $\mathcal{X}$ in the least squares sense. We consider the approach based on using *radial basis functions* (RBFs) providing the traditional tools for multivariate scattered data interpolation. To that end, the radial basis function (RBF) interpolation approach deals with a class of interpolants P_N in the form

$$P_N(x) = \sum_{j=1}^{N} c_j p(\|x - x_j\|) + Q(x), \quad Q \text{ is some smooth function,} \tag{15.20}$$

where $p : [0, \infty) \to \mathbb{R}$ is a fixed radial function, and $\|\cdot\|$ is the Euclidean norm on $\mathbb{R}^d$. To fix the idea, here we consider the particular version of (15.20) by setting $Q = 0$. Notice that the interpolation ansatz P_N in (15.20) has the same form as the multi-particle interaction potential in (15.7). This observation indicates that the numerical treatment of various problems based on the use of interpolant P_N can be handled by using the same tools of model reduction via rank-structured RS tensor approximation.

The particular choice of RBFs described in [42, 141] includes functions $p(r)$ in the form

$$r^\nu, \quad (1 + r^2)^\nu, \ (\nu \in \mathbb{R}), \quad \exp(-r^2), \quad r^2 \log(r).$$

For our tensor-based approach, the common feature of all these function classes is the existence of low-rank tensor approximations to the grid-based discretization of the RBF $p(\|x\|) = p(x_1, \dots, x_d)$, $x \in \mathbb{R}^d$, where we set $r = \|x\|$. We can add to the above examples a few examples of traditional RBFs functions commonly used in quantum chemistry, such as the Coulomb potential $1/r$, Slater function $\exp(-\lambda r)$, Yukawa potential $\exp(-\lambda r)/r$, and the class of Matérn RBFs, traditionally applied in stochastic modeling [219, 206]. Other examples are given by the Lennard-Jones (Van der Waals), dipole–dipole interaction, and Stokeslet potentials (see [205]), given by $p(r) = 4\epsilon[(\frac{\sigma}{r})^{12} - (\frac{\sigma}{r})^6]$, $p(r) = \frac{1}{r^3}$, and 3×3 matrix $P(\|x\|) = I/r + (xx^T)/r^3$ for $x \in \mathbb{R}^3$, respectively.

In the context of numerical data modeling, we shall focus on the following computational tasks:

(A) Fixed coefficient vector $\mathbf{c} = (c_1, \dots, c_N)^T \in \mathbb{R}^N$: the efficient representation and storage of the interpolant in (15.20), sampled on fine tensor grid in $\mathbb{R}^d$, that allows the $O(1)$-fast point evaluation of P_N in the whole volume Ω and computation of various integral-differential operations on that interpolant, such as gradients, forces, scalar products, convolution integrals, etc.

(B) Finding the coefficient vector $\mathbf{c}$ that solves the interpolation problem (15.19).

We look on the problems (A) and (B) with the intent to apply the RS tensor representation to the interpolant $P_N(x)$. The point is that the representation (15.20) can be viewed as the many-particle interaction potential (with charges c_j) considered in the previous sections. Hence, the RS tensor approximation can be successfully applied if the d-dimensional tensor approximating the RBF $p(\|x\|)$, $x \in \mathbb{R}^d$, on tensor grid allows the low-rank canonical representation that can be split into the short- and long-range parts. This can be proven for functions listed above (see the example in Section 6.1 for the Newton kernel $1/\|x\|$). Notice that the Gaussian is already the rank-1 separable function.

Problem (A). We consider the particular choice of the set $\mathcal{X} \subset [0,1]^d$, which can be represented by using the nearly optimal point sampling. The so-called optimal point sets give rise to the trade-off between the separation distance $q_{\mathcal{X}} = \min_{s\in\mathcal{X}} \min_{s_\nu\in\mathcal{X}\setminus s} d(s_\nu, s)$ (see (15.9)) and the fill distance $h_{\mathcal{X},\Omega} = \max_{y\in\Omega} d(\mathcal{X}, y)$ thereby solving the problem (see [42])

$$q_{\mathcal{X}}/h_{\mathcal{X},\Omega} \to \max .$$

We choose the set of points $\mathcal{X}$ as a subset of the $n^{\otimes}$ square grid Ω_h with the mesh-size $h = 1/(n-1)$, such that the separation distance satisfies $\sigma_* = q_{\mathcal{X}} \geq \alpha h$, $\alpha \geq 1$. Here, $N \leq N_0 = n^d$. The square grid Ω_h is an example of the almost optimal point set (see the discussion in [141]). The construction below also applies to nonuniform rectangular grids.

Now, we are in a position to apply the RS tensor representation to the total interpolant P_N. Let $\mathbf{P}_R$ be the $n \times n \times n$ (say, for $d = 3$) rank-R tensor representing the RBF $p(\|\cdot\|)$, which allows the RS splitting by (15.3) generating the global RS representation (15.10). Then P_N can be represented by the tensor $\mathbf{P}_N$ in the RS-Tucker (15.17) or RS-canonical (15.16) formats. The storage cost scales linear in both N and n, $O(N + dR_l n)$.

Problem (B). The interpolation problem (15.19) reduces to solve the linear system of equations for unknown coefficient vector $\mathbf{c} = (c_1, \dots, c_N)^T \in \mathbb{R}^N$,

$$A_{p,\mathcal{X}}\mathbf{c} = \mathbf{f}, \quad \text{where } A_{p,\mathcal{X}} = [p(\|x_i - x_j\|)]_{1\leq i,j\leq N} \in \mathbb{R}^{N\times N} \tag{15.21}$$

with the symmetric matrix $A_{p,\mathcal{X}}$. Here, without loss of generality, we assume that the RBF $p(\|\cdot\|)$ is continuous. The solvability conditions for the linear system (15.21) with the matrix $A_{p,\mathcal{X}}$ are discussed, for example, in [42]. We consider two principal cases.

Case (A). We assume that the point set $\mathcal{X}$ coincides with the set of grid-points in Ω_h, that is, $N = n^d$. Introducing the d-tuple multi-index $\mathbf{i} = (i_1, \dots, i_d)$ and $\mathbf{j} = (j_1, \dots, j_d)$, we reshape the matrix $A_{p,\mathcal{X}}$ into the tensor form

$$A_{p,\mathcal{X}} \mapsto \mathbf{A} = [a(i_1, j_1, \dots, i_d, j_d)] \in \bigotimes_{\ell=1}^{d} \mathbb{R}^{n\times n},$$

which corresponds to folding of an N-vector to a d-dimensional $n^{\otimes d}$ tensor. This d-level Toeplitz matrix is generated by the tensor $\mathbf{P}_R$ obtained by collocation of the RBF $p(\|\cdot\|)$

on the grid Ω_h. Splitting the rank-R canonical tensor $\mathbf{P}_R$ into a sum of short- and long-range terms

$$\mathbf{P}_R = \mathbf{P}_{R_s} + \mathbf{P}_{R_l} \quad \text{with } \mathbf{P}_{R_l} = \sum_{k=1}^{R_l} \mathbf{p}_k^{(1)} \otimes \cdots \otimes \mathbf{p}_k^{(d)}$$

allows representing the matrix $\mathbf{A}$ in the RS form as a sum of low-rank canonical tensors $\mathbf{A} = \mathbf{A}_{R_s} + \mathbf{A}_{R_l}$. Here, the first one corresponds to the diagonal (nearly diagonal in the case of "soft" separation strategy) matrix by assumption on the locality of $\mathbf{P}_{R_s}$. The second matrix takes the form of R_l-term Kronecker product sum

$$\mathbf{A}_{R_l} = \sum_{k=1}^{R_l} A_k^{(1)} \otimes \cdots \otimes A_k^{(d)},$$

where each "univariate" matrix $A_k^{(\ell)} \in \mathbb{R}^{n \times n}$, $\ell = 1, \ldots, d$, takes the symmetric Toeplitz form generated by the first column vector $\mathbf{p}_k^{(\ell)}$. The storage complexity of the resultant RS representation to the matrix $\mathbf{A}$ is estimated by $O(N + dR_l n)$.

Now, we let the coefficient vector $\mathbf{c} \in \mathbb{R}^N$ be represented as the d-dimensional $n^{\otimes d}$ tensor $\mathbf{c} \mapsto \mathbf{C} \in \mathbb{R}^{n^{\otimes d}}$. Then the matrix vector multiplication $\mathbf{AC} = (\mathbf{A}_{R_s} + \mathbf{A}_{R_l})\mathbf{C}$ implemented in tensor formats can be accomplished in $O(cN + dR_l N \log n)$ operations, that is, with the asymptotically optimal cost in the number of sampling points N. The reason is that the matrix $\mathbf{A}_{R_s}$ has the diagonal form, whereas the matrix–vector multiplication between Toeplitz matrices $A_k^{(\ell)}$ constituting the Kronecker factors $\mathbf{A}_{R_l}$ and the corresponding n-columns (fibers) of the tensor $\mathbf{C}$ can be implemented by 1D FFT in $O(n \log n)$ operations. One can customarily enhance this scheme by introducing the low-rank tensor structure for the target vector (tensor) $\mathbf{C}$.

Case (B). This construction can be generalized to the situation when $\mathcal{X}$ is a subset of Ω_h, i. e., $N < n^d$. In this case, the complexity again scales linearly in N if $N = O(n^d)$. When $N \ll n^d$, the matrix–vector operation applies to the vector $\mathbf{C}$ that vanishes beyond the small set $\mathcal{X}$. In this case, the corresponding block-diagonal sub-matrices in $A_k^{(\ell)}$ loose the Toeplitz form, thus resulting in the slight increase in the overall cost $O(N^{1+1/d})$.

In both cases (A) and (B) the new rank-structured matrix construction can be applied within any favorable preconditioned iteration for solving the linear system (15.21).

15.3.2 Interaction energy for many-particle systems

Consider the calculation of the interaction energy (IE) for a charged multiparticle system. In the case of lattice-structured systems, the fast tensor-based computation scheme for IE was described in [152]. Here we follow [24].

Recall that the interaction energy of the total electrostatic potential generated by the system of N charged particles located at $x_k \in \mathbb{R}^3$ ($k = 1, \ldots, N$) is defined by the weighted sum

$$E_N = E_N(x_1, \ldots, x_N) = \frac{1}{2} \sum_{j=1}^{N} z_j \sum_{k=1, k \neq j}^{N} \frac{z_k}{\|x_j - x_k\|}, \tag{15.22}$$

where z_k denotes the particle charge. Letting $\sigma > 0$ be the minimal physical distance between the centers of particles, we arrive at the σ-separable systems (see Definition 15.1). The double sum in (15.22) applies only to the particle positions $\|x_j - x_k\| \geq \sigma$. Hence, the quantity in (15.22) is computable also for singular kernels such as $p(r) = 1/r$.

We observe that the quantity of interest E_N can be recast in terms of the interconnection matrix $A_{p,\mathcal{X}}$ defined by (15.21) with $p(r) = 1/r$, $\mathcal{X} = \{x_1, \ldots, x_N\}$,

$$E_N = \frac{1}{2} \langle (A_{p,\mathcal{X}} - \operatorname{diag} A_{p,\mathcal{X}}) \mathbf{z}, \mathbf{z} \rangle, \quad \text{where } \mathbf{z} = (z_1, \ldots, z_N)^T. \tag{15.23}$$

Hence, E_N can be calculated by using the approach already addressed in the previous section.

To fix the idea, we recall that the reference canonical tensor $\mathbf{P}_R$ approximating the single Newton kernel on an $n \times n \times n$ tensor grid Ω_h in the computational box $\Omega = [-b, b]^3$ is represented by (6.6), where $h > 0$ is the fine mesh size. For ease of exposition, we further assume that the particle centers x_k are located exactly at some grid points in Ω_h (otherwise, an additional approximation error may be introduced) such that each point x_k inherits some multi-index $\mathbf{i}_k \in \mathcal{I}$, and the origin $x = 0$ corresponds to the central point $\mathbf{n}_0 = (n/2, n/2, n/2)$ on the grid. In turn, the canonical tensor $\mathbf{P}_0$ approximating the total interaction potential $P_N(x)$ ($x \in \Omega$) for the N-particle system,

$$P_N(x) = \sum_{k=1}^{N} \frac{z_k}{\|x - x_k\|} \rightsquigarrow \mathbf{P}_0 = \mathbf{P}_s + \mathbf{P}_l \in \mathbb{R}^{n \times n \times n},$$

is represented by (15.10) as a sum of short- and long-range tensor components. Now, the tensor $\mathbf{P}_0 = \mathbf{P}_0(x^h)$ can be defined as a function of discrete variable x^h at each point $x^h \in \Omega_h$ and, in particular, in the vicinity of each particle center x_k, that is, at the grid-points $x_k + h\mathbf{e}$, where the directional vector $\mathbf{e} = (e_1, e_2, e_3)^T$ is specified by some choice of 3D coordinates $e_\ell \in \{-1, 0, 1\}$ for $\ell = 1, 2, 3$. This allows introducing the useful notations $\mathbf{P}_0(x_k + h\mathbf{e})$, which can be applied to all tensors living on Ω_h.

The following lemma describes the tensor scheme for calculating E_N by utilizing the long-range part $\mathbf{P}_l$ only in the tensor representation of $P_N(x)$.

Lemma 15.15 ([24]). *Let the effective support of the short-range components in the reference potential $\mathbf{P}_R$ not exceed $\sigma > 0$. Then the interaction energy E_N of the N-particle system can be calculated by using only the long-range part in the total potential sum*

$$E_N = E_N(x_1, \ldots, x_N) = \frac{1}{2} \sum_{j=1}^{N} z_j(\mathbf{P}_l(x_j) - z_j \mathbf{P}_{R_l}(x = 0)) \tag{15.24}$$

in $O(dR_l N)$ operations, where R_l is the canonical rank of the long-range component.

Proof. Similarly to [152], where the case of lattice-structured systems was analyzed, we show that the interior sum in (15.22) can be obtained from the tensor $\mathbf{P}_0$ traced onto the centers of particles x_k, where the term corresponding to $x_j = x_k$ is removed:

$$\sum_{k=1, k \neq j}^{N} \frac{z_k}{\|x_j - x_k\|} \rightsquigarrow \mathbf{P}_0(x_j) - z_j \mathbf{P}_R(x = 0).$$

Here, the value of the reference canonical tensor $\mathbf{P}_R$ (see (6.6)) is evaluated at the origin $x = 0$, i. e., corresponding to the multi-index $\mathbf{n}_0 = (n/2, n/2, n/2)$. Hence, we arrive at the tensor approximation

$$E_N \rightsquigarrow \frac{1}{2} \sum_{j=1}^{N} z_j(\mathbf{P}_0(x_j) - z_j \mathbf{P}_R(x = 0)). \tag{15.25}$$

Now, we split $\mathbf{P}_0$ into the long-range part (15.11) and the remaining short-range potential to obtain $\mathbf{P}_0(x_j) = \mathbf{P}_s(x_j) + \mathbf{P}_l(x_j)$, and the same for the reference tensor $\mathbf{P}_R$. By assumption, the short-range part $\mathbf{P}_s(x_j)$ at point x_j in (15.25) consists only of the local term $P_{R_s}(x = 0) = z_j \mathbf{P}_R(x = 0)$. Due to the corresponding cancellations in the right-hand side of (15.25), we find that E_N depends only on $\mathbf{P}_l$, leading to the final tensor representation in (15.24).

We arrive at the linear complexity scaling $O(dR_l N)$ taking into account the $O(dR_l)$ cost of the point evaluation for the canonical tensor $\mathbf{P}_l$. □

Table 15.4 presents the error of energy computation by (15.25) by using the RS tensor format with $R_l = 14$ and $R_s = 13$.

Table 15.4: Absolute and relative errors in the interaction energy of N-particle clusters computed by RS-tensor approximation with $R_l = 14$ ($R_s = 13$).

grid size, h	N	100	200	400	782
	Exact E_N	−8.4888	−18.1712	−35.9625	−90.2027
$8192^3, 6.8 \cdot 10^{-3}$	$(E_N - E_{N,T})/E_N$	10^{-4}	$2 \cdot 10^{-4}$	$2 \cdot 10^{-4}$	10^{-4}
$16\,384^3, 3.4 \cdot 10^{-3}$	$(E_N - E_{N,T})/E_N$	$2 \cdot 10^{-4}$	10^{-4}	10^{-4}	10^{-5}

Table 15.5 represents the approximation error in E_N computed by RS tensor representation (15.24) for the different values of system size. The grid size is fixed to $n^3 = 4096^3$ and $h = 0.0137$; the canonical rank for the reference tensor is $R = 29$. The short-range part of the RS tensor is taken as $R_s = 10$.

Table 15.5: Error in the interaction energy of clusters of N particles computed by the RS tensor approach ($R_s = 10$).

N	200	300	400	500	600	700
Exact E_N	−17.91	−26.47	−35.56	−47.1009	−62.32	−77.47
$(E_N - E_{N,T})/E_N$	$6 \cdot 10^{-5}$	$9 \cdot 10^{-7}$	$3.8 \cdot 10^{-5}$	$2.4 \cdot 10^{-4}$	$3.0 \cdot 10^{-4}$	$2.0 \cdot 10^{-4}$

Table 15.6 shows the results for several clusters of particles generated by random assignment of charges z_j to finite lattices of sizes 8^3, 12^3, $16 \times 16 \times 8$, and 16^3. Newton kernel is approximated with $\varepsilon_N = 10^{-4}$ on the grid of size 4096^3 with the rank $R = 25$. Computation of the interaction energy was performed using the only long-range part with $R_l = 12$. For the rank reduction, the multigrid C2T algorithm is applied [174], with the rank truncation parameters $\varepsilon_{\mathrm{C2T}} = 10^{-5}$, and $\varepsilon_{\mathrm{T2C}} = 10^{-6}$. The box size is about $40 \times 40 \times 40$ atomic units with mesh size $h = 0.0098$.

Table 15.6: Errors in the interaction energy of clusters of N particles computed by RS tensor approximation with the long-range rank parameter $R_l = 12$ ($R_s = 13$).

N of particles	512	1728	2048	4096
Exact E_N	51.8439	−133.9060	−138.5562	−207.8477
$(E_N - E_{N,T})/E_N$	0.0022	0.001	0.0016	0.001

Table 15.6 illustrates that the relative accuracy of energy calculations by using the RS tensor format remains of order 10^{-3} almost independent of the cluster size. Tucker ranks only slightly increase with the system size N. The computation time for the tensor $\mathbf{P}_l$ remains almost constant, whereas the point evaluations time for this tensor (with pre-computed data) increases linearly in N (see Lemma 15.15).

15.3.3 Gradients and forces

Computation of electrostatic forces and gradients of interaction potential in multiparticle systems is a computationally extensive problem. The algorithms based on Ewald summation technique was discussed in [63, 133]. We describe the alternative approach using RS tensor format proposed in [24].

First, we consider the computation of gradients. Given an RS-canonical tensor $\mathbf{A}$ as in (15.16) with the width parameter $\gamma > 0$, the discrete gradient $\nabla_h = (\nabla_1, \ldots, \nabla_d)^T$ applied to the long-range part in $\mathbf{A}$ at the grid points of Ω_h can be calculated simultaneously as the R-term canonical tensor by applying the simple one-dimensional finite-difference (FD) operations to the long-range part of $\mathbf{A} = \mathbf{A}_s + \mathbf{A}_l$,

$$\nabla_h \mathbf{A}_l = \sum_{k=1}^{R} \xi_k (\mathbf{G}_k^{(1)}, \ldots, \mathbf{G}_k^{(d)})^T, \tag{15.26}$$

with tensor entries

$$\mathbf{G}_k^{(\ell)} = \mathbf{u}_k^{(1)} \otimes \cdots \otimes \nabla_\ell \mathbf{u}_k^{(\ell)} \otimes \cdots \otimes \mathbf{u}_k^{(d)},$$

where ∇_ℓ ($\ell = 1, \ldots, d$) is the univariate FD differentiation scheme (by using backward or central differences). Numerical complexity of the representation (15.26) can be estimated by $O(dRn)$, provided that the canonical rank is almost uniformly bounded in the number of particles. The gradient operator applies locally to each short-range term in (15.16), which amounts to complexity $O(dR_0\gamma N)$.

The gradient of an RS-Tucker tensor can be calculated in a completely similar way.

Furthermore, in the setting of Section 15.3.2, the force vector $\mathbf{F}_j$ on the particle j is obtained by differentiating the electrostatic potential energy $E_N(x_1, \ldots, x_N)$ with respect to x_j,

$$\mathbf{F}_j = -\frac{\partial}{\partial x_j} E_N = -\nabla_{|x_j} E_N,$$

which can be calculated explicitly (see [133]) in the form

$$\mathbf{F}_j = \frac{1}{2} z_j \sum_{k=1, k \neq j}^{N} z_k \frac{x_j - x_k}{\|x_j - x_k\|^3}.$$

The Ewald summation technique for force calculations was presented in [64, 133]. In principle, it is possible to construct the RS tensor representation for this vector field directly by using the radial basis function $p(r) = 1/r^2$.

However, here we describe the alternative approach based on numerical differentiation of the energy functional by using RS tensor representation of the N-particle interaction potential on fine spacial grid. The differentiation in RS-tensor format with respect to x_j is based on the explicit representation (15.24), which can be rewritten in the form

$$E_N(x_1, \ldots, x_N) = \widehat{E}_N(x_1, \ldots, x_N) - \frac{1}{2}\left(\sum_{j=1}^{N} z_j^2\right) \mathbf{P}_{R_l}(x = 0), \tag{15.27}$$

where $\widehat{E}_N(x_1, \ldots, x_N) = \frac{1}{2} \sum_{j=1}^{N} z_j \mathbf{P}_l(x_j)$ denotes the "non-calibrated" interaction energy with the long-range tensor component $\mathbf{P}_l$. In the following discussion, for definiteness, we set $j = N$. Since the second term in (15.27) does not depend on the particle positions, it can be omitted in calculation of variations in E_N with respect to x_N. Hence, we arrive at the representation for the first difference in direction $\mathbf{e}_i$, $i = 1, 2, 3$,

$$E_N(x_1, \ldots, x_N) - E_N(x_1, \ldots, x_N - h\mathbf{e}_i) = \widehat{E}_N(x_1, \ldots, x_N) - \widehat{E}_N(x_1, \ldots, x_N - h\mathbf{e}_i).$$

The straightforward implementation of the above relation for three different values of $\mathbf{e}_1 = (1,0,0)^T$, $\mathbf{e}_2 = (0,1,0)^T$, and $\mathbf{e}_3 = (0,0,1)^T$ is reduced to the four calls of the basic procedure for computation the tensor $\mathbf{P}_l$ corresponding to *four* different dispositions of points $x_1, \ldots, x_N$ leading to the cost of order $O(dRn)$.

However, the factor *four* can be reduced to merely one, taking into account that the two canonical/Tucker tensors $\mathbf{P}_l$ computed for particle positions $(x_1, \ldots, x_{N-1}, x_N)$ and $(x_1, \ldots, x_{N-1}, x_N - h\mathbf{e})$ differ in a small part (since the positions $x_1, \ldots, x_{N-1}$ remain fixed). This requires only minor modifications compared with repeating the full calculation of $\widehat{E}_N(x_1, \ldots, x_N)$.

15.3.4 Regularization scheme for the Poisson–Boltzmann equation

Following [24], we describe the application scheme to the Poisson–Boltzmann equation (PBE) commonly used for numerical modeling of the electrostatic potential of proteins [135, 209].

Consider a solvated biomolecular system modeled by dielectrically separated domains with singular Coulomb potentials distributed in the molecular region. For schematic representation, we consider the system occupying a rectangular domain Ω with boundary $\partial\Omega$ (see Figure 15.11). The solute (molecule) region is represented by Ω_m, and the solvent region by Ω_s.

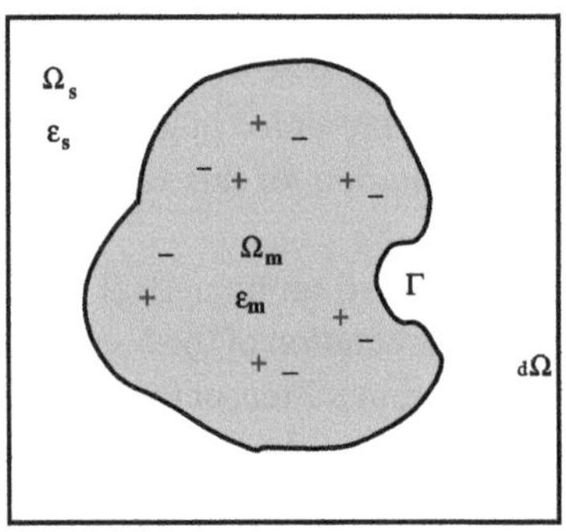

Figure 15.11: Computational domain for PBE.

The linearized Poisson–Boltzmann equation takes the form (see [209])

$$-\nabla \cdot (\epsilon \nabla u) + \kappa^2 u = \rho_f \quad \text{in } \Omega, \tag{15.28}$$

where u denotes the target electrostatic potential of a protein, and $\rho_f = \sum_{k=1}^{N} z_k \delta(\|x - x_k\|)$ is the scaled singular charge distribution supported at points x_k in Ω_m, where δ is the Dirac delta. Here, $\epsilon = 1$ and $\kappa = 0$ in Ω_m, whereas in the solvent region Ω_s, we have $\kappa \geq 0$ and $\epsilon \leq 1$. The boundary conditions on the external boundary $\partial\Omega$ can be specified depending on the particular problem setting. For definiteness, we impose the simplest Dirichlet boundary condition $u_{|\partial\Omega} = 0$. The interface conditions on the interior boundary $\Gamma = \partial\Omega_m$ arise from the dielectric theory:

$$[u] = 0, \quad \left[\epsilon \frac{\partial u}{\partial n}\right] \quad \text{on } \Gamma. \tag{15.29}$$

The practically useful solution methods for the PBE are based on regularization schemes aimed at removing the singular component from the potentials in the governing equation. Among others, we consider one of the most commonly used approaches based on the additive splitting of the potential only in the molecular region Ω_m (see [209]). To that end, we introduce the additive splitting

$$u = u^r + u^s, \quad \text{where } u^s = 0 \quad \text{in } \Omega_s,$$

where the singular component satisfies the equation

$$-\epsilon_m \Delta u^s = \rho_f \quad \text{in } \Omega_m; \quad u^s = 0 \quad \text{on } \Gamma. \tag{15.30}$$

Now, equation (15.28) can be transformed to that for the regular potential u^r:

$$\begin{aligned} &-\nabla \cdot (\epsilon \nabla u^r) + \kappa^2 u^r = \rho_f \quad \text{in } \Omega, \\ &[u^r] = 0, \quad \left[\epsilon \frac{\partial u^r}{\partial n}\right] = -\epsilon_m \frac{\partial u^s}{\partial n} \quad \text{on } \Gamma. \end{aligned} \tag{15.31}$$

To facilitate solving equation (15.30) with singular data, we define the singular potential U in the free space by

$$\epsilon_m \Delta U = \rho_f \quad \text{in } \mathbb{R}^3$$

and introduce its restriction U^s onto Ω_m,

$$U^s = U_{|\overline{\Omega}_m} \quad \text{in } \overline{\Omega}_m; \quad U^s = 0 \quad \text{in } \Omega_s.$$

Then we have $u^s = U^s + u^h$, where a harmonic function u^h compensates the discontinuity of U^s on Γ,

$$\Delta u^h = 0 \quad \text{in } \Omega_m; \quad u^h = -U^s \quad \text{on } \Gamma.$$

The advantage of this formulation is due to:

(a) the absence of singularities in the solution u^r and

(b) the localization of the solution splitting only on the domain Ω_m.

Calculating the singular potential U, which may include a sum of hundreds or even thousands of single Newton kernels in 3D, leads to a challenging computational problem. In the considered approach, it can be represented on large tensor grids with controlled precision by using the range separated tensor formats described above. The long-range component in the formatted parametrization remains smooth and allows global low-rank representation.

Notice that the short-range part in the tensor representation of U does not contribute to the right-hand side in the interface conditions on Γ in equation (15.31). This crucial simplification is possible since the physical distance between the atomic centers in protein modeling is bounded from below by the fixed constant $\sigma > 0$, whereas the effective support of the localized parts in the tensor representation of U can be chosen as the half of σ.

Moreover, all normal derivatives can be easily calculated by differentiation of univariate canonical vectors in the long-range part of the electrostatic potential U precomputed a on fine tensor grid in $\mathbb{R}^3$ (see Section 15.3.3). Hence, the numerical cost to build up the interface conditions in (15.31) becomes negligible compared with the solution of the equation (15.31). We conclude with the following:

Proposition 15.16. *Let the effective support of the short-range components in the reference potential $\mathbf{P}_R$ be not larger than $\sigma/2$. Then the interface conditions in the regularized formulation* (15.31) *of the PBE depend only on the low-rank long-range component in the free-space electrostatic potential of the system. The numerical cost to build up the interface conditions on Γ in* (15.31) *does not depend on the number of particles N.*

An important characterization of a protein molecule is the electrostatic solvation energy [209], which is the difference between the electrostatic-free energy in the solvated state (described by the PBE) and the electrostatic-free energy in the absence of solvent, that is, E_N. Now, the electrostatic solvation energy can be computed in the framework of the new regularized formulation (15.31) of PBE.

The particular numerical schemes for solving the PBE by using the RS tensor format are considered in [26]. An accurate tensor representation of the right-hand side in the PBE can be modelled by using the range-separated splitting of the Dirac delta introduced in [172].

Bibliography

[1] P.-A. Absil, R. Mahoni, and R. Sepulchre. *Optimization Algorithms on Matrix Manifolds*. Princeton University Press, Princeton, 2008.

[2] E. Acar, T. G. Kolda, and D. M. Dunlavy. *A scalable optimization approach for fitting canonical tensor decompositions*. J. Chemom., 25 (2), 67–86, 2011.

[3] J. Almlöf. *Direct methods in electronic structure theory*. In D. R. Yarkony , ed., *Modern Electronic Structure Theory, vol. II*, World Scientific, Singapore, pp. 110–151, 1995.

[4] F. Aquilante, L. Gagliardi, T. B. Pedersen, and R. Lindh. *Atomic Cholesky decompositions: a root to unbiased auxiliary basis sets for density fitting approximation with tunable accuracy and efficiency*. J. Chem. Phys., 130, 154107, 2009.

[5] D. Z. Arov and I. P. Gavrilyuk. *A method for solving initial value problems for linear differential equations in Hilbert space based on the Cayley transform*. Numer. Funct. Anal. Optim., 14 (5–6), 456–473, 1993.

[6] P. Y. Ayala and G. E. Scuseria. *Linear scaling second-order Møller–Plesset theory in the atomic orbital basis for large molecular systems*. J. Chem. Phys., 110 (8), 3660–3671, 1999.

[7] M. Bachmayr. *Adaptive Low-Rank Wavelet Methods and Applications to Two-Electron Schrödinger Equations*. PhD dissertation, RWTH Aachen, 2012.

[8] M. Bachmayr and W. Dahmen. *Adaptive near-optimal rank tensor approximation for high-dimensional operator equations*. Found. Comput. Math., 15 (4), 2015.

[9] B. W. Bader and T. G. Kolda. *Algorithm 862: MATLAB tensor classes for fast algorithm prototyping*. ACM Trans. Math. Softw., 32 (4), 2006.

[10] J. Ballani and L. Grasedyck. *A projection method to solve linear systems in tensor format*. Numer. Linear Algebra Appl., 20 (1), 27–43, 2013.

[11] J. Ballani, L. Grasedyck, and M. Kluge. *Black box approximation of tensors in hierarchical Tucker format*. Linear Algebra Appl., 428, 639–657, 2013.

[12] M. Barrault, E. Cancés, W. Hager, and C. Le Bris. *Multilevel domain decomposition for electronic structure calculations*. J. Comput. Phys., 222, 86–109, 2007.

[13] P. Baudin, J. Marin, I. G. Cuesta, and A. M. S.de Meras. *Calculation of excitation energies from the CC2 linear response theory using Cholesky decomposition*. J. Chem. Phys., 140, 104111, 2014.

[14] M. Bebendorf. *Adaptive cross approximation of multivariate functions*. Constr. Approx. 34 (2), 149–179, 2011.

[15] M. Bebendorf and S. Rjasanow. *Adaptive low-rank approximation of collocation matrices*. Computing, 70 (1), 1–24, 2003.

[16] T. Beck. *Real-space mesh techniques in density-functional theory*. Rev. Mod. Phys., 72, 1041–1080, 2000.

[17] N. H. F. Beebe and J. Linderberg. *Simplifications in the generation and transformation of two-electron integrals in molecular calculations*. Int. J. Quant. Chem., 12 (7), 683–705, 1977.

[18] R. E. Bellman. *Dynamic Programming*. Princeton University Press, Princeton, 1957.

[19] P. Benner, V. Mehrmann, and H. Xu. *A new method for computing the stable invariant subspace of a real Hamiltonian matrix*. J. Comput. Appl. Math., 86, 17–43, 1997.

[20] P. Benner, H. Faßbender, and M. Stoll. *Solving large-scale quadratic eigenvalue problems with Hamiltonian eigenstructure using a structure-preserving Krylov subspace method*. Electron. Trans. Numer. Anal., 29, 212–229, 2008.

[21] P. Benner, A. Onwunta, and M. Stoll. *Low-rank solution of unsteady diffusion equations with stochastic coefficients*. SIAM/ASA J. Uncertain. Quantificat., 3 (1), 622–649, 2015.

https://doi.org/10.1515/9783110365832-016

[22] P. Benner, H. Faßbender, and C. Yang. *Some remarks on the complex J-symmetric eigenproblem*. Preprint, Max Planck Institute Magdeburg, MPIMD/15-12, July 2015, http://www2.mpi-magdeburg.mpg.de/preprints/2015/12/
[23] P. Benner, V. Khoromskaia, and B. N. Khoromskij. *A reduced basis approach for calculation of the Bethe–Salpeter excitation energies using low-rank tensor factorizations*. Mol. Phys., 114 (7–8), 1148–1161, 2016.
[24] P. Benner, V. Khoromskaia, and B. N. Khoromskij. *Range-separated tensor formats for numerical modeling of many-particle interaction potentials*. arXiv:1606.09218 (39 pp.), 2016.
[25] P. Benner, S. Dolgov, V. Khoromskaia, and B. N. Khoromskij. *Fast iterative solution of the Bethe–Salpeter eigenvalue problem using low-rank and QTT tensor approximation*. J. Comput. Phys., 334, 221–239, 2017.
[26] P. Benner, V. Khoromskaia, B. N. Khoromskij, C. Kweyu, and M. Stein. *Application of the range-separated tensor format in solution of the Poisson–Boltzmann equation*. Manuscript, 2017.
[27] P. Benner, V. Khoromskaia, B. N. Khoromskij, and C. Yang. *Computing the density of states for optical spectra by low-rank and QTT tensor approximation*. arXiv:1801.03852, 2017.
[28] P. Benner, V. Khoromskaia, and B. N. Khoromskij. *Range-separated tensor format for many-particle modeling*. SIAM J. Sci. Comput., 40 (2), A1034–A1062, 2018.
[29] A. Bensoussan, J.-L. Lions, and G. Papanicolaou. *Asymptotic Analysis for Periodic Structures*. North-Holland, Amsterdam, 1978.
[30] C. Bertoglio, and B. N. Khoromskij. *Low-rank quadrature-based tensor approximation of the Galerkin projected Newton/Yukawa kernels*. Comput. Phys. Commun., 183 (4), 904–912, 2012.
[31] G. Beylkin and M. J. Mohlenkamp. *Numerical operator calculus in higher dimensions*. Proc. Natl. Acad. Sci. USA, 99, 10246–10251, 2002.
[32] G. Beylkin and M. J. Mohlenkamp. *Algorithms for numerical analysis in high dimension*. SIAM J. Sci. Comput., 26 (6), 2133–2159, 2005.
[33] G. Beylkin, M. J. Mohlenkamp, and F. Pérez. *Approximating a wavefunction as an unconstrained sum of Slater determinants*. J. Math. Phys., 49, 032107, 2008.
[34] G. Beylkin, J. Garcke, and M. J. Mohlenkamp, *Multivariate regression and machine learning with sums of separable functions*. SIAM J. Sci. Comput., 31 (3), 1840–1857, 2009.
[35] F. A. Bischoff, E. F. Valeev. *Computing molecular correlation energies with guaranteed precision*. J. Chem. Phys., 139 (11), 114106, 2013.
[36] T. Blesgen, V. Gavini, and V. Khoromskaia. *Tensor product approximation of the electron density of large aluminium clusters in OFDFT*. J. Comput. Phys., 231 (6), 2551–2564, 2012.
[37] A. Bloch. *Les theoremes de M. Valiron sur les fonctions entieres et la theorie de l'uniformisation*. Ann. Fac. Sci. Univ. Toulouse, 17 (3), 1–22, 1925, ISSN 0240-2963.
[38] S. F. Boys, G. B. Cook, C. M. Reeves, and I. Shavitt. *Automatic fundamental calculations of molecular structure*. Nature, 178, 1207–1209, 1956.
[39] D. Braess. *Nonlinear Approximation Theory*. Springer-Verlag, Berlin, 1986.
[40] D. Braess. *Asymptotics for the approximation of wave functions by exponential-sums*. J. Approx. Theory, 83, 93–103, 1995.
[41] S. Brenner and R. Scott. *The Mathematical Theory of Finite Element Methods*. Springer, Berlin, 1994.
[42] M. D. Buhmann. *Radial Basis Functions*. Cambridge University Press, Cambridge, 2003.
[43] H. J. Bungartz, and M. Griebel. *Sparse grids*. Acta Numer., 1–123, 2004.
[44] E. Cancés and C. Le Bris. *On the convergence of SCF algorithms for the Hartree–Fock equations*. ESAIM: M2AN, 34 (4), 749–774, 2000.
[45] E. Cancés and C. Le Bris. *Mathematical modeling of point defects in materials science*. Math. Models Methods Appl. Sci., 23, 1795–1859, 2013.

[46] E. Cancés, A. Deleurence, and M. Lewin. *A new approach to the modeling of local defects in crystals: the reduced Hartree–Fock case*. Commun. Math. Phys., 281, 129–177, 2008.

[47] E. Cancés, V. Ehrlacher, and Y. Maday. *Periodic Schrödinger operator with local defects and spectral pollution*. SIAM J. Numer. Anal., 50 (6, 3016–3035, 2012.

[48] E. Cancés, V. Ehrlacher, and T. Leliévre. *Greedy algorithms for high-dimensional non-symmetric linear problems*. ESAIM Proc. 41, 95–131, 2013.

[49] J. D. Carrol and J. Chang. *Analysis of individual differences in multidimensional scaling via an N-way generalization of 'Eckart–Young' decomposition*. Psychometrika 35, 283–319, 1970.

[50] J. D. Carrol, S. Pruzansky, and J. B. Kruskal. *CANDELINC: A general approach to multidimensional analysis of many-way arrays with linear constraints on parameters*. Psychometrika, 45, 3–24, 1980.

[51] M. E. Casida. *Time-dependent density-functional response theory for molecules*. In D. P. Chong, ed., *Recent Advances in Density Functional Methods, Part I*, World Scientific, Singapore, 155–192, 1995.

[52] S. R. Chinnamsetty, M. Espig, W. Hackbusch, B. N. Khoromskij, and H.-J. Flad. *Kronecker tensor product approximation in quantum chemistry*. J. Chem. Phys., 127, 084110, 2007.

[53] P. G. Ciarlet and C. Le Bris, eds. *Handbook of Numerical Analysis, vol. X, Computational Chemistry*. Elsevier, Amsterdam, 2003.

[54] A. Cichocki and Sh. Amari. *Adaptive Blind Signal and Image Processing: Learning Algorithms and Applications*. Wiley, New York, 2002.

[55] A. Cichocki, N. Lee, I. Oseledets, A. H. Pan, Q. Zhao, and D. P. Mandic. *Tensor networks for dimensionality reduction and large-scale optimization: Part 1 low-rank tensor decompositions*. Found. Trends Mach. Learn., 9 (4–5), 249–429, 2016.

[56] C. Cramer and D. Truhlar. *Density functional theory for transition metals and transition metal chemistry*. Phys. Chem. Chem. Phys., 11 (46), 10757–10816, 2009.

[57] W. Dahmen, R. Devore, L. Grasedyck, E. Süli. *Tensor-sparsity of solutions to high-dimensional elliptic partial differential equations*. Found. Comput. Math., 16 (4), 813–874, 2016.

[58] T. Darten, D. York, and L. Pedersen. *Particle mesh Ewald: an $O(N \log N)$ method for Ewald sums in large systems*. J. Chem. Phys., 98, 10089–10091, 1993.

[59] L. De Lathauwer. *Signal Processing Based on Multilinear Algebra*. PhD thesis, Katholeke Universiteit Leuven, 1997.

[60] L. De Lathauwer, B. De Moor, and J. Vandewalle. *On the best rank-1 and rank-$(R_1, \dots, R_N)$ approximation of higher-order tensors*. SIAM J. Matrix Anal. Appl., 21, 1324–1342, 2000.

[61] L. De Lathauwer, B. De Moor, and J. Vandewalle. *A multilinear singular value decomposition*. SIAM J. Matrix Anal. Appl., 21, 1253–1278, 2000.

[62] V. De Silva and L.-H. Lim. *Tensor rank and the ill-posedness of the best low-rank approximation problem*. SIAM J. Matrix Anal. Appl., 30 (3), 1084–1127, 2008.

[63] M. Deserno and C. Holm. *How to mesh up Ewald sums. I. A theoretical and numerical comparison of various particle mesh routines*. J. Chem. Phys., 109 (18), 7678–7693, 1998.

[64] M. Deserno and C. Holm. *How to mesh up Ewald sums. II. A theoretical and numerical comparison of various particle mesh routines*. J. Chem. Phys., 109 (18), 7694–7701, 1998.

[65] S. Dolgov. *Tensor Product Methods in Numerical Simulation of High-Dimensional Dynamical Problems*. PhD thesis, University of Leipzig, 2014. http://nbn-resolving.de/urn:nbn:de:bsz:15-qucosa-151129

[66] S. V. Dolgov, and B. N. Khoromskij. *Two-level Tucker-TT-QTT format for optimized tensor calculus*. SIAM J. Matrix Anal. Appl., 34 (2),593–623, 2013.

[67] S. Dolgov, and B. N. Khoromskij. *Simultaneous state-time approximation of the chemical master equation using tensor product formats*. Numer. Linear Algebra Appl., 22 (2), 197–219, 2015.

[68] S. V. Dolgov, B. N. Khoromskij, and I. Oseledets. *Fast solution of multi-dimensional parabolic problems in the TT/QTT formats with initial application to the Fokker–Planck equation.* SIAM J. Sci. Comput., 34 (6), A3016–A3038, 2012.

[69] S. V. Dolgov, B. N. Khoromskij, and D. Savostyanov. *Superfast Fourier transform using QTT approximation.* J. Fourier Anal. Appl., 18 (5), 915–953, 2012.

[70] S. Dolgov, B. N. Khoromskij, D. Savostyanov, and I. Oseledets. *Computation of extreme eigenvalues in higher dimensions using block tensor train format.* Comput. Phys. Commun., 185 (4), 1207–1216, 2014.

[71] S. Dolgov, B. N. Khoromskij, A. Litvinenko, and H. G. Matthies. *Computation of the response surface in the tensor train data format.* SIAM J. Uncertain. Quantificat., 3, 1109–1135, 2015.

[72] R. Dovesi, R. Orlando, C. Roetti, C. Pisani, and V. R. Sauders. *The periodic Hartree–Fock method and its implementation in the CRYSTAL code.* Phys. Status Solidi (b), 217, 63, 2000.

[73] D. A. Drabold and O. F. Sankey. *Maximum entropy approach for linear scaling in the electronic structure problem.* Phys. Rev. Lett., 70, 3631–3634, 1993.

[74] F. Ducastelle and F. Cyrot-Lackmann. *Moments developments and their application to the electronic charge distribution of d bands.* J. Phys. Chem. Solids, 31, 1295–1306, 1970.

[75] T. H. Dunning, Jr. *Gaussian basis sets for use in correlated molecular calculations. I. The atoms boron through neon and hydrogen.* J. Chem. Phys., 90, 1007–1023, 1989.

[76] A. Durdek, S. R. Jensen, J. Juselius, P. Wind, T. Flå, and L. Frediani. *Adaptive order polynomial algorithm in a multi-wavelet representation scheme.* Appl. Numer. Math., 92, 40–53, 2015.

[77] V. Ehrlacher, C. Ortner, and A. V. Shapeev. *Analysis of boundary conditions for crystal defect atomistic simulations.* Arch. Ration. Mech. Anal., 222 (3), 1217–1268, 2016.

[78] L. Elden and B. Savas. *A Newton–Grassmann method for computing the best multilinear rank-(r_1, r_2, r_3) approximation of a tensor.* SIAM J. Matrix Anal. Appl., 31 (2), 248–271, 2009.

[79] P. P. Ewald. *Die Berechnung optische und elektrostatischer Gitterpotentiale.* Ann. Phys., 64, 253, 1921.

[80] H.-J. Flad, W. Hackbusch, and R. Schneider. *Best N-term approximation in electronic structure calculations: I. One-electron reduced density matrix.* ESAIM: M2AN 40, 49–61, 2006.

[81] H.-J. Flad, B. N. Khoromskij, D. V. Savostyanov, and E. E. Tyrtyshnikov. *Verification of the cross 3d algorithm on quantum chemistry data.* Russ. J. Numer. Anal. Math. Model., 4, 1–16, 2008.

[82] H.-J. Flad, R. Schneider, and B.-W. Schulze. *Asymptotic regularity of solutions to Hartree–Fock equations with Coulomb potential.* Math. Methods Appl. Sci., 31 (18), 2172–2201, 2008.

[83] H.-J. Flad, W. Hackbusch, B. N. Khoromskij, and R. Schneider. *Concepts of data-sparse tensor-product approximation in many-particle modeling.* In V. Olshevsky, E. Tyrtyshnikov, eds., *Matrix Methods: Theory, Algorithms, Applications (Dedicated to the Memory of Gene Golub)*, World Scientific Publishing, Singapore, pp. 313–347, 2010.

[84] B. Fornberg and N. Flyer. *A Primer on Radial Basis Functions with Applications to the Geosciences.* CBMS–NSF Regional Conference Series in Applied Mathematics, vol. 87, SIAM, Philadelphia, 2015.

[85] L. Frediani and D. Sundholm. *Real-space numerical grid methods in quantum chemistry.* Phys. Chem. Chem. Phys., 17, 31357–31359, 2015.

[86] L. Frediani, E. Fossgaard, T. Flå, and K. Ruud. *Fully adaptive algorithms for multivariate integral equations using the non-standard form and multiwavelets with applications to the Poisson and bound-state Helmholtz kernels in three dimensions.* Mol. Phys., 111 (9–11), 1143–1160, 2013.

[87] S. Friedland, V. Mehrmann, A. Miedlar, and M. Nkengla. *Fast low rank approximation of matrices and tensors.* Electron. J. Linear Algebra, 22, 1031–1048, 2011.

[88] M. J. Frisch, G. W. Trucks, H. B. Schlegel, G. E. Scuseria, M. A. Robb, J. R. Cheeseman, G. Scalmani, V. Barone, B. Mennucci, G. A. Petersson, et al. *Gaussian Development Version Revision H.1*, Gaussian Inc., Wallingford, CT, 2009.

[89] I. P. Gavrilyuk, *Super exponentially convergent approximation to the solution of the Schrödinger equation in abstract setting*. Comput. Methods Appl. Math., 10 (4), 345–358, 2010.

[90] I. V. Gavrilyuk and B. N. Khoromskij. *Quantized-TT-Cayley transform to compute dynamics and spectrum of high-dimensional Hamiltonians*. Comput. Methods Appl. Math., 11 (3), 273–290, 2011.

[91] I. P. Gavrilyuk, W. Hackbusch, and B. N. Khoromskij. *$\mathcal{H}$-matrix approximation for the operator exponential with applications*. Numer. Math., 92, 83–111, 2002.

[92] I. P. Gavrilyuk, W. Hackbusch, and B. N. Khoromskij. *Data-sparse approximation to operator-valued functions of elliptic operator*. Math. Comput., 73, 1297–1324, 2003.

[93] I. P. Gavrilyuk, W. Hackbusch, and B. N. Khoromskij. *Hierarchical tensor-product approximation to the inverse and related operators in high-dimensional elliptic problems*. Computing, 74, 131–157, 2005.

[94] I. P. Gavrilyuk, W. Hackbusch, and B. N. Khoromskij. *Tensor-product approximation to elliptic and parabolic solution operators in higher dimensions*. Computing, 74, 131–157, 2005.

[95] I. P. Gavrilyuk, W. Hackbusch, and B. N. Khoromskij. *Data-sparse approximation to a class of operator-valued functions*. Math. Comput., 74, 681–708, 2005.

[96] I. P. Gavrilyuk, W. Hackbusch, and B. N. Khoromskij. *Data-sparse approximation of a class of operator-valued functions*. Math. Comput. 74, 681–708, 2005.

[97] L. Genovese, A. Neelov, S. Goedecker, T. Deutsch, S. A. Ghasemi, A. Willand, D. Caliste, O. Zilberberg, M. Rayson, A. Bergman, and R. Schneider. *Daubechies wavelets as a basis set for density functional pseudopotential calculations*. J. Chem. Phys., 129, 014109, 2008.

[98] G. H. Golub, C. F. Van Loan. *Matrix Computations*, 4th edn. Johns Hopkins University Press, Baltimore, 2013.

[99] S. A. Goreinov, E. E. Tyrtyshnikov, and N. L. Zamarashkin. *A theory of pseudoskeleton approximations*. Linear Algebra Appl., 261, 1–21, 1997.

[100] L. Grasedyck. *Hierarchical singular value decomposition of tensors*. SIAM J. Matrix Anal. Appl., 31, 2029–2054, 2010.

[101] L. Grasedyck. *Polynomial approximation in hierarchical Tucker format by vector tensorization*. Preprint 43, DFG/SPP1324, RWTH Aachen, 2010.

[102] L. Grasedyck, D. Kressner, and C. Tobler. *A literature survey of low-rank tensor approximation techniques*. GAMM-Mitt., 36 (1), 53–78, 2013.

[103] L. Greengard and V. Rochlin. *A fast algorithm for particle simulations*. J. Comput. Phys., 73, 325, 1987.

[104] W. H. Greub. *Multilinear Algebra*, 2nd edn. Springer, Berlin, 1978.

[105] M. Griebel and J. Hamaekers. *Sparse grids for the Schroedinger equation*. Modél. Math. Anal. Numér., 41, 215–247, 2007.

[106] M. Griebel and J. Hamaekers. *Tensor product multiscale many-particle spaces with finite-order weights for the electronic Schrödinger equation*. Z. Phys. Chem., 224, 527–543, 2010.

[107] E. Gross and W. Kohn. *Time-dependent density-functional theory*. Adv. Quantum Chem., 21, 255–291, 1990.

[108] W. Hackbusch. *Efficient convolution with the Newton potential in d dimensions*. Numer. Math., 110 (4), 449–489, 2008.

[109] W. Hackbusch. *Convolution of hp-functions on locally refined grids*. IMA J. Numer. Anal., 29, 960–985, 2009.

[110] W. Hackbusch. *Tensor Spaces and Numerical Tensor Calculus*. Springer, Berlin, 2012.
[111] W. Hackbusch and B. N. Khoromskij. *Low-rank Kronecker product approximation to multi-dimensional nonlocal operators. Part I. Separable approximation of multi-variate functions*. Computing, 76, 177–202, 2006.
[112] W. Hackbusch and B. N. Khoromskij. *Low-rank Kronecker-product approximation to multi-dimensional nonlocal operators. Part II. HKT representation of certain operators*. Computing, 76, 203–225, 2006.
[113] W. Hackbusch and B. N. Khoromskij. *Tensor-product approximation to operators and functions in high dimension*. J. Complex., 23, 697–714, 2007.
[114] W. Hackbusch and B. N. Khoromskij. *Tensor-product approximation to multi-dimensional integral operators and Green's functions*. SIAM J. Matrix Anal. Appl., 30 (3), 1233–1253, 2008.
[115] W. Hackbusch, and S. Kühn. *A new scheme for the tensor representation*. J. Fourier Anal. Appl., 15, 706–722, 2009.
[116] W. Hackbusch, B. N. Khoromskij, and E. E. Tyrtyshnikov. *Hierarchical Kronecker tensor-product approximations*. J. Numer. Math., 13, 119–156, 2005.
[117] W. Hackbusch, B. N. Khoromskij, and E. Tyrtyshnikov. *Approximate iteration for structured matrices*. Numer. Math., 109, 365–383, 2008.
[118] W. Hackbusch, B. N. Khoromskij, S. Sauter, and E. Tyrtyshnikov. *Use of tensor formats in elliptic eigenvalue problems*. Numer. Linear Algebra Appl., 19 (1), 133–151, 2012.
[119] W. Hackbusch, and R. Schneider. *Tensor spaces and hierarchical tensor representations*. In S. Dahlke, W. Dahmen, et al., eds, *Lecture Notes in Computer Science and Engineering*, vol. 102, Springer, Berlin, 2014.
[120] N. Hale and L. N. Trefethen. *Chebfun and numerical quadrature*. Sci. China Math., 55 (9), 1749–1760, 2012.
[121] H. Harbrecht, M. Peters, and R. Schneider. *On the low-rank approximation by the pivoted Cholesky decomposition*. Appl. Numer. Math., 62 (4), 428–440, 2012.
[122] R. J. Harrison, G. I. Fann, T. Yanai, Z. Gan, and G. Beylkin. *Multiresolution quantum chemistry: basic theory and initial applications*. J. Chem. Phys., 121 (23), 11587–11598, 2004.
[123] D. R. Hartree. *The Calculation of Atomic Structure*. Wiley, New York, 1957.
[124] R. Haydock, V. Heine, and M. J. Kelly. *Electronic structure based on the local atomic environment for tight-binding bands*. J. Phys. C, Solid State Phys., 5, 2845–2858, 1972.
[125] M. Head-Gordon, J. A. Pople, and M. Frisch. *MP2 energy evaluation by direct methods*. Chem. Phys. Lett., 153 (6), 503–506, 1988.
[126] L. Hedin. *New method for calculating the one-particle Green's function with application to the electron–gas problem*. Phys. Rev. 139, A796, 1965.
[127] T. Helgaker, P. Jørgensen, and N. Handy. *A numerically stable procedure for calculating Møller–Plesset energy derivatives, derived using the theory of Lagrangians*. Theor. Chim. Acta, 76, 227–245, 1989.
[128] T. Helgaker, P. Jørgensen, and J. Olsen. *Molecular Electronic-Structure Theory*. Wiley, New York, 1999.
[129] J. S. Hesthaven, G. Rozza, and B. Stamm. *Certified Reduced Basis Methods for Parametrized Partial Differential Equations*. Springer, Berlin, 2016.
[130] N. Higham. *Analysis of the Cholesky decomposition of a semi-definite matrix*. In M. G. Cox and S. J. Hammarling, eds, *Reliable Numerical Computations*, Oxford University Press, Oxford, pp. 161–185, 1990.
[131] F. L. Hitchcock. *The expression of a tensor or a polyadic as a sum of products*. J. Math. Phys., 6, 164–189, 1927.
[132] F. L. Hitchcock. *Multiple invariants and generalized rank of a p-way matrix or tensor*. J. Math. Phys., 7, 39–79, 1927.

[133] R. W. Hockney and J. W. Eastwood. *Computer Simulation Using Particles*. IOP, Bristol, 1988.

[134] E. G. Hohenstein, R. M. Parrish, and T. J. Martinez. *Tensor hypercontraction density fitting. Quartic scaling second- and third-order Møller–Plesset perturbation theory*. J. Chem. Phys., 137, 044103, 2012.

[135] M. Holst, N. Baker, and F. Wang. *Adaptive multilevel finite element solution of the Poisson–Boltzmann equation: algorithms and examples*. J. Comput. Chem., 21, 1319–1342, 2000.

[136] S. Holtz, T. Rohwedder, and R. Schneider. *On manifold of tensors of fixed TT-rank*. Numer. Math., 120 (4), 701–731, 2012.

[137] S. Holtz, T. Rohwedder, and R. Schneider. *The alternating linear scheme for tensor optimization in the tensor train format*. SIAM J. Sci. Comput., 34 (2), A683–A713, 2012.

[138] T. Huckle, K. Waldherr, and T. Schulte-Herbrüggen. *Computations in quantum tensor networks*. Linear Algebra Appl., 438, 750–781, 2013.

[139] P. H. Hünenberger. *Lattice-sum methods for computing electrostatic interactions in molecular simulations*. AIP Conf. Proc., 492, 17, 1999.

[140] M. Ishteva, L. De Lathauwer, P.-A. Absil, and S. Van Huffel. *Differential-geometric Newton method for the best rank-(R_1, R_2, R_3) approximation of tensors*. Numer. Algorithms, 51 (2), 179–194, 2009.

[141] A. Iske. *Multiresolution Methods in Scattered Data Modeling*. Springer, Berlin, 2004.

[142] V. Kazeev, and B. N. Khoromskij. *Explicit low-rank QTT representation of Laplace operator and its inverse*. SIAM J. Matrix Anal. Appl., 33 (3), 2012, 742–758.

[143] V. Kazeev, B. N. Khoromskij, and E. E. Tyrtyshnikov. *Multilevel Toeplitz matrices generated by tensor-structured vectors and convolution with logarithmic complexity*. SIAM J. Sci. Comput. 35 (3), A1511–A1536, 2013.

[144] V. Kazeev, M. Khammash, M. Nip, and Ch. Schwab. *Direct solution of the chemical master equation using quantized tensor trains*. PLoS Comput. Biol. 10 (3), 2014.

[145] V. Khoromskaia. *Computation of the Hartree–Fock exchange in the tensor-structured format*. Comput. Methods Appl. Math., 10 (2), 1–16, 2010.

[146] V. Khoromskaia. *Numerical Solution of the Hartree–Fock Equation by Multilevel Tensor-Structured Methods*. PhD dissertation, TU Berlin, 2010. https://depositonce.tu-berlin.de/handle/11303/3016

[147] V. Khoromskaia. *Black-box Hartree–Fock solver by tensor numerical methods*. Comput. Methods Appl. Math., 14 (1), 89–111, 2014.

[148] V. Khoromskaia and B. N. Khoromskij. *Grid-based lattice summation of electrostatic potentials by assembled rank-structured tensor approximation*. Comput. Phys. Commun., 185, 3162–3174, 2014.

[149] V. Khoromskaia and B. N. Khoromskij. *Tucker tensor method for fast grid-based summation of long-range potentials on 3D lattices with defects*. arXiv:1411.1994, 2014.

[150] V. Khoromskaia and B. N. Khoromskij. *Møller–Plesset (MP2) energy correction using tensor factorizations of the grid-based two-electron integrals*. Comput. Phys. Commun., 185, 2–10, 2014.

[151] V. Khoromskaia and B. N. Khoromskij. *Tensor approach to linearized Hartree–Fock equation for lattice-type and periodic systems. Preprint 62/2014, Max-Planck Institute for Mathematics in the Sciences, Leipzig*. arXiv:1408.3839, 2014.

[152] V. Khoromskaia and B. N. Khoromskij. *Tensor numerical methods in quantum chemistry: from Hartree–Fock to excitation energies*. Phys. Chem. Chem. Phys., 17 (47), 31491–31509, 2015.

[153] V. Khoromskaia and B. N. Khoromskij. *Fast tensor method for summation of long-range potentials on 3D lattices with defects*. Numer. Linear Algebra Appl., 23, 249–271, 2016.

[154] V. Khoromskaia and B. N. Khoromskij. *Block circulant and Toeplitz structures in the linearized Hartree–Fock equation on finite lattices: tensor approach*. Comput. Methods Appl. Math., 17 (3), 431–455, 2017.

[155] V. Khoromskaia, B. N. Khoromskij, and R. Schneider. *QTT representation of the Hartree and exchange operators in electronic structure calculations*. Comput. Methods Appl. Math., 11 (3), 327–341, 2011.

[156] V. Khoromskaia, D. Andrae, and B. N. Khoromskij. *Fast and accurate 3D tensor calculation of the Fock operator in a general basis*. Comput. Phys. Commun., 183, 2392–2404, 2012.

[157] V. Khoromskaia, B. N. Khoromskij, and R. Schneider. *Tensor-structured factorized calculation of two-electron integrals in a general basis*. SIAM J. Sci. Comput., 35 (2), A987–A1010, 2013.

[158] V. Khoromskaia, B. N. Khoromskij, and F. Otto. *A numerical primer in 2D stochastic homogenization: CLT scaling in the representative volume element*. Preprint 47/2017, Max-Planck Institute for Math. in the Sciences, Leipzig 2017.

[159] B. N. Khoromskij. *Data-sparse elliptic operator inverse based on explicit approximation to the Green function*. J. Numer. Math., 11 (2), 135–162, 2003.

[160] B. N. Khoromskij. *An Introduction to Structured Tensor-Product Representation of Discrete Nonlocal Operators*. Lecture Notes, vol. 27, Max-Planck Institute for Mathematics in the Sciences, Leipzig, 2005.

[161] B. N. Khoromskij. *Structured rank-*$(r_1,\dots,r_d)$ *decomposition of function-related tensors in* $\mathbb{R}^d$. Comput. Methods Appl. Math., 6 (2), 194–220, 2006.

[162] B. N. Khoromskij. *Structured data-sparse approximation to high order tensors arising from the deterministic Boltzmann equation*. Math. Comput., 76, 1292–1315, 2007.

[163] B. N. Khoromskij. *On tensor approximation of Green iterations for Kohn–Sham equations*. Comput. Vis. Sci., 11, 259–271, 2008.

[164] B. N. Khoromskij. *Tensor-structured preconditioners and approximate inverse of elliptic operators in* $\mathbb{R}^d$. Constr. Approx., 30, 599–620, 2009.

[165] B. N. Khoromskij. $O(d \log N)$*-quantics approximation of N-d tensors in high-dimensional numerical modeling*. Preprint 55/2009, Max-Planck Institute for Mathematics in the Sciences, Leipzig 2009.
http://www.mis.mpg.de/publications/preprints/2009/prepr2009-55.html

[166] B. N. Khoromskij. *Fast and accurate tensor approximation of a multivariate convolution with linear scaling in dimension*. J. Comput. Appl. Math., 234, 3122–3139, 2010.

[167] B. N. Khoromskij. $O(d \log N)$*-quantics approximation of N-d tensors in high-dimensional numerical modeling*. Constr. Approx., 34 (2), 257–289, 2011.

[168] B. N. Khoromskij. *Introduction to tensor numerical methods in scientific computing*. Lecture Notes, Preprint 06-2011, University of Zuerich, Institute of Mathematics, 2011, pp. 1–238, http://www.math.uzh.ch/fileadmin/math/preprints/06_11.pdf

[169] B. N. Khoromskij. *Tensors-structured numerical methods in scientific computing: survey on recent advances*. Chemom. Intell. Lab. Syst., 110, 1–19, 2012.

[170] B. N. Khoromskij. *Tensor numerical methods for high-dimensional PDEs: basic theory and initial applications*. ESAIM, 48, 1–28, 2014.

[171] B. N. Khoromskij. *Tensor Numerical Methods in Scientific Computing*. Research Monograph, De Gruyter Verlag, Berlin, 2018.

[172] B. N. Khoromskij. *Operator-Dependent Approximation of the Dirac Delta by Using Range-Separated Tensor Format*. Manuscript, 2018.

[173] B. N. Khoromskij and V. Khoromskaia. *Low rank Tucker-type tensor approximation to classical potentials*. Cent. Eur. J. Math., 5 (3), 523–550, 2007 (Preprint 105/2006 Max-Planck Institute for Mathematics in the Sciences, Leipzig 2006).

[174] B. N. Khoromskij and V. Khoromskaia. *Multigrid tensor approximation of function related arrays*. SIAM J. Sci. Comput., 31 (4), 3002–3026, 2009.

[175] B. N. Khoromskij and S. Miao. *Superfast wavelet transform using QTT approximation. I: Haar wavelets*. Comput. Methods Appl. Math., 14 (4), 537–553, 2014.

[176] B. N. Khoromskij and I. Oseledets. *Quantics-TT collocation approximation of parameter-dependent and stochastic elliptic PDEs*. Comput. Methods Appl. Math., 10 (4), 34–365, 2010.

[177] B. N. Khoromskij and I. Oseledets. *DMRG+QTT approach to the computation of ground state for the molecular Schrödinger operator*. Preprint 68/2010, Max-Planck Institute for Mathematics in the Sciences, Leipzig, 2010.

[178] B. N. Khoromskij, and I. Oseledets. *Quantics-TT approximation of elliptic solution operators in higher dimensions*. Russ. J. Numer. Anal. Math. Model., 26 (3), 303–322, 2011.

[179] B. N. Khoromskij and I. Oseledets. *Quantics-TT approximation of elliptic solution operators in higher dimensions*. Preprint 79/2009, Max-Planck Institute for Mathematics in the Sciences, Leipzig 2009.

[180] B. N. Khoromskij and S. Repin. *A fast iteration method for solving elliptic problems with quasiperiodic coefficients*. Russ. J. Numer. Anal. Math. Model., 30 (6), 329–344, 2015.

[181] B. N. Khoromskij and S. Repin. *Rank structured approximation method for quasi-periodic elliptic problems*. Comput. Methods Appl. Math. 17 (3), 457–477, 2017.

[182] B. N. Khoromskij, and Ch. Schwab. *Tensor approximation of multi-parametric elliptic problems in SPDEs*. SIAM J. Sci. Comput., 33 (1), 364–385, 2011.

[183] B. Khoromskij and A Veit. *Efficient computation of highly oscillatory integrals by using QTT tensor approximation*. Comput. Methods Appl. Math., 16 (1), 145–159, 2016.

[184] B. N. Khoromskij and G. Wittum. *Numerical Solution of Elliptic Differential Equations by Reduction to the Interface*. Research Monograph, LNCSE, vol. 36, Springer-Verlag, Berlin, 2004.

[185] B. N. Khoromskij, A. Litvinenko, and H. G. Matthies. *Application of hierarchical matrices for computing the Karhunen–Loéve expansion*. Computing, 84, 49–67, 2009.

[186] B. N. Khoromskij, V. Khoromskaia, S. R. Chinnamsetty, and H.-J. Flad. *Tensor decomposition in electronic structure calculations on 3D Cartesian grids*. J. Comput. Phys., 228, 5749–5762, 2009.

[187] B. N. Khoromskij, V. Khoromskaia, and H.-J. Flad. *Numerical solution of the Hartree–Fock equation in multilevel tensor-structured format*. SIAM J. Sci. Comput., 33 (1), 45–65, 2011.

[188] B. N. Khoromskij, S. Sauter, and A. Veit. *Fast quadrature techniques for retarded potentials based on TT/QTT tensor approximation*. Comput. Methods Appl. Math., 11 (3), 342–362, 2011.

[189] B. N. Khoromskij, K. K. Naraparaju, and J. Schneider. *Quantized-CP approximation and sparse tensor interpolation of function generated data*. arXiv:1707.04525, 2017.

[190] A. V. Knyazev. *Toward the optimal preconditioned eigensolver: locally optimal block preconditioned conjugate gradient method*. SIAM J. Sci. Comput., 23 (2), 517–541, 2001.

[191] O. Koch and Ch. Lubich. *Dynamical low rank approximation*. SIAM J. Matrix Anal. Appl., 29 (2), 434–454, 2007.

[192] T. Kolda. *Orthogonal tensor decompositions*. SIAM J. Matrix Anal. Appl., 23, 243–255, 2001.

[193] T. G. Kolda and B. W. Bader. *Tensor decompositions and applications*. SIAM Rev., 51 (3), 455–500, 2009.

[194] S. Körbel, P. Boulanger, I. Duchemin, X. Blase, M. AL Marques, and S. Botti. *Benchmark many-body GW and Bethe–Salpeter calculations for small transition metal molecules*. J. Chem. Theory Comput., 10 (9), 3934–3943, 2014.

[195] D. Kressner and C. Tobler. *Preconditioned low-rank methods for high-dimensional elliptic PDE eigenvalue problems*. Comput. Methods Appl. Math., 11 (3), 363–381, 2011.

[196] D. Kressner, M. Steinlechner, and A. Uschmajew. *Low-rank tensor methods with subspace correction for symmetric eigenvalue problems*. SIAM J. Sci. Comput., 36 (5), A2346–A2368, 2014.

[197] P. M. Kroonenberg and J. De Leeuw. *Principal component analysis of three-mode data by means of alternating least squares algorithms*. Psychometrika, 45, 69–97, 1980.

[198] J. B. Kruskal. *Three-way arrays: rank and uniqueness of trilinear decompositions, with applications to arithmetic complexity and statistics*. Linear Algebra Appl., 18, 95–138, 1977.

[199] K. N. Kudin, and G. E. Scuseria, *Revisiting infinite lattice sums with the periodic Fast Multipole Method*, J. Chem. Phys. 121, 2886–2890, 2004.

[200] L. Laaksonen, P. Pyykkö, and D. Sundholm. *Fully numerical Hartree–Fock methods for molecules*, Comput. Phys. Rep., 4, 313–344, 1986.

[201] J. M. Landsberg. *Tensors: Geometry and Applications*. American Mathematical Society, Providence, RI, 2012.

[202] S. Lang. *Linear Algebra*, 3rd edn. Springer, Berlin, 1987.

[203] C. Le Bris, *Computational chemistry from the perspective of numerical analysis*. Acta Numer., 363–444, 2005.

[204] L. Lin, Y. Saad, and Ch. Yang. *Approximating spectral densities of large matrices*. SIAM Rev., 58, 34, 2016.

[205] D. Lindbo and A.-K. Tornberg. *Fast and spectrally accurate Ewald summation for 2-periodic electrostatic systems*. J. Chem. Phys., 136, 164111, 2012.

[206] A. Litvinenko, D. Keyes, V. Khoromskaia, B. Khoromskij, and H. Matthies. *Tucker Tensor analysis of Matérn functions in spatial statistics*. arXiv:1711.06874, 2017.

[207] M. Lorenz, D. Usvyat, and M. Schütz. *Local ab initio methods for calculating optical band gaps in periodic systems. I. Periodic density fitted local configuration interaction singles method for polymers*. J. Chem. Phys., 134, 094101, 2011.

[208] S. A. Losilla, D. Sundholm, and J. Juselius. *The direct approach to gravitation and electrostatics method for periodic systems*. J. Chem. Phys., 132 (2), 024102, 2010.

[209] B. Z. Lu, Y. C. Zhou, M. J. Holst, and J. A. McCammon. *Recent progress in numerical methods for Poisson–Boltzmann equation in biophysical applications*. Commun. Comput. Phys., 3 (5), 973–1009, 2008.

[210] Ch. Lubich. *On variational approximations in quantum molecular dynamics*. Math. Comput., 74, 765–779, 2005.

[211] Ch. Lubich. *From Quantum to Classical Molecular Dynamics: Reduced Models and Numerical Analysis*. Zurich Lectures in Advanced Mathematics, EMS, Zurich, 2008.

[212] Ch. Lubich and I. V. Oseledets. *A projector-splitting integrator for dynamical low-rank approximation*. BIT Numer. Math., 54 (1), 171–188, 2014.

[213] Ch. Lubich, T. Rohwedder, R. Schneider, and B. Vandereycken. *Dynamical approximation of hierarchical Tucker and tensor-train tensors*. SIAM J. Matrix Anal. Appl., 34 (2), 470–494, 2013.

[214] C Lubich, I. V. Oseledets, and B. Vandereycken. *Time integration of tensor trains*. SIAM J. Numer. Anal., 53 (2), 917–941, 2015.

[215] J. Lund and K. L. Bowers. *Sinc Methods for Quadrature and Differential Equations*. SIAM, Philadelphia, 1992.

[216] F. R. Manby. *Density fitting in second-order linear-r_{12} Møller–Plesset perturbation theory*. J. Chem. Phys., 119 (9), 4607–4613, 2003.

[217] F. R. Manby, P. J. Knowles, and A. W. Lloyd. *The Poisson equation in density fitting for the Kohn–Sham Coulomb problem*. J. Chem. Phys., 115, 9144–9148, 2001.

[218] G. I. Marchuk and V. V. Shaidurov. *Difference Methods and Their Extrapolations*. Applications of Mathematics, Springer, New York, 1983.

[219] H. G. Matthies, A. Litvinenko, O. Pajonk, B. L. Rosic, and E. Zander. *Parametric and uncertainty computations with tensor product representations*. In: *Uncertainty Quantification in Scientific Computing*, Springer, Berlin, pp. 139–150, 2012.

[220] V. Mazyja, G. Schmidt. *Approximate Approximations*. Mathematical Surveys and Monographs, vol. 141, AMS, Providence, 2007.

[221] H.-D. Meyer, F. Gatti, and G. A. Worth. *Multidimensional Quantum Dynamics: MCTDH Theory and Applications*. Willey–VCH, Wienheim, 2009.

[222] C. Møller and M. S. Plesset. *Note on an approximation treatment for many-electron systems*. Phys. Rev., 46, 618, 1934.

[223] K. K. Naraparaju and J. Schneider. *Generalized cross approximation for 3d-tensors*. Comput. Vis. Sci., 14 (3), 105–115, 2011.

[224] G. Onida, L. Reining, A. Rubio. *Electronic excitations: density-functional versus many-body Green's-function approaches*. Rev. Mod. Phys., 74 (2), 601, 2002.

[225] I. V. Oseledets. *Approximation of $2^d \times 2^d$ matrices using tensor decomposition*. SIAM J. Matrix Anal. Appl., 31 (4), 2130–2145, 2010.

[226] I. V. Oseledets. *Tensor-train decomposition*. SIAM J. Sci. Comput., 33 (5), 2295–2317, 2011.

[227] I. V. Oseledets. *Constructive representation of functions in low-rank tensor formats*. Constr. Approx., 37 (1), 1–18, 2013.

[228] I. V. Oseledets and S. V. Dolgov. *Solution of linear systems and matrix inversion in the TT-format*. SIAM J. Sci. Comput., 34 (5), A2718–A2739, 2012.

[229] I. V. Oseledets, and E. E. Tyrtyshnikov, *Breaking the curse of dimensionality, or how to use SVD in many dimensions*. SIAM J. Sci. Comput., 31 (5), 3744–3759, 2009.

[230] I. Oseledets and E. E. Tyrtyshnikov. *TT-cross approximation for multidimensional arrays*. Linear Algebra Appl., 432 (1), 70–88, 2010.

[231] I. V. Oseledets, D. V. Savostyanov, and E. E. Tyrtyshnikov. *Tucker dimensionality reduction of three-dimensional arrays in linear time*. SIAM J. Matrix Anal. Appl., 30 (3), 939–956, 2008.

[232] I. V. Oseledets et al. *Tensor Train Toolbox*, 2014. https://github.com/oseledets/TT-Toolbox

[233] R. Parrish, E. G. Hohenstein, T. J. Martinez, and C. D. Sherrill. *Tensor hypercontraction. II. Least-squares renormalization*. J. Chem. Phys., 137, 224106, 2012.

[234] K. A. Peterson, D. E. Woon, and T. H. Dunning, Jr. *Benchmark calculations with correlated molecular wave functions. IV. The classical barrier height of the $H + H_2 \rightarrow H_2 + H$ reaction*. J. Chem. Phys., 100, 7410–7415, 1994.

[235] C. Pisani, M. Schütz, S. Casassa, D. Usvyat, L. Maschio, M. Lorenz, and A. Erba. *CRYSCOR: a program for the post-Hartree–Fock treatment of periodic systems*. Phys. Chem. Chem. Phys., 14, 7615–7628, 2012.

[236] E. L. Pollock and J. Glosli. *Comments on p(3)m, fmm and the Ewald method for large periodic Coulombic systems*. Comput. Phys. Commun., 95, 93–110, 1996.

[237] R. Polly, H.-J. Werner, F. R. Manby, and P. J. Knowles. *Fast Hartree–Fock theory using density fitting approximations*. Mol. Phys., 102, 2311–2321, 2004.

[238] P. Pulay. *Improved SCF convergence acceleration*. J. Comput. Chem., 3, 556–560, 1982.

[239] M. Rakhuba and I. Oseledets. *Fast multidimensional convolution in low-rank tensor formats via cross approximation*. SIAM J. Sci. Comput., 37 (2), A565–A582, 2015.

[240] M. Rakhuba and I. Oseledets. *Grid-based electronic structure calculations: the tensor decomposition approach*. J. Comput. Phys., 312, 19–30, 2016.

[241] G. Rauhut, P. Pulay, H.-J. Werner. *Integral transformation with low-order scaling for large local second-order Mollet–Plesset calculations*. J. Comput. Chem., 19, 1241–1254, 1998.

[242] H. Rauhut, R. Schneider, and Z. Stojanac. *Low rank tensor recovery via iterative hard thresholding*. Linear Algebra Appl., 523, 220–262, 2017.

[243] E. Rebolini, J. Toulouse, and A. Savin. *Electronic excitation energies of molecular systems from the Bethe–Salpeter equation: Example of H_2 molecule*. In S. Ghosh, P. Chattaraj, eds, *Concepts and Methods in Modern Theoretical Chemistry, vol. 1: Electronic Structure and Reactivity*, p. 367, 2013.

[244] E. Rebolini, J. Toulouse, and A. Savin. *Electronic excitations from a linear-response range-separated hybrid scheme*. Mol. Phys., 111, 1219, 2013.

[245] E. Rebolini, J. Toulouse, A. M. Teale, T. Helgaker, and A. Savin. *Calculating excitation energies by extrapolation along adiabatic connections*. Phys. Rev. A, 91, 032519, 2015.

[246] M. Reed and B. Simon. *Functional Analysis*. Academic Press, San Diego, 1972.

[247] S. Reine, T. Helgaker, and R. Lindh. *Multi-electron integrals*. WIREs Comput. Mol. Sci., 2, 290–303, 2012.

[248] L. Reining, V. Olevano, A. Rubio, and G. Onida. *Excitonic effects in solids described by time-dependent density-functional theory*. Phys. Rev. Lett., 88 (6), 66404, 2002.

[249] T. Rohwedder and R. Schneider. *Error estimates for the coupled cluster method*. ESAIM: M2AN, 47 (6), 1553–1582, 2013.

[250] T. Rohwedder and A. Uschmajew. *On local convergence of alternating schemes for optimization of convex problems in the tensor train format*. SIAM J. Numer. Anal., 51 (2), 1134–1162, 2013.

[251] E. Runge and E. K. U. Gross. *Density-functional theory for time-dependent systems*. Phys. Rev. Lett., 52 (12), 997, 1984.

[252] E. E. Salpeter and H. A. Bethe. *A relativistic equation for bound-state problems*. Phys. Rev., 84 (6), 1951.

[253] G. Sansone, B. Civalleri, D. Usvyat, J. Toulouse, K. Sharkas, and L. Maschio. *Range-separated double-hybrid density-functional theory applied to periodic systems*. J. Chem. Phys., 143, 102811, 2015.

[254] B. Savas and L.-H. Lim. *Quasi-Newton methods on Grassmanians and multilinear approximations of tensors*. SIAM J. Sci. Comput., 32 (6), 3352–3393, 2010.

[255] D. V. Savostianov. *Fast revealing of mode ranks of tensor in canonical form*. Numer. Math., Theory Methods Appl., 2 (4), 439–444, 2009.

[256] D. V. Savostyanov and I. V. Oseledets. *Fast adaptive interpolation of multi-dimensional arrays in tensor train format*. In *Multidimensional (ND) Systems, 7th International Workshop*, University of Poitiers, France, 2011, doi:10.1109/nDS.2011.6076873

[257] D. V. Savostyanov, S. V. Dolgov, J. M. Werner, and I. Kuprov. *Exact NMP simulation of protein-size spin systems using tensor train formalism*. Phys. Rev. B, 90, 085139, 2014.

[258] G. Schaftenaar and J. H. Noordik. *Molden: a pre- and post-processing program for molecular and electronic structures*. J. Comput.-Aided Mol. Des., 14, 123–134, 2000.

[259] W. G. Schmidt, S. Glutsch, P. H. Hahn, and F. Bechstedt. *Efficient O(N 2) method to solve the Bethe–Salpeter equation*. Phys. Rev. B, 67, 085307, 2003.

[260] R. Schneider. *Analysis of the projected coupled cluster method in electronic structure calculation*. Numer. Math., 113, (3), 433–471, 2009.

[261] R. Schneider and A. Uschmajew. *Approximation rates for the hierarchical tensor format in periodic Sobolev spaces*. J. Complex., 30 (2), 56–71, 2014.

[262] R. Schneider and A. Uschmajew. *Convergence results for projected line-search methods on varieties of low-rank matrices via Lojasiewicz inequality*. SIAM J. Optim., 25 (1), 622–646, 2015.

[263] R. Schneider, Th. Rohwedder, J. Blauert, and A. Neelov. *Direct minimization for calculating invariant subspaces in density functional computations of the electronic structure.* J. Comput. Math., 27 (2–3), 360–387, 2009.

[264] U. Schollwöck. *The density-matrix renormalization group in the age of matrix product states,* Ann. Phys., 326 (1), 96–192, 2011.

[265] K. L. Schuchardt, B. T. Didier, T. Elsethagen, L. Sun, V. Gurumoorthi, J. Chase, J. Li, and T. L. Windus. *Basis set exchange: a community database for computational sciences,* J. Chem. Inf. Model., 47, 1045–1052, 2007.

[266] C. Schwab and R.-A. Todor, *Karhunen–Loéve approximation of random fields by generalized fast multipole methods.* J. Comput. Phys., 217, 100–122, 2006.

[267] H. Sekino, Y. Maeda, T. Yanai, and R. J. Harrison. *Basis set limit Hartree Fock and density functional theory response property evaluation by multiresolution multiwavelet basis.* J. Chem. Phys., 129, 034111, 2008.

[268] Y. Shao, L. F. Molnar, Y. Jung, J. Kussmann, C. Ochsenfeld, S. T. Brown, et al. *Advances in methods and algorithms in a modern quantum chemistry program package.* Phys. Chem. Chem. Phys., 8 (27), 3172–3191.

[269] J. Sherman, W. Morrison. *Adjustment of an inverse matrix corresponding to a change in one element of a given matrix.* Ann. Math. Stat., 21 (1), 124–127, 1950.

[270] A. Smilde, R. Bro, and P. Geladi. *Multi-Way Analysis.* Wiley, New York, 2004.

[271] F. Stenger. *Numerical Methods Based on Sinc and Analytic Functions.* Springer-Verlag, Berlin, 1993.

[272] G. Strang. *Introduction to Linear Algebra,* 5th edn. Wellesley–Cambridge Press, Wellesley, 2016.

[273] G. Strang and G. J. Fix. *An Analysis of the Finite Element Method.* Prentice-Hall, Inc., NJ, 1973.

[274] R. E. Stratmann, G. E. Scuseria, and M. J. Frisch. *An efficient implementation of time-dependent density-functional theory for the calculation of excitation energies of large molecules.* J. Chem. Phys., 109, 8218, 1998.

[275] E. Süli and D. F. Mayers. *An Introduction to Numerical Analysis.* Cambridge University Press, Cambridge, 2003.

[276] D. Sundholm, P. Pyykkö, and L. Laaksonen. *Two-dimensional fully numerical molecular calculations. X. Hartree–Fock results for* He_2, Li_12, Be_2, *HF,* OH^-, N_2, *CO, BF,* NO^+, *and* CN^-. Mol. Phys., 56, 1411–1418, 1985.

[277] A. Szabo and N. Ostlund. *Modern Quantum Chemistry.* Dover Publication, New York, 1996.

[278] A. Y. Toukmaji, and J. Board Jr. *Ewald summation techniques in perspective: a survey.* Comput. Phys. Commun., 95, 73–92, 1996.

[279] J. Toulouse, A. Savin. *Local density approximation for long-range or for short-range energy functionals?* J. Mol. Struct., Theochem, 762, 147, 2006.

[280] J. Toulouse, F. Colonna, and A. Savin. *Long-range – short-range separation of the electron–electron interaction in density-functional theory.* Phys. Rev. A, 70, 062505, 2004.

[281] L. N. Trefethen. *Spectral Methods in MATLAB.* SIAM, Philadelphia, 2000.

[282] L. N. Trefethen and D Bau III. *Numerical Linear Algebra.* SIAM, Philadelphia, 1997.

[283] L. N. Trefethen and M. Embree. *Spectra and Pseudospectra: The Behavior of Nonnormal Matrices and Operators.* Princeton University Press, Princeton and Oxford, 2005.

[284] L. R. Tucker. *Some mathematical notes on three-mode factor analysis.* Psychometrika, 31, 279–311, 1966.

[285] I. Turek. *A maximum-entropy approach to the density of states within the recursion method.* J. Phys. C, 21, 3251–3260, 1988.

[286] E. E. Tyrtyshnikov. *Mosaic-skeleton approximations.* Calcolo, 33, 47–57, 1996.

[287] E. E. Tyrtyshnikov. *Incomplete cross approximation in the mosaic-skeleton method.* Computing, 64, 367–380, 2000.

[288] E. E. Tyrtyshnikov. *Tensor approximations of matrices generated by asymptotically smooth functions.* Sb. Math., 194 (5–6), 941–954, 2003 (translated from *Mat. Sb.*, 194 (6), 146–160, 2003).

[289] E. E. Tyrtyshnikov. *Kronecker-product approximations for some function-related matrices.* Linear Algebra Appl., 379, 423–437, 2004.

[290] J. L. M. Van Dorsselaer and M. E. Hoschstenbach. *Computing probabilistic bounds for extreme eigenvalues of symmetric matrices with the Lanczos method.* SIAM J. Matrix Anal. Appl., 22, 837–852, 2000.

[291] C. F. Van Loan and J. P. Vokt. *Approximating matrices with multiple symmetries.* SIAM J. Matrix Anal. Appl., 36 (3), 974–993, 2015.

[292] J. VandeVondele, M. Krack, F. Mohamed, M. Parinello, Th. Chassaing, and J. Hutter. *QUICKSTEP: fast and accurate density functional calculations using a mixed Gaussian and plane waves approach.* Comput. Phys. Commun., 167, 103–128, 2005.

[293] F. Verstraete, D. Porras, and J. I. Cirac. *DMRG and periodic boundary conditions: a quantum information perspective.* Phys. Rev. Lett., 93 (22), 227205, 2004.

[294] G. Vidal. *Efficient classical simulation of slightly entangled quantum computations.* Phys. Rev. Lett. 91 (14), 147902, 2003.

[295] E. Voloshina, D. Usvyat, M. Schütz, Y. Dedkov, and B. Paulus. *On the physisorption of water on graphene: a CCSD(T) study.* Phys. Chem. Chem. Phys., 13, 12041–12047, 2011.

[296] L.-W. Wang. *Calculating the density of states and optical-absorption spectra of large quantum systems by the plane-wave moments method.* Phys. Rev. B, 49, 10154–10158, 1994.

[297] H. Wang, and M. Thoss. *Multilayer formulation of the multiconfiguration time-dependent Hartree theory.* J. Chem. Phys., 119, 1289–1299, 2003.

[298] H.-J. Werner, F. R. Manby, and P. J. Knowles. *Fast linear scaling second order Møller–Plesset perturbation theory (MP2) using local and density fitting approximations.* J. Chem. Phys., 118, 8149–8160, 2003.

[299] H.-J. Werner, P. J. Knowles, G. Knozia, F. R. Manby, and M. Schuetz. *Molpro: a general-purpose quantum chemistry program package.* WIREs Comput. Mol. Sci., 2, 242–253, 2012.

[300] H.-J. Werner, P. J. Knowles, et al. *MOLPRO, Version 2002.10, A Package of Ab Initio Programs for Electronic Structure Calculations.*

[301] J. C. Wheeler and C. Blumstein. *Modified moments for harmonic solids.* Phys. Rev. B, 6, 4380–4382, 1972.

[302] S. R. White. *Density-matrix algorithms for quantum renormalization groups.* Phys. Rev. B, 48 (14), 10345–10356, 1993.

[303] S. Wilson. *Universal basis sets and Cholesky decomposition of the two-electron integral matrix.* Comput. Phys. Commun., 58, 71–81, 1990.

[304] P. Wind, W. Klopper, and T. Helgaker. *Second order Møller–Plesset perturbation theory with terms linear in interelectronic coordinates and exact evaluation of three-electron integrals.* Theor. Chem. Acc., 107, 173–179, 2002.

[305] T. Yanai, G. Fann, Z. Gan, R. Harrison, and G. Beylkin. *Multiresolution quantum chemistry: Hartree–Fock exchange.* J. Chem. Phys., 121 (14), 6680–6688, 2004.

[306] Y. Yang, Y. Kurashige, F. R. Manby, and G. K. L. Chan. *Tensor factorizations of local second-order Møller–Plesset theory.* J. Chem. Phys., 134, 044123, 2011.

[307] H. Yserentant. *The hyperbolic cross space approximation of electronic wavefunctions.* Numer. Math., 105, 659–690, 2007.

[308] H. Yserentant. *Regularity and Approximability of Electronic Wave Functions.* Lecture Notes in Mathematics Series, Springer-Verlag, Berlin, 2010.

[309] E. Zeidler. *Applied Functional Analysis: Applications to Mathematical Physics*. Springer, Berlin, 1995.
[310] T. Zhang and G. Golub. *Rank-One approximation to high order tensors*. SIAM J. Matrix Anal. Appl., 23, 534–550, 2001.
[311] J. Zienau, L. Clin, B. Doser, and C. Ochsenfeld. *Cholesky-decomposed densities in Laplace-based second-order Møller–Plesset perturbation theory*. J. Chem. Phys., 130, 204112, 2009.
[312] M. Zuzovski, A. Boag, and A. Natan. *An auxilliary grid method for the calculation of electrostatic terms in density functional theory on a real-space grid*. Phys. Chem. Chem. Phys., 17, 31550–31557, 2015.

Index

Bei Fragen zur Produktsicherheit wenden Sie sich bitte an:
If you have any questions regarding product safety,
please contact:

Walter de Gruyter GmbH
Genthiner Straße 13
10785 Berlin
productsafety@degruyterbrill.com